# Simulation for Designing Clinical Trials

# DRUGS AND THE PHARMACEUTICAL SCIENCES

## DRUGS AND THE PHARMACEUTICAL SCIENCES

A Series of Textbooks and Monographs

1. Pharmacokinetics, *Milo Gibaldi and Donald Perrier*
2. Good Manufacturing Practices for Pharmaceuticals: A Plan for Total Quality Control, *Sidney H. Willig, Murray M. Tuckerman, and William S. Hitchings IV*
3. Microencapsulation, *edited by J. R. Nixon*
4. Drug Metabolism: Chemical and Biochemical Aspects, *Bernard Testa and Peter Jenner*
5. New Drugs: Discovery and Development, *edited by Alan A. Rubin*
6. Sustained and Controlled Release Drug Delivery Systems, *edited by Joseph R. Robinson*
7. Modern Pharmaceutics, *edited by Gilbert S. Banker and Christopher T. Rhodes*
8. Prescription Drugs in Short Supply: Case Histories, *Michael A. Schwartz*
9. Activated Charcoal: Antidotal and Other Medical Uses, *David O. Cooney*
10. Concepts in Drug Metabolism (in two parts), *edited by Peter Jenner and Bernard Testa*
11. Pharmaceutical Analysis: Modern Methods (in two parts), *edited by James W. Munson*
12. Techniques of Solubilization of Drugs, *edited by Samuel H. Yalkowsky*
13. Orphan Drugs, *edited by Fred E. Karch*
14. Novel Drug Delivery Systems: Fundamentals, Developmental Concepts, Biomedical Assessments, *Yie W. Chien*
15. Pharmacokinetics: Second Edition, Revised and Expanded, *Milo Gibaldi and Donald Perrier*
16. Good Manufacturing Practices for Pharmaceuticals: A Plan for Total Quality Control, Second Edition, Revised and Expanded, *Sidney H. Willig, Murray M. Tuckerman, and William S. Hitchings IV*
17. Formulation of Veterinary Dosage Forms, *edited by Jack Blodinger*
18. Dermatological Formulations: Percutaneous Absorption, *Brian W. Barry*
19. The Clinical Research Process in the Pharmaceutical Industry, *edited by Gary M. Matoren*
20. Microencapsulation and Related Drug Processes, *Patrick B. Deasy*
21. Drugs and Nutrients: The Interactive Effects, *edited by Daphne A. Roe and T. Colin Campbell*
22. Biotechnology of Industrial Antibiotics, *Erick J. Vandamme*

23. Pharmaceutical Process Validation, *edited by Bernard T. Loftus and Robert A. Nash*
24. Anticancer and Interferon Agents: Synthesis and Properties, *edited by Raphael M. Ottenbrite and George B. Butler*
25. Pharmaceutical Statistics: Practical and Clinical Applications, *Sanford Bolton*
26. Drug Dynamics for Analytical, Clinical, and Biological Chemists, *Benjamin J. Gudzinowicz, Burrows T. Younkin, Jr., and Michael J. Gudzinowicz*
27. Modern Analysis of Antibiotics, *edited by Adjoran Aszalos*
28. Solubility and Related Properties, *Kenneth C. James*
29. Controlled Drug Delivery: Fundamentals and Applications, Second Edition, Revised and Expanded, *edited by Joseph R. Robinson and Vincent H. Lee*
30. New Drug Approval Process: Clinical and Regulatory Management, *edited by Richard A. Guarino*
31. Transdermal Controlled Systemic Medications, *edited by Yie W. Chien*
32. Drug Delivery Devices: Fundamentals and Applications, *edited by Praveen Tyle*
33. Pharmacokinetics: Regulatory • Industrial • Academic Perspectives, *edited by Peter G. Welling and Francis L. S. Tse*
34. Clinical Drug Trials and Tribulations, *edited by Allen E. Cato*
35. Transdermal Drug Delivery: Developmental Issues and Research Initiatives, *edited by Jonathan Hadgraft and Richard H. Guy*
36. Aqueous Polymeric Coatings for Pharmaceutical Dosage Forms, *edited by James W. McGinity*
37. Pharmaceutical Pelletization Technology, *edited by Isaac Ghebre-Sellassie*
38. Good Laboratory Practice Regulations, *edited by Allen F. Hirsch*
39. Nasal Systemic Drug Delivery, *Yie W. Chien, Kenneth S. E. Su, and Shyi-Feu Chang*
40. Modern Pharmaceutics: Second Edition, Revised and Expanded, *edited by Gilbert S. Banker and Christopher T. Rhodes*
41. Specialized Drug Delivery Systems: Manufacturing and Production Technology, *edited by Praveen Tyle*
42. Topical Drug Delivery Formulations, *edited by David W. Osborne and Anton H. Amann*
43. Drug Stability: Principles and Practices, *Jens T. Carstensen*
44. Pharmaceutical Statistics: Practical and Clinical Applications, Second Edition, Revised and Expanded, *Sanford Bolton*
45. Biodegradable Polymers as Drug Delivery Systems, *edited by Mark Chasin and Robert Langer*
46. Preclinical Drug Disposition: A Laboratory Handbook, *Francis L. S. Tse and James J. Jaffe*
47. HPLC in the Pharmaceutical Industry, *edited by Godwin W. Fong and Stanley K. Lam*
48. Pharmaceutical Bioequivalence, *edited by Peter G. Welling, Francis L. S. Tse, and Shrikant V. Dinghe*

49. Pharmaceutical Dissolution Testing, *Umesh V. Banakar*
50. Novel Drug Delivery Systems: Second Edition, Revised and Expanded, *Yie W. Chien*
51. Managing the Clinical Drug Development Process, *David M. Cocchetto and Ronald V. Nardi*
52. Good Manufacturing Practices for Pharmaceuticals: A Plan for Total Quality Control, Third Edition, *edited by Sidney H. Willig and James R. Stoker*
53. Prodrugs: Topical and Ocular Drug Delivery, *edited by Kenneth B. Sloan*
54. Pharmaceutical Inhalation Aerosol Technology, *edited by Anthony J. Hickey*
55. Radiopharmaceuticals: Chemistry and Pharmacology, *edited by Adrian D. Nunn*
56. New Drug Approval Process: Second Edition, Revised and Expanded, *edited by Richard A. Guarino*
57. Pharmaceutical Process Validation: Second Edition, Revised and Expanded, *edited by Ira R. Berry and Robert A. Nash*
58. Ophthalmic Drug Delivery Systems, *edited by Ashim K. Mitra*
59. Pharmaceutical Skin Penetration Enhancement, *edited by Kenneth A. Walters and Jonathan Hadgraft*
60. Colonic Drug Absorption and Metabolism, *edited by Peter R. Bieck*
61. Pharmaceutical Particulate Carriers: Therapeutic Applications, *edited by Alain Rolland*
62. Drug Permeation Enhancement: Theory and Applications, *edited by Dean S. Hsieh*
63. Glycopeptide Antibiotics, *edited by Ramakrishnan Nagarajan*
64. Achieving Sterility in Medical and Pharmaceutical Products, *Nigel A. Halls*
65. Multiparticulate Oral Drug Delivery, *edited by Isaac Ghebre-Sellassie*
66. Colloidal Drug Delivery Systems, *edited by Jörg Kreuter*
67. Pharmacokinetics: Regulatory • Industrial • Academic Perspectives, Second Edition, *edited by Peter G. Welling and Francis L. S. Tse*
68. Drug Stability: Principles and Practices, Second Edition, Revised and Expanded, *Jens T. Carstensen*
69. Good Laboratory Practice Regulations: Second Edition, Revised and Expanded, *edited by Sandy Weinberg*
70. Physical Characterization of Pharmaceutical Solids, *edited by Harry G. Brittain*
71. Pharmaceutical Powder Compaction Technology, *edited by Göran Alderborn and Christer Nyström*
72. Modern Pharmaceutics: Third Edition, Revised and Expanded, *edited by Gilbert S. Banker and Christopher T. Rhodes*
73. Microencapsulation: Methods and Industrial Applications, *edited by Simon Benita*
74. Oral Mucosal Drug Delivery, *edited by Michael J. Rathbone*
75. Clinical Research in Pharmaceutical Development, *edited by Barry Bleidt and Michael Montagne*

76. The Drug Development Process: Increasing Efficiency and Cost Effectiveness, *edited by Peter G. Welling, Louis Lasagna, and Umesh V. Banakar*
77. Microparticulate Systems for the Delivery of Proteins and Vaccines, *edited by Smadar Cohen and Howard Bernstein*
78. Good Manufacturing Practices for Pharmaceuticals: A Plan for Total Quality Control, Fourth Edition, Revised and Expanded, *Sidney H. Willig and James R. Stoker*
79. Aqueous Polymeric Coatings for Pharmaceutical Dosage Forms: Second Edition, Revised and Expanded, *edited by James W. McGinity*
80. Pharmaceutical Statistics: Practical and Clinical Applications, Third Edition, *Sanford Bolton*
81. Handbook of Pharmaceutical Granulation Technology, *edited by Dilip M. Parikh*
82. Biotechnology of Antibiotics: Second Edition, Revised and Expanded, *edited by William R. Strohl*
83. Mechanisms of Transdermal Drug Delivery, *edited by Russell O. Potts and Richard H. Guy*
84. Pharmaceutical Enzymes, *edited by Albert Lauwers and Simon Scharpé*
85. Development of Biopharmaceutical Parenteral Dosage Forms, *edited by John A. Bontempo*
86. Pharmaceutical Project Management, *edited by Tony Kennedy*
87. Drug Products for Clinical Trials: An International Guide to Formulation • Production • Quality Control, *edited by Donald C. Monkhouse and Christopher T. Rhodes*
88. Development and Formulation of Veterinary Dosage Forms: Second Edition, Revised and Expanded, *edited by Gregory E. Hardee and J. Desmond Baggot*
89. Receptor-Based Drug Design, *edited by Paul Leff*
90. Automation and Validation of Information in Pharmaceutical Processing, *edited by Joseph F. deSpautz*
91. Dermal Absorption and Toxicity Assessment, *edited by Michael S. Roberts and Kenneth A. Walters*
92. Pharmaceutical Experimental Design, *Gareth A. Lewis, Didier Mathieu, and Roger Phan-Tan-Luu*
93. Preparing for FDA Pre-Approval Inspections, *edited by Martin D. Hynes III*
94. Pharmaceutical Excipients: Characterization by IR, Raman, and NMR Spectroscopy, *David E. Bugay and W. Paul Findlay*
95. Polymorphism in Pharmaceutical Solids, *edited by Harry G. Brittain*
96. Freeze-Drying/Lyophilization of Pharmaceutical and Biological Products, *edited by Louis Rey and Joan C. May*
97. Percutaneous Absorption: Drugs–Cosmetics–Mechanisms–Methodology, Third Edition, Revised and Expanded, *edited by Robert L. Bronaugh and Howard I. Maibach*

98. Bioadhesive Drug Delivery Systems: Fundamentals, Novel Approaches, and Development, *edited by Edith Mathiowitz, Donald E. Chickering III, and Claus-Michael Lehr*
99. Protein Formulation and Delivery, *edited by Eugene J. McNally*
100. New Drug Approval Process: Third Edition, The Global Challenge, *edited by Richard A. Guarino*
101. Peptide and Protein Drug Analysis, *edited by Ronald E. Reid*
102. Transport Processes in Pharmaceutical Systems, *edited by Gordon L. Amidon, Ping I. Lee, and Elizabeth M. Topp*
103. Excipient Toxicity and Safety, *edited by Myra L. Weiner and Lois A. Kotkoskie*
104. The Clinical Audit in Pharmaceutical Development, *edited by Michael R. Hamrell*
105. Pharmaceutical Emulsions and Suspensions, *edited by Francoise Nielloud and Gilberte Marti-Mestres*
106. Oral Drug Absorption: Prediction and Assessment, *edited by Jennifer B. Dressman and Hans Lennernäs*
107. Drug Stability: Principles and Practices, Third Edition, Revised and Expanded, *edited by Jens T. Carstensen and C. T. Rhodes*
108. Containment in the Pharmaceutical Industry, *edited by James P. Wood*
109. Good Manufacturing Practices for Pharmaceuticals: A Plan for Total Quality Control from Manufacturer to Consumer, Fifth Edition, Revised and Expanded, *Sidney H. Willig*
110. Advanced Pharmaceutical Solids, *Jens T. Carstensen*
111. Endotoxins: Pyrogens, LAL Testing, and Depyrogenation, Second Edition, Revised and Expanded, *Kevin L. Williams*
112. Pharmaceutical Process Engineering, *Anthony J. Hickey and David Ganderton*
113. Pharmacogenomics, *edited by Werner Kalow, Urs A. Meyer, and Rachel F. Tyndale*
114. Handbook of Drug Screening, *edited by Ramakrishna Seethala and Prabhavathi B. Fernandes*
115. Drug Targeting Technology: Physical • Chemical • Biological Methods, *edited by Hans Schreier*
116. Drug–Drug Interactions, *edited by A. David Rodrigues*
117. Handbook of Pharmaceutical Analysis, *edited by Lena Ohannesian and Anthony J. Streeter*
118. Pharmaceutical Process Scale-Up, *edited by Michael Levin*
119. Dermatological and Transdermal Formulations, *edited by Kenneth A. Walters*
120. Clinical Drug Trials and Tribulations: Second Edition, Revised and Expanded, *edited by Allen Cato, Lynda Sutton, and Allen Cato III*
121. Modern Pharmaceutics: Fourth Edition, Revised and Expanded, *edited by Gilbert S. Banker and Christopher T. Rhodes*
122. Surfactants and Polymers in Drug Delivery, *Martin Malmsten*
123. Transdermal Drug Delivery: Second Edition, Revised and Expanded, *edited by Richard H. Guy and Jonathan Hadgraft*

124. Good Laboratory Practice Regulations: Second Edition, Revised and Expanded, *edited by Sandy Weinberg*
125. Parenteral Quality Control: Sterility, Pyrogen, Particulate, and Package Integrity Testing: Third Edition, Revised and Expanded, *Michael J. Akers, Daniel S. Larrimore, and Dana Morton Guazzo*
126. Modified-Release Drug Delivery Technology, *edited by Michael J. Rathbone, Jonathan Hadgraft, and Michael S. Roberts*
127. Simulation for Designing Clinical Trials: A Pharmacokinetic-Pharmacodynamic Modeling Perspective, *edited by Hui C. Kimko and Stephen B. Duffull*
128. Affinity Capillary Electrophoresis in Pharmaceutics and Biopharmaceutics, *edited by Reinhard H. H. Neubert and Hans-Hermann Rüttinger*

**ADDITIONAL VOLUMES IN PREPARATION**

Pharmaceutical Process Validation: An International Third Edition, Revised and Expanded, *edited by Robert A. Nash and Alfred H. Wachter*

Ophthalmic Drug Delivery Systems: Second Edition, Revised and Expanded, *edited by Ashim K. Mitra*

Pharmaceutical Gene Delivery Systems, *edited by Alain Rolland and Sean M. Sullivan*

Biomarkers in Clinical Drug Development, *edited by John Bloom*

Pharmaceutical Inhalation Aerosol Technology: Second Edition, Revised and Expanded, *edited by Anthony J. Hickey*

Pharmaceutical Extrusion Technology, *edited by Isaac Ghebre-Sellassie and Charles Martin*

# Simulation for Designing Clinical Trials

## A Pharmacokinetic-Pharmacodynamic Modeling Perspective

Hui C. Kimko
***Johnson & Johnson Pharmaceutical Research & Development Raritan, New Jersey, U.S.A.***

Stephen B. Duffull
***University of Queensland Brisbane, Queensland, Australia***

CRC Press
Taylor & Francis Group
Boca Raton London New York

CRC Press is an imprint of the
Taylor & Francis Group, an **informa** business

CRC Press
Taylor & Francis Group
6000 Broken Sound Parkway NW, Suite 300
Boca Raton, FL 33487-2742

First issued in paperback 2019



ISBN-13: 978-0-8247-0862-7 (hbk)
ISBN-13: 978-0-367-39560-5 (pbk)



**Library of Congress Cataloging-in-Publication Data**
A catalog record for this book is available from the Library of Congress.

**Visit the Taylor & Francis Web site at**
**http://www.taylorandfrancis.com**

**and the CRC Press Web site at**
**http://www.crcpress.com**

*To Dennis, Edwin, Jason & Aynsley, and Ella & Holly*

# Foreword

Computer simulations of clinical trials, employing realistic virtual subjects and typical trial conditions, based on both experimentally informed disease progress and drug intervention models, originated within the last decade. Previously, clinical trials were designed using ad hoc empirical approaches, unaided by a systematic clinical pharmacology orientation or a quantitative pharmacokinetic-pharmacodynamic framework, leading to highly inefficient drug development programs. Stimulated by research and educational contributions from academia, and encouragement from regulatory agencies, expert simulation teams have recently deployed hundreds of clinical trial simulation projects. The advent of modern clinical trial simulation is transforming clinical drug development from naïve empiricism to a mechanistic scientific discipline.

The editors and contributors have provided more than just a comprehensive history, critical vocabulary, insightful compilation of motivations, and clear explanation of the state of the art of modern clinical trial simulation. This book advances a rigorous framework for employing simulation as an experiment, according to a predefined simulation plan that reflects good simulation practices.* We describe attributes of the multidisciplinary simulation team that position it to achieve benefits of enhanced communication and collaboration during the development and employment of the simulation.

* Holford NHG, Hale M, Ko, HC, Steimer J-L, Sheiner LB, Peck CC. Simulation in Drug Development: Good Practices, 1999. http://cdds.georgetown.edu/SDDGP.html.

While the far future of scientific drug development is difficult to predict, successful advancement and integration of clinical trial simulation lead to a daring prediction: in the not so distant future, most clinical trials will be virtual—only a few actual trials will be undertaken. These few human trials will be designed to inform simulation models and to confirm model predictions. The academic, pharmaceutical, and regulatory scientists who have articulated the state of the art of clinical trial simulations in this book provide the first comprehensive description of a breakthrough technology that is enabling this bold departure from inefficient past practices.

*Carl Peck, M.D.*
*Center for Drug Development Science*
*Georgetown University, Washington D.C., U.S.A.*

# Preface

Simulation has been used widely in various disciplines such as engineering, physics, and economics to support the development and testing of the performance of systems. Some of the most notable examples arise from the wealth of experience in the aerospace industry, in which over the past 30 years it has become routine practice to simulate the performance of aircraft before production and launch. The use of simulations in these industries has been shown to reduce costs and shorten development time. Some of the experience gained from these highly technical systems has pervaded the highly stochastic area of biological systems. Within the pharmaceutical industry, this has culminated in the growth of modeling and simulation in the drug development process, where computer simulation is gaining popularity as a tool for the design of clinical trials.

The integration of modeling and simulation into drug development has been a gradual process in spite of the long-term use of stochastic simulations by biostatisticians for exploration and analyses of data and deterministic simulations by pharmacologists for descriptive purposes. However, simulation has considerable potential for design of trials as evidenced by the rapid growth in discussion groups, conferences, and publications. The basis of the use of simulation lies in the argument that if, in theory, a virtual trial could be constructed that incorporated all relevant influences (controllable and uncontrollable, and deterministic and stochastic) and their related outcomes, the researcher could then explore the influence of changes in the design on the performance of the trial. If this theoretical construct were a reality, then it is equally conceivable that trial designs could be selected based on their probability for success. While this seems a straightforward task, albeit computationally intensive, its introduction has been at the mercy of the availability of powerful computing. Since fast computing machines are now available on almost every office desk, it is no surprise that both the design of trials and the necessary knowledge and understanding of the time course of drug effects that underpins the design process have shown a dramatic upsurge. The parallel development of more complex and mechanistic models for drug effects and design of trials is not coincidental, since an understanding of the effects of drugs is paramount for design of trials and the more complex models themselves rely heavily on computational methods for their solution.

This book describes the background and lays the foundation for simulation as a tool for the design of clinical trials. The target audience is any researcher or practitioner who is involved in the design, implementation, analysis, or regulatory decisions concerning clinical trials. This book does not embrace all aspects of trial design, nor is it intended as a recipe for using computers to design trials. Rather, it is a source of information that enables the reader to gain a better understanding of the theoretical background and knowledge of the practical applications of simulation for design. It is assumed that the reader has a working understanding of pharmacokinetics and pharmacodynamics, modeling, and the drug development process. In addition, some knowledge of types and practicalities of designs commonly used for clinical trials is assumed.

The book is divided into parts that describe model development, model evaluation, execution of simulation, choice of design, and applications. It is useful to partition the simulation model into submodels (e.g., input-output model, covariate distribution model, execution model) in order to describe specific aspects of the process. The input-output model (Chapter 2) describes the relationship between dosing schedule and response in an explanatory manner for any given patient. This model itself usually comprises a number of submodels: the pharmacokinetic and pharmacodynamic models, disease progression models, and (patho)physiological models of homeostatic systems in the body. The covariate distribution model (Chapter 3) describes the characteristics of the virtual patient that affect the input-output models and execution models. The execution model (Chapter 4) describes the deviation from the nominal design, thereby mimicking the actual trial conduct.

Details of model evaluation methods are provided in Chapter 5. The mechanics required for simulation of a trial, including replications, random number generation, and the implementation of numerical integration are outlined in Chapter 6. Analysis of replicates of the subsequent virtual trial is discussed in Chapter 7. Chapter 8 addresses the important issue of the sensitivity of the trial design to assumptions in the development of the models that underpin the response of the virtual patient. Finally, in this section discussion is raised about how a sufficient design might be selected from all possible designs (Chapter 9).

While simulation as an investigation tool has proven conceptually to be straightforward, complete acceptance by regulatory authorities and the pharmaceutical industry remains elusive. Details of perspectives by regulatory authorities, academia, and the pharmaceutical industry are provided by Chapters 10, 11, and 12, respectively. In addition to these perspectives, an overview and history of mechanism-based model development for physiological/pharmacological processes are presented in Chapter 13.

We have also included a part devoted to applications of simulation for trial design and evaluation (Chapters 14 to 18). These include a wide range of practical applications, including optimization of sampling strategies, dose selection, integration of optimal design with simulation, prediction of efficacy, and side effects.

We accept that our current knowledge of predicting clinical responses for the individual patient pales beside that imagined by science fiction writers. Our way of exploring the nature of drug activity is limited to conducting clinical trials in the hope of learning about clinical responses and confirming these findings using methods that are often empirical. Since it is recognized that the analysis of data is dependent on the quality of the data, and the quality of the data dependent on the quality of the study design, we are dependent on adequately designed clinical trials to pave the way for better treatments. In due course, it is expected that methods that promote more informative and rigorous designs such as those based on modeling and simulation will provide the same benefits for drug development that have been seen in other industries.

We thank the authors of the chapters and Marcel Dekker, Inc., for providing the opportunity to publish this book.

*Hui C. Kimko*
*Stephen B. Duffull*

# Contents

# Contributors

**Leon Aarons** School of Pharmacy and Pharmaceutical Sciences, University of Manchester, Manchester, United Kingdom

**Jeffrey S. Barrett** Aventis Pharmaceuticals, Bridgewater, New Jersey, U.S.A.

**Peter L. Bonate*** Quintiles Transnational Corp., Kansas City, Missouri, U.S.A.

**Stephen B. Duffull** School of Pharmacy, University of Queensland, Brisbane, Queensland, Australia

**Ene I. Ette** Vertex Pharmaceuticals, Inc., Cambridge, Massachusetts, U.S.A.

**Bärbel Fotteler** Modeling and Simulation Team, Clinical Pharmacology and Biostatistics, Pharma Development, F. Hoffmann-La Roche, Basel, Switzerland

**Patrice Francheteau†** Sandoz Pharmaceuticals, Rueil-Malmaison, France

**Marc R. Gastonguay** School of Pharmacy, University of Connecticut, Farmington, Connecticut, U.S.A.

* *Current affilation*: ILEX Oncology, San Antonio, Texas, U.S.A.

† Retired.

**Ekaterina Gibiansky*** GloboMax LLC, Hanover, Maryland, U.S.A.

**Leonid Gibiansky** The Emmes Corporation, Rockville, Maryland, U.S.A.

**Ronald Gieschke** Modeling and Simulation Team, Clinical Pharmacology and Biostatistics, Pharma Development, F. Hoffmann-La Roche, Basel, Switzerland

**Pascal Girard** Pharsight, Lyon, France

**Christopher J. Godfrey** Vertex Pharmaceuticals, Inc., Cambridge, Massachusetts, U.S.A.

**Timothy Goggin†** Modeling and Simulation Team, Clinical Pharmacology and Biostatistics, Pharma Development, F. Hoffmann-La Roche, Basel, Switzerland

**Gordon Graham** Centre for Applied Pharmacokinetic Research, School of Pharmacy and Pharmaceutical Sciences, University of Manchester, Manchester, United Kingdom

**Madeleine Guerret‡** Sandoz Pharmaceuticals, Rueil-Malmaison, France

**Nicholas H. G. Holford** Division of Pharmacology and Clinical Pharmacology, University of Auckland, Auckland, New Zealand

**Matthew M. Hutmacher** Pharmacia Corp, Skokie, Illinois, U.S.A.

**Paul Jordan** Modeling and Simulation Team, Clinical Pharmacology and Biostatistics, Pharma Development, F. Hoffmann-La Roche, Basel, Switzerland

**Mats O. Karlsson** Division of Pharmacokinetics and Drug Therapy, University of Uppsala, Uppsala, Sweden

**Helen Kastrissios§** School of Pharmacy, University of Queensland, Brisbane, Queensland, Australia

---

* *Current affiliation*: Guilford Pharmaceuticals, Baltimore, Maryland, U.S.A.

† *Current affiliation*: Human Pharmacology Group, Serono International S.A., Geneva, Switzerland.

‡ Retired.

§ *Current affiliation*: GloboMax LLC, Hanover, Maryland, U.S.A.

**Hui C. Kimko*** Center for Drug Development Science, Georgetown University, Washington, D.C., U.S.A.

**Stéphane Kirkesseli†** Sandoz Pharmaceuticals, Rueil-Malmaison, France

**Kenneth G. Kowalski‡** Pfizer, Inc., Ann Arbor, Michigan, U.S.A.

**Peter I. D. Lee** Office of Clinical Pharmacology and Biopharmaceutics, Center for Drug Evaluation and Research, Food and Drug Administration, Rockville, Maryland, U.S.A.

**Lawrence J. Lesko** Office of Clinical Pharmacology and Biopharmaceutics, Center for Drug Evaluation and Research, Food and Drug Administration, Rockville, Maryland, U.S.A.

**Jaap W. Mandema** Pharsight Corporation, Mountain View, California, U.S.A.

**Henri Merdjan§** Sandoz Pharmaceuticals, Rueil-Malmaison, France

**Jonathan P. R. Monteleone** Division of Pharmacology and Clinical Pharmacology, School of Medicine, University of Auckland, Auckland, New Zealand

**Diane R. Mould**** Center for Drug Development Science, Georgetown University, Washington, D.C., U.S.A.

**Ivan A. Nestorov††** Centre for Applied Pharmacokinetic Research, School of Pharmacy and Pharmaceutical Sciences, University of Manchester, Manchester, United Kingdom

**Stephan Ogenstad** Vertex Pharmaceutical, Inc., Cambridge, Massachusetts, U.S.A.

---

* *Current affiliation*: Department of Advanced Pharmacokinetic/Pharmacodynamic Modeling & Simulation, Johnson & Johnson Pharmaceutical Research & Development, LLC, Raritan, New Jersey, U.S.A.

† *Current affiliation*: Clinical Pharmacology, Aventis Pharmaceuticals, Bridgewater, New Jersey, U.S.A.

‡ *Current affiliation*: Pfizer, Inc., Ann Arbor, Michigan, U.S.A.

§ *Current affiliation*: Laboratory of Pharmacokinetics and Metabolism, Servier Research and Development, Fulmer, Slough, United Kingdom.

** *Current affiliation*: Projections Research, Inc., Phoenixville, Pennsylvania, U.S.A.

†† *Current affiliation*: Amgen, Inc., Seattle, Washington, U.S.A.

**John C. Pezzullo** Departments of Pharmacology and Biostatistics, Georgetown University, Washington, D.C., U.S.A.

**Goonaseelan (Colin) Pillai*** Modeling and Simulation Team, Clinical Pharmacology and Biostatistics, Pharma Development, F. Hoffmann-La Roche, Basel, Switzerland

**Paolo Sassano†** Sandoz Pharmaceuticals, Rueil-Malmaison, France

**Jean-Louis Steimer‡** Modeling and Simulation Team, Clinical Pharmacology and Biostatistics, Pharma Development, F. Hoffmann-La Roche, Basel, Switzerland

**John Urquhart** Chief Scientist, AARDEX Ltd./APREX Corp., Zug, Switzerland, and Union City, California, U.S.A.; Department of Epidemiology, Maastricht University, Maastricht, the Netherlands; and Department of Biopharmaceutical Sciences, University of California, San Francisco, California, U.S.A.

**Wenping Wang** Pharsight Corporation, Mountain View, California, U.S.A.

**Paul J. Williams** Trials by Design, LLC, La Jolla and Stockton, and School of Pharmacy & Allied Health Sciences, University of the Pacific, Stockton, California, U.S.A.

* *Current affiliation*: Modeling and Simulation, Clinical Pharmacology, Novartis Pharma AG, Basel, Switzerland.

† *Current affiliation*: Novartis Pharmaceuticals, Rueil-Malmaison, France.

‡ *Current affiliation*: Stochastic Modeling and Related Technologies, Biostatistics, Novartis Pharma AG, Basel, Swizerland.

# Simulation for Designing Clinical Trials

# 1

# Introduction to Simulation for Design of Clinical Trials

**Hui C. Kimko***
Georgetown University, Washington, D.C., U.S.A.

**Stephen B. Duffull**
University of Queensland, Brisbane, Queensland, Australia

## 1.1 BACKGROUND

The decision processes of designing clinical trials have followed a largely ad hoc manner, driven by empiricism and embracing concepts such as "what was done previously" and "it has always been done this way." In contrast, other disciplines have been effectively designing experiments using statistical techniques to aid design for many years, but it is only recently that these methods have filtered into the clinical pharmacological arena. Simulation has become a powerful tool for the practitioner due to its generality of application to a wide array of potential problems. In many cases, it can be employed without the practitioner being required to derive closed form solutions to complex statistical problems that would otherwise remain inaccessible. In addition, simulation gains credibility with the nonscientist since it can be explained in essentially nonscientific terms, which allows its transparency to be grasped with ease. It is not surprising, therefore,

* *Current affiliation*: Johnson & Johnson Pharmaceutical Research and Development, Raritan, New Jersey, U.S.A.

that clinical trial simulation (CTS) has been used in designing clinical trials in drug development (1–3).

Perhaps the biggest single advance in the design of clinical trials has been the union of clinicians and mathematicians (we use this term broadly to include those scientists with predominantly mathematical, statistical, or engineering backgrounds) toward a common goal. This has led to the evolution of a position that is between the two disciplines, with the emergence of practitioners who are able to comfortably converse with both clinicians and mathematicians. These practitioners are often referred to as pharmacometricians, and they tend to have backgrounds in pharmaceutical science, biostatistics, and/or clinical pharmacology. The emergence of the discipline of pharmacometrics and the interaction of clinicians and mathematicians have paved the way for significant advances in the design of trials, particularly in the area of simulation for design.

### 1.1.1 History of Clinical Trial Simulation

Simulation has been used widely in many disciplines, including engineering, statistics, astronomy, economy, and marketing to name but a few. In the field of drug development, it is a relatively new concept. The term CTS was perhaps first coined in the 1970s with its appearance in Ref. 4, where a game called Instant Experience was used as an aid to teach doctors and scientists interested in learning about practical difficulties and sources of error in designing and executing clinical trials. While developing the game rules, the authors created a cohort of virtual subjects with characteristics that were appropriate for the fictive drugs in question. The drug efficacies were unknown to the game participants. The participants then allocated treatments to the virtual patients by a method of their choice, e.g., randomization, and the organizers provided the results of their trial, which the participant would analyze. Ultimately, the participant would be able to determine what factors affected the virtual subject's response and design the trial accordingly.

Since its inception in a teaching environment, there have been several success stories recounting the value of simulation for design of clinical trials. These include reports where simulations proved helpful in explaining proposed trial designs not only to their own internal members but also to regulatory authorities, such as the Food and Drug Administration (FDA) (5). The simulation of a proposed randomized concentration controlled trial of mycophenolate mofetil was the first reported demonstration of the practical utility of stochastic clinical trial simulation (6). This simulation helped to determine trial feasibility, and influenced design of an actual trial. It is important to note that this simulation project was motivated by a specific request by the FDA, implying interest and receptivity of the regulatory agency to this technology. The main supporter at the FDA was Carl Peck, who was then the director of the Center for Drug Evaluation and

Research. After the formation of the Center for Drug Development Science (CDDS) at Georgetown University in 1994, the CDDS has been championing the use of simulation to accelerate the drug development process (7). Two software companies (Mitchell & Gauthier Associates Corporation and Pharsoft Corporation) were interested in developing clinical trial simulation programs initially, and Pharsight Corporation (previously known as Pharsoft Corporation) (8) integrated the two programs into one. More detailed history of CTS can be found in the review by Holford et al. (1).

### 1.1.2 FDA Opinion on Clinical Trial Simulation

The interest in performing clinical trial simulations has been boosted by the support of government authorities. The Center for Drug Evaluation and Research (CDER) and Center for Biologics Evaluation and Research (CBER) at the Food and Drug Administration (FDA) of the U.S. Department of Health and Human Services published the Guidance for Industry on Population Pharmacokinetics in February 1999 (9). It defines simulation as "the generation of data with certain types of mathematical and probabilistic models describing the behavior of the system under study."

Section V, "Study Design and Execution," of the Guidance succinctly summarizes the usefulness of simulations in their statement: "Obtaining preliminary information on variability from pilot studies makes it possible through simulation to anticipate certain fatal study designs, and to recognize informative ones." This section includes a subsection, "Simulation," which states:

> Simulation is a useful tool to provide convincing objective evidence of the merits of a proposed study design and analysis (10). Simulating a planned study offers a potentially useful tool for evaluating and understanding the consequences of different study designs. Shortcomings in study design result in the collection of uninformative data. Simulation can reveal the effect of input variables and assumptions on the results of a planned population PK study. Simulation allows study designers to assess the consequences of the design factors chosen and the assumptions made. Thus, simulation enables the pharmacometrician to better predict the results of a population PK study and to choose the study design that will best meet the study objectives (2, 11–15). A simulation scheme should entail repetitive simulation and appropriate analysis of data sets to control for the effect of sampling variability on parameter estimates. Alternative study designs may be simulated to determine the most informative design.

In subsection G, Application of Results of Section IX, Population PK Study Report, further support for simulation is provided: "In addition, the use of graphics, often based on simulations of potential responses under the final fitted model, to communicate the application of a population model (e.g., for dosage adjustments) is recommended."

Further documentation, made in the "Points to Consider" document issued by the Committee for Proprietary Medicinal Products, supported the use of CTS in drug development: "Through simulation, the influence of certain aspects of the planned Phase III trial can be assessed, and, the design (for example, with respect to dose or dosing interval) subsequently modified if needed" (16).

The FDA has cosponsored several meetings to discuss the value of modeling and simulation (7). The subgroup of pharmacometricians among the reviewers in the division of Biopharmaceutics of the FDA (personal communication with Jogarao Gobburu at the FDA) has advised pharmaceutical companies to perform modeling and simulations when deciding the best trial design. Detailed discussion on a regulatory perspective of CTS can be found in Chapter 10.

## 1.2 WHAT IS CLINICAL TRIAL SIMULATION?

A technical definition of clinical trial simulations is the generation of a response for a virtual subject by approximating (a) trial design, (b) human behavior, (c) disease progress, and (d) drug behavior using mathematical models and numerical methods. The trial design provides dosing algorithms, subject selection criteria, and demography. Human behavior includes trial execution characteristics such as adherence in drug administration (pertaining to subjects) and missing records (pertaining to the investigators). Disease status may change during a trial, for which a disease progress model may need to be developed. The drug behavior in the body is generally characterized by pharmacokinetic (PK) and pharmacodynamic (PD) models. These models are developed from prior experience and from prior data.

Figure 1 summarizes the factors and models to be understood in the performance of Monte Carlo simulation, applied in drug development. Factors other than controllable factors, such as number of subjects or doses, are subject to stochastic components. The stochastic element is presumed to arise from some predefined statistical distribution that is characterized from modeling previous data, the simulator generates sets of each element that provide the characteristics and expected responses of one virtual subject. Repetition of this process with a random number generator yields a population of different virtual subjects that can be included in a virtual trial.

Simulation for design of trials is, however, not without difficulty. While the process of developing simulation models allows a complex biological system(s) to be represented by a series of linked mathematical expressions, the full complexity of the biological system is necessarily condensed and information lost. The process of developing these models is itself associated with significant

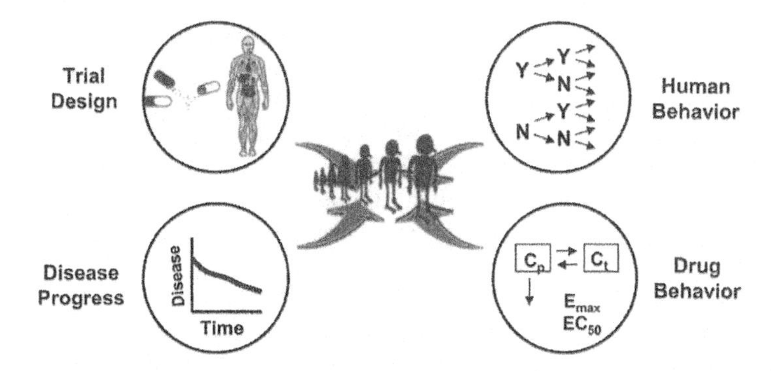


**Figure 1** Factors and models to be understood in performing Monte Carlo simulation.

cost and time, although these pale in comparison to the resources required to conduct the actual trial. The loss of information associated with the gross simplification of biological processes requires assumptions to be made about both the models and the systems that they represent. In this circumstance, the model must adequately describe the system as the system is currently observed (i.e., descriptive) and that the system under usual regulatory conditions can be adequately described by the model (i.e., predictive). Further difficulty may be encountered since many of the assumptions may not be obvious to all members of the design team. Some assumptions may be purely statistical and others may be wholly biological. In all cases, they need to be identified and assessed in terms of their effect on the outcomes of the virtual trial. The greater the simplification of the system, the more assumptions are necessary. A balance must then be sought between what is a livable "amount" of assumptions and the level of complexity of the model. Complex models may also be difficult to solve numerically due to a lack of a closed form solution. In general, overly complex models become both difficult to interpret and summarize and may falsely convey confidence in the study findings due to the sheer scale of the undertaking and the volume of numbers provided per simulation.

## 1.3 WHY SIMULATE CLINICAL TRIALS?

Simulation allows the evaluation of competing trial designs, e.g., comparison of dosage regimens, to be compared in terms of their efficacy to produce the desired outcomes prior to conducting the actual trial. The act of performing a simulation requires the development of a series of linked simulation models. This model development process has innate educational benefits since it allows clear identification of what is known about the drug in question. The corollary of this process is that model development identifies areas that are not known or are poorly characterized. It should not be surprising that many characteristics of a drug that were thought to have been known are found to be lacking when collation of available information is formalized. The act of model development in itself is therefore valuable for identifying areas that may require further development. This may initiate the necessity for additional early phase studies prior to undertaking the larger virtual or actual trial in question.

In addition to formalizing the available background data, model development provides a useful construct to draw out the ideas and experiences of those involved in the design team. Traditionally, drug development has been based on intuitive decision-making processes that reflected personal experience, judgment, and effort rather than rigorous quantitative scientific evidence. However, problems can arise due to differences in the personal experiences of those involved in designing the trial. It can be difficult for all design team members to appreciate all aspects of the trial design, and therefore it is challenging for a team leader to integrate all the ideas and expertise from all the design team members. The introduction of more formal methods of design require a higher level of interaction at the outset and provides the necessary framework for compiling these individual initiatives which should in itself lead to more powerful trial designs that reflect the ideas of the whole design team. An essential and nontrivial aspect of trial design is the requirement to describe quantitatively the relationship(s) between dose schedule of a particular drug and likely outcomes for any given subject. It is not a requirement, however, to describe every aspect of the drug-host interaction explicitly as long as a sufficient mathematical model is developed that can be used for descriptive and in theory predictive purposes. Predictions about the outcomes for different patient groups or dosing schedules are often the goal of trial simulations but are prone to errors due to the assumptions in the model-building process. This leaves the trial designer in the unenviable position of making predictions that are outside the scope of the data from which the model was originally defined.

Although these predictions risk possible errors, the more formal nature of simulation for trial design allows many of the assumptions inherent in the process to be evaluated in a rigorous manner. The beauty of a more formal approach to designing clinical trials is the ability to ask "what if?" questions: "What if the model parameters were a little different?" "What if the structural aspects of the model were changed?" "What if more subjects dropped out prior to assessment of trial outcomes?" and so on. There are of course circumstances when a descriptive model is sufficient in itself, and can be used to answer questions such as, "What went wrong?" "Can the outcome of the actual trial be trusted, i.e., is the outcome of the trial similar to what might have been expected given what is known about the drug?"

## 1.4 CLINICAL TRIAL SIMULATION PLAN

A simulation plan refers to a "protocol" for the simulation of virtual trials. The pharmaceutical industry relies heavily on the use of standard operating procedures and protocols in the drug development process. These processes are vital in order that standards are met and all outcomes are recorded and can be tracked. Preparation of a simulation plan is not exceptional to this rule. By preparing a simulation plan, the simulation team can focus on the problem to be solved and during the procedure may discover missing information needed to describe the problem.

A simulation plan may work as a communication tool, especially in the model-building procedures where many assumptions are inevitable. In all cases, any assumptions should be listed in the plan. It should convey a level of transparency that allows any or all of the work to be reproduced or continued by a newly joined person in the simulation team. In addition, the simulation plan can provide a pro forma for the development of similar drugs or similar types of trial designs. It is essential that the simulation plan be approved by the drug development team, which in itself adds credibility and acceptance of the CTS process.

Figure 2 summarizes suggested steps to follow in performing CTS. The process should be initiated with a clear objective statement, followed by data collation and model development. The program code is then constructed and verified by pilot runs to assess the behavior of the code. This also allows for some rudimentary, preliminary investigations of the model. The code can then be generalized to allow several proposed clinical trial designs to be simulated,

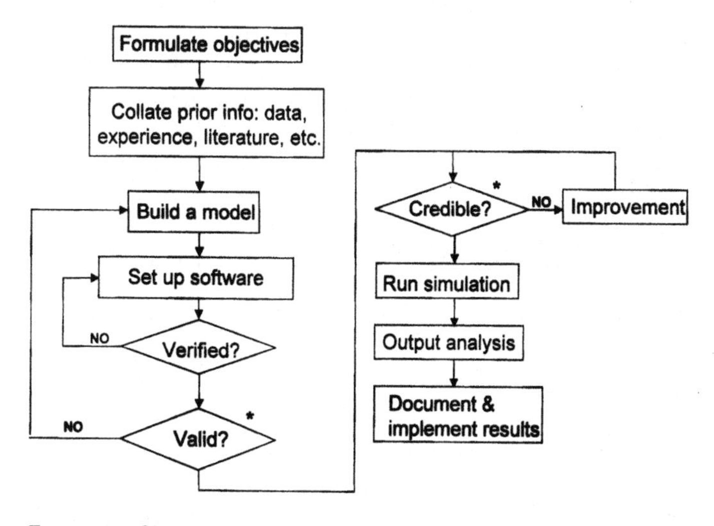


**FIGURE 2** Clinical trial simulation procedures. *Indicates model qualification steps (20).

analyzed, and documented. In developing a simulation plan, these steps are designed to concord with the documentation of the plan.

There are three guiding principles in performing a simulation experiment (Simulation in Drug Development: Good Practices [http://cdds.georgetown.edu/research/sddgp723.html]): clarity, completeness and parsimony. For each principle, it is stated that the report of the simulation should be understandable in terms of scope and conclusions by the intended client, such as those responsible for committing resources to a clinical trial; the assumptions, methods, and critical results should be described in sufficient detail to be reproduced by an independent team; the complexity of the models and simulation procedures should be no more than necessary to meet the objectives of the simulation project. With these principles in mind, the simulation plan should include the following sections.

### 1.4.1 Objectives

The primary focus in the preparation of a simulation plan is to identify the question that the project team wants to answer by running a simulation experiment. Consider that the objectives of the simulation are different from those of the actual trial. The objectives of the trial are either to provide knowledge about (i.e., learn) or to confer clinical utility to (i.e., confirm knowledge about) a drug, whereas a simulation experiment may be based on prospective (e.g., to assess what might happen in a trial yet to be conducted) or retrospective analyses (e.g., to assess what happened in a trial that has been completed). It is therefore important that the objectives of the simulation experiment can be considered independently of the actual trial. Objectives should give a clear high-level view of what the experiment intends to achieve. In practice, most of the time a drug development team will want the simulation team to come up with a design that is sufficiently powered to show desirable outcomes. However, sometimes, they may want to design a special population study where some model parameters can be precisely and accurately estimated. In simulation, the "true" parameters of virtual subjects are known by the simulation team, making it possible to assess how precisely and accurately those parameters can be estimated from simulated data.

The specific aims should be sufficiently detailed to evaluate the "deliverables" of the simulation experiment and indicate the scope of the experiment. The specific aims may be to find the dosage regimen, number of subjects, sampling times, inclusion/exclusion criteria, and study period, to name a few potential examples.

### 1.4.2 Simulation Team

A simulation experiment should be undertaken as part of a team effort. In the simulation plan, the members of the simulation team should be listed. The mission of the members is to build models, execute simulations, interpret the simulation outcomes, choose designs, and apply the ensuing knowledge and understanding to a real clinical trial. To accomplish these goals, each member's expertise should be reflected appropriately. Preclinical scientists should be involved especially in early phase study designs. Clinicians are vital in the provision of information about the disease of interest, disease progression, prognosis, and current treatment modalities. Those with expertise in clinical trial execution are central to the whole process of assessment of the feasibility of a trial (design). Constraints on the study design are of similar importance, and in many circumstances, the marketing requirements may dictate which doses or dosing regimens are likely to be successful when the drug becomes registered. Pharmacometricians and statisticians are key in integrating this information in the form of various models and designs and analyzing simulated outcomes. The simulation team is reconvened to consider sufficiency of designs that result from the process.

### 1.4.3 Simulation Models

A simulation model can be categorized into three submodels: input-output (IO), covariate distribution, and execution models. These models are generally developed from previous data sets that may include preclinical data. However, the values of the model parameters (both structural and statistical elements) and the structure used in the simulation of a proposed trial may be different from those that were originally derived from the analysis of the previous data. This may be due to perceived differences between the design of previous trials and how the intended trial may be performed, e.g., patients versus healthy volunteers. In all cases, justification must be provided for the choice of any particular model and parameter estimates.

#### 1.4.3.1 Input-Output Model

Input-output (IO) models include pharamacokinetic/pharmacodynamic (PKPD) models and disease progress models. The structural and statistical models and their parameter estimates to be used in simulation should be specified. Building PKPD models has been discussed extensively in the literature [see Gabrielsson and Weiner (17) for a recent discussion]. In contrast, disease progress models are less abundant, but may be derived from the placebo group data; in some (rare) circumstances databases may be available that provide the natural course data of disease progress (18). The statistical models should include those levels of random effects that are pertinent to the drug in question; these will usually include between-subject variability and residual unexplained variability. See Chapter 2 for more detail.

#### 1.4.3.2 Covariate Distribution Models

Covariate distribution models describe the characteristics of the subjects that might affect the drug behavior in the body. The model may include demographic, physiologic, and pathophysiologic aspects that are both representative of the patient type that is likely to be enrolled in the actual trial and pertinent to the simulation model. The distributions may be assumed to be similar to those of previous trials or based on clinical experience. The relevant covariates such as body weight, height, frequency of concomitant drug use, and baseline measurements are identified in the model development process. In addition, correlation between covariates should be considered, where appropriate, in order to avoid generating unrealistic representations of virtual patients. The distribution function of covariates may be altered in the what-if scenarios of simulation. See Chapter 3 for more detail.

#### 1.4.3.3 Execution Models

Deviations from the protocol will occur inevitably during execution of a real clinical trial. Similarly, such deviations should be introduced into the virtual trial in order to make the virtual trial as representative of reality as possible. To simulate a clinical trial that accounts for trial protocol deviations, an execution model that describes these deviations must be developed. Typically, execution model(s) will include patient-specific effects such as dropouts and adherence of the patient to the dosing regimen, and investigator effects such as measurement error. Ideally, models need to be developed from prior data that describe these uncontrollable factors so that the probability of such occurrences can be mimicked for the virtual trial.

In some circumstances, execution models may not be able to be developed from prior data or extrapolated from similar drugs. In these circumstances, it is possible to either assign a probability of adherence and dropout rate, etc. based on prior experience or to design a trial that is relatively insensitive to likely protocol deviations. See Chapter 4 for more detail.

### 1.4.4 Model Evaluation Methods

The FDA has provided guidelines for the assessment of the performance of models (9). They state, "The choice of a validation approach depends on the objective of the analysis because the model is both unknown and complex, subject to multiplicity of unknown covariate effects and nonlinearity." This sentiment is well deserved and is the basis for discussion about the assumptions in model building.

However, the primary question is not whether the model is right or wrong, but whether deficiencies in the model have a noticeable effect on the substantive inferences (19). The goal, therefore, should be the attainment of a "reasonable" model. Defining a "reasonable" model is, however, the subject of much debate and a large body of literature has been published in this area. The detailed description on evaluation methods is discussed in Chapter 5.

### 1.4.5 Computation Methods for Model Building

The details about computational methods for simulation can be found in Chapter 6.

#### 1.4.5.1 Software

The software and details of versions and release that were used to develop the model and to execute simulations must be stated explicitly. This includes any language compilers that may be required. Where appropriate, inbuilt algorithms and options selected within the software should be identified. In addition, it is useful to record the operating system under which the software was run. The hardware platform is quantitatively less important unless a comparison of computational time is of importance.

#### 1.4.5.2 Random Number Generator

Random numbers used in computer simulations are always generated by means of a computation that produces a *pseudo-random* sequence of numbers that have most of the statistical characteristics of truly random numbers. That is, individually they do not appear to follow any pattern that could be used to guess the next value in the sequence, but collectively they tend to be distributed according to some mathematically defined function. Depending on the size of the models, number of subjects, and trial replications, care may be required in selecting software. In general, however, most modern mathematical software programs have excellent random number generators with a suitably long repetition period. In a simulation plan, it is recommended to include the size of random number repetition period. In addition, the seed number for the random number generation should be recorded so that the same sequence of random numbers can be regenerated for reproducibility of the simulation outcome.

#### 1.4.5.3 Differential Equation Solver

The choice of differential equation solver represents a trade-off between the desired accuracy and computational efficiency. Since solving a differential equation for each measure for each virtual subject for each virtual trial for each virtual design is a very numerically intensive process that is prone to errors, it should be emphasized that, where possible, the algebraic solution of the simulation model should be used in preference to differential equations. However, in some cases the luxury of a closed-form solution may not be available and the simulation model may need to be represented as a series of differential equations. In this scenario, the analyst should be aware of the limitations of the various numerical integrators.

#### 1.4.5.4 Replications

It is interesting that as trial designers we would almost never be satisfied with a single mathematical replication of our design, preferring in many cases to assess the outcomes of several hundred replications. Yet as practitioners we are prepared to make recommendations concerning populations of patients on the outcomes of a few trials undertaken as singletons. Hence, it is not surprising that actual trials can provide conflicting results. The contradictory trial outcomes may be due to subtle differences in design aspects, e.g., slight variations in inclusion and exclusion criteria, or simply because each trial is an independent realization of a distribution of outcomes and deviations might be caused by random fluctuations that have combined in a particular way to produce results that may seem quite unrepresentative of the expected outcome. In such cases, it is not uncommon for these "actual" trials to be considered as "replicates" from a distribution of possible outcomes and combined in a single analysis, termed a meta-analysis. This process is similar in spirit to replicating a virtual trial that has been constructed by the trial designer. The main difference being that the true result is known for each virtual trial and the trial designer can exclude sources of heterogeneity between replications, neither are possible for the meta-analysis.

### 1.4.6 Statistical Analysis of Simulation Results

The analysis of a series of replicates of a virtual trial will require the same statistical consideration as the analysis of a single occurrence of an actual trial (see Chapter 7 for details). There may be three levels of statistical analyses: analyses of each virtual trial, each design scenario, and each simulation experiment (i.e., if there are many design scenarios in each simulation experiment). In addition, since the expected outcome of the virtual trial is known, analysis of the virtual trial needs to not only determine the power of the trial but also consider the power of the statistical test to provide the appropriate answer. For example, does setting the alpha error to 0.05 for each replication actually result in 5% of all virtual trials inappropriately rejecting the null hypothesis?

### 1.4.7 Model-Building Assumptions

Typically, assumptions in modeling and simulation can be classified into mechanism-based and empirical assumptions. The need for assumptions is inherent in the predictive nature of simulation for designing trials that are yet to be undertaken. It is ironic that the very assumptions that are necessary to design a trial form the basis for skepticism about the usefulness of the CTS process. Fortunately, many of the assumptions inherent in the process may be avoided if well-defined and "mechanistic"-type models are chosen to form the basis of the simulation model. However, in many cases components of these models may be partly or even completely empirical, since a fully mechanistic model may be impractical to implement or sometimes even poorly understood due to its complexity. In these scenarios, it is incumbent on the pharmacometrician to justify the modeling approach and rigorously evaluate the assumptions in terms of model stability, reliability, and the sensitivity of the simulated trial.

In addition, the development of models that link IO model outputs to clinically relevant endpoints may ultimately be empirical and/or be of a probabilistic nature. This will ultimately lead to a divergence in the opinion of experts, who may well agree on the implementation of mechanistic-type models but are likely to disagree on the empirical assumptions. The simulation plan therefore provides the forum for all experts in the simulation team to discuss, debate, and finally (compromise to) agree on various assumptions. The virtual trial will then provide a testing ground for assessment of the assumptions inherent in its design, therefore enabling the simulation team to advance their knowledge about the drug and improve subsequent trial designs.

The inclusion of assumptions in the model-building process, e.g., choice of parameter values, structural, statistical and probabilistic models, leads naturally to the assessment of the influence of these assumptions on the final design; or the development of a final design that is robust to these assumptions. This is usually addressed formally by way of a sensitivity analysis where the sensitivity of the trial outcome to perturbations in parameter and model specification is gauged. See Chapter 8 for more detail.

### 1.4.8 Clinical Trial Design Scenarios

A simulation project can involve several scenarios with various combinations of trial design features. It follows that once a design has been simulated and the results analyzed, the trial designer may wish to consider other designs, but this raises the question: "What design should be considered next?" "How does one arrive at a sufficient design?" Despite the more formal approach afforded by the design of trials by simulation compared to empiricism, there remains no easily accessible method for determining the process of selecting a sufficient, or, preferably, the best design from the myriad of possible designs. In many cases, practitioners might rely on experience to lead the way, but this may be both a blessing and a hindrance. In the first instance, the experience of the practitioner is invaluable in setting up and evaluating designs and determining boundaries within which sufficient designs and (also) the best design is likely to lie. However, searching within the boundaries can be a daunting task even for the more experienced practitioner and is likely to lead to inefficient designs when compared with search techniques that are directed mathematically. See Chapter 9 for more detail.

### 1.4.9 Plan for Comparison of the Simulated Versus Actual Results

Once the results from the actual trial have become available, it is always good practice to return to the comparable virtual trial to evaluate how accurately the virtual trial described the actual trial. An exact prediction of the actual trial is not expected, and indeed would be extraordinarily unlikely. The goal is to determine whether the actual trial conferred outcomes that are encompassed within the distribution of possible outcomes generated by the many replications of the virtual trial. If the actual trial outcomes are not encompassed within reasonable bounds of the distribution of simulated outcomes, then this suggests that either the underlying models of the simulated trial were not representative of how the actual data arose, or that there may be some other processes involved that were not accounted for—such as some error in the manner in which the actual trial outcomes were analyzed. This retrospective comparison is reported in Chapters 15, 16, and 19) and in Refs. 5 and 21. From this comparative assessment step, the simulation team can learn the predictability of the simulation models and techniques in order to improve the process on subsequent simulation projects.

## 1.5 CONCLUSION

It is still too early to generalize about the acceptance of this technique by the pharmaceutical industry. This is in part due to the time required to develop an active modeling and simulation team, the complexity of building credible models for simulation of a trial, and the relatively limited number of trials that have been designed by simulation from which a test case can be made. However, given the soaring costs of drug development, it is likely that economic implications of failed trials, i.e., those that do not contribute to the registration documentation, will provide the necessary impetus for the industry to seriously consider more formal methods of design. In this light, design by simulation has a favorable profile due to its increasing use and its "apparent" simplicity when compared with other numerical methods, e.g., those based on design of experiments for optimizing designs. The success of modeling and simulation in designing clinical trials will eventually be decided in pharmacoeconomic evaluations.

Several groups have conducted design projects using modeling and simulation. These projects are often discussed at meetings such as East Coast Population Analysis group, Midwest Users for Population Analysis Discussion group, the Population Approach Group in Europe, and the Population Approach Group in Australia and New Zealand. In addition, less specialized meetings, such as the American Association of Pharmaceutical Scientists in 2000 and 2001, included a number of symposia on clinical trial simulation to optimize resources to meet unmet medical needs. These meetings indicate the current interest of the modeling and simulation technique to the decision-making process of selecting clinical trial designs.

Modeling has its roots firmly entrenched in the drug development process and simulation as a method for exploring models and potential experimental designs is a natural next step. The role of simulation, even though it is still in its infancy, has already shown significant benefits in drug development and has the capability to advance the understanding of drug actions and improve efficiency and rigor of trial designs.

## REFERENCES


1. NHG Holford, HC Kimko, JPR Monteleone, CC Peck. Simulation of clinical trials. Annual Review Pharmacol Therap 40:209–234, 2000.
2. M Hale, DW Gillespie, SK Gupta, NH Holford. Clinical trial simulation streamlining your drug development process. Appl Clin Trials 15:35–40, 1996.
3. C Peck, R Desjardins. Simulation of clinical trials—encouragement and cautions. Appl Clin Trials 5:30–32, 1996.
4. C Maxwell, JG Domenet, CRR Joyce. Instant experience in clinical trials: a novel aid to teaching by simulation. J Clin Pharmacol 11(5):323–331, 1971.
5. MD Hale. Using population pharmacokinetics for planning a randomized concentration-controlled trial with a binary response. In: L Aarons, LP Balant, M Danhof, eds. European Cooperation in the Field of Scientific and Technical Research, Geneva: Eur. Comm., 1997, pp 227–235.
6. MD Hale, AJ Nicholls, RES Bullingham RH Hene, A Hoitsman, JP Squifflet, et al. The pharmacokinetic-pharmacodynamic relationship for mycophenolate mofetil in renal transplantation. Clin Pharmacol Ther 64:672–683, 1998.
7. Center for Drug Development Science, Georgetown University, http://cdds.georgetown.edu.
8. Pharsight Corporation, http://www.pharsight.com.
9. Guidance for Industry on Population Pharmacokinetics, FDA, http://www.fda.gov/cder/guidance/1852fnl.pdf.
10. MK Al-Banna, AW Kelman, B Whiting. Experimental design and efficient parameter estimation in population pharmacokinetics. J Pharmacokinet Biopharm 18:347–360, 1990.
11. Y Hashimoto, LB Sheiner. Designs for population pharmacodynamics: value of pharmacokinetic data and population analysis. J Pharmacokinet Biopharm 19:333–353, 1991.
12. EI Ette, H Sun, TM Ludden. Design of population pharmacokinetic studies, Proc Am Stat Assoc (Biopharmaceutics Section) 487–492, 1994.
13. NE Johnson, JR Wade, MO Karlsson. Comparison of some practical sampling strategies for population pharmacokinetic studies. J Pharmacokinet Biopharm 24:245–272, 1996.
14. H Sun, EI Ette, TM Ludden. On the recording of sampling times and parameter estimation from repeated measures pharmacokinetic data. J Pharmacokinet Biopharm 24:635–648, 1996.

15. MO Karlsson, LB Sheiner. The importance of modeling interoccasion variability in population pharmacokinetic analyses. J Pharmacokinet Biopharm 21:735–750, 1993.
16. Points to consider on pharmacokinetics and pharmacodynamics in the development of antibacterial medicinal products, Committee for Proprietary Medicinal Products. http://www.emea.eu.int/pdfs/human/ewp/265599en.pdf.
17. J Gabrielsson, D Weiner. Pharmacokinetic and Pharmacodynamic Data Analysis: Concepts and Applications, 3rd ed. Apotekarsocieteten, 2001.
18. PLS Chan, NHG Holford. Drug treatment effects on disease progression. Ann Rev Pharmacol Toxicol 41:625–659, 2001.
19. A Gelman, JB Carlin, HS Stern, DB Rubin. Bayesian Data Analysis, CRC Press, 1995.
20. HC Kimko, P Ma, SB Duffull. Qualification of simulation models in designing clinical trials. Clin Pharmacokinet, 2002 (submitted).
21. HC Kimko, SS Reele, NH Holford, CC Peck. Prediction of the outcome of a phase 3 clinical trial of an antischizophrenic agent (quetiapine fumarate) by simulation with a population pharmacokinetic and pharmacodynamic model. Clin Pharmacol Ther 68(5):568–577, 2000.

# 2

# Input-Output Models

**Nicholas H. G. Holford**
University of Auckland, Auckland, New Zealand

## 2.1 CLINICAL TRIAL SIMULATION MODELS

Clinical trial simulation (CTS) depends fundamentally on a set of models to simulate observations that might arise in a clinical trial. Three distinct categories of model have been proposed (1):

- Covariate distribution
- Input-output
- Execution

They are presented in this sequence because the first decision that must be made when designing a clinical trial is what kind of subjects will be enrolled. The covariate distribution model defines the population of subjects in terms of their characteristics, such as weight, renal function, sex, etc. Next, the input-output model can be developed to predict the observations expected in each subject using that individual's characteristics defined by the covariate distribution model. Finally, deviations from the clinical trial protocol may arise during execution of the trial. These may be attributed to subject withdrawal, incomplete adherence to dosing, lost samples, etc. The execution model will modify the output of the input-output model to simulate these sources of variability in actual trial performance.

This chapter discusses the structure of input-output (IO) models. A single pharmacokinetic model is used to illustrate features of IO models, but it should be understood that IO models are quite general and the principles of IO models described below can be applied to any process which might describe the occurrence of an observation in a clinical trial.

## 2.2 SIMULATION AND ANALYSIS MODELS

It is a common aphorism that all models are wrong but some are useful (2). The usefulness of models for simulating observations that could arise in a clinical trial is directly dependent on the complexity of the model. In general, all the levels of the model hierarchy (Section 2.3.3) should be implemented for the purposes of clinical trial simulation in order to make the predicted observations as realistic as possible.

Analysis of clinical trial observations, however, can be useful with much less complexity. One of the purposes of clinical trial simulation is to evaluate alternative analysis models by applying them to simulated data that may arise from a much more complex but mechanistically plausible model. The following description of input-output models is oriented toward the development of models for simulation. Similar models could be used for analysis of actual data or simulated data, but this is usually not required to satisfy the objectives of many clinical trial simulation experiments, e.g., an analysis of variance may be all that is required to evaluate a simulated data set.

## 2.3 INPUT-OUTPUT MODEL

The input-output (IO) model is responsible for predicting the observations in each subject. The simplest IO models are nonstochastic; i.e., they do not include any random effects such as residual unexplained variability or between subject variability. More complex IO models may include one or both of these random effect components.

### 2.3.1 IO Model Anatomy

Equation (1) is a model for predicting the time course of concentration $C(t)$ using a one-compartment first-order elimination model with bolus input. The left-hand side of the equation $C(t)$ is the *dependent variable*. The symbol $t$ is usually the *independent variable* in the right-hand side of the equation. The symbols $V$ (volume of distribution) and CL (clearance) are constants that reflect drug disposition in an individual. The symbol dose is also a model constant. In contrast to $V$ and CL, the value of dose is under experimental control and is part of the design of a clinical trial. It is helpful to refer to such controllable experimental factors as *properties* to distinguish them from uncontrollable factors such as $V$ and CL that are usually understood as the *parameters* of the model. In a more general sense all constants of the model are parameters.

$$C(t) = \frac{\text{dose}}{V} \cdot \exp\left(-\frac{\text{CL}}{V} \cdot t\right) \tag{1}$$

### 2.3.2 IO Model Hierarchy

IO models can be ordered in a hierarchy that makes predictions about populations, groups, individuals and observations. Each level of model is dependent on its predecessor. The simplest IO model is at the population level and the most complex is at the level of an observation. It is the observation IO model prediction that is the foundation of clinical trial simulation.

### 2.3.3 Population IO Model

The predictions of IO models that do not account for either systematic or apparently random differences between individuals are referred to here as population IO models.*

#### 2.3.3.1 Population Parameter Model

Population models are based on parameter values that represent the population. They may have been estimated without consideration of covariates such as weight, etc. and simply reflect the characteristics of the obvserved population. These parameters can be considered naive population parameters (e.g., $V_{\text{pop}}$, $\text{CL}_{\text{pop}}$).

For the purposes of comparing population parameters obtained from different studies, population parameters need to be standardized to a common set of covariates (3), e.g., male, weight 70 kg, age 40 years, creatinine clearance 6 liters/h. Standardized population parameter values can be estimated using group IO models (see below) and should be distinguished from naïve population parameters. All examples shown below refer to standardized parameters, e.g., $V_{\text{std}}$, $\text{CL}_{\text{std}}$ in Eq. (2).

#### 2.3.3.2 Population IO Model Simulation

Equation (2) illustrates the use of population standardized parameters for population IO model simulation. A population IO model simulation based on this equation is shown in Figure 1.

* Others may use this term to encompass a model including what are defined below as group IO, individual IO, and observation IO models. However, it seems clearer to define the model based on the source of its parameters.

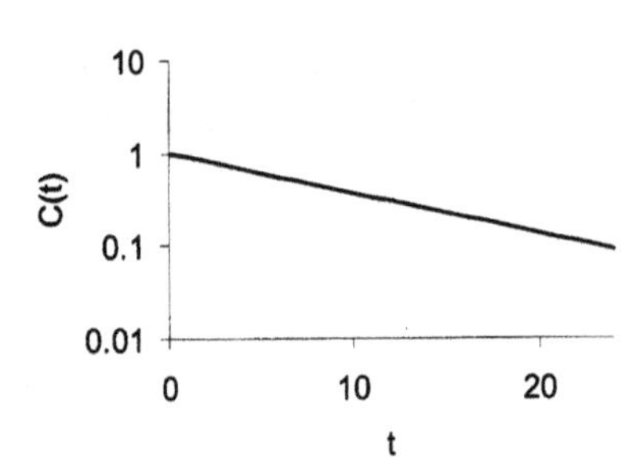


**FIGURE 1** Population IO simulation. Solid line is population IO model prediction.

$$C_{\text{pop}(t)} = \frac{\text{dose}}{V_{\text{std}}} \cdot \exp\left(-\frac{\text{CL}_{\text{std}}}{V_{\text{std}}} \cdot t\right) \tag{2}$$

### 2.3.4 Group IO Model

The group IO model is used to simulate nonstochastic variation in the model predictions. Statisticians refer to a model for this kind of variation as a fixed effects model. Note that "effects" has nothing to do with pharmacological drug effects. It is a statistical term referring to the source of variability—in this case a predictable or "fixed" source. The group IO model uses the same functional form as the population IO model but, instead of population parameters, group parameters are used.

#### 2.3.4.1 Group Parameter Model

If the covariate distribution model includes values that distinguish individuals by their physical characteristics, e.g., weight, then the model parameters may be predicted from that particular combination of covariate values. Equations (3) to (8) illustrate models that may be used to predict values of $V$ and CL in subjects with a particular weight or age. These predicted parameters are typical of individuals with that weight or age and are sometimes known as the typical value parameters but are more clearly identified as group parameters, $V_{\text{grp}}$ and $\text{CL}_{\text{grp}}$, because they are representative of a group with similar covariates. The group parameter model includes the population parameter and usually a constant that standardizes the population parameter ($\text{Wt}_{\text{std}}$, $\text{Age}_{\text{std}}$). These normalizing constants may reflect a central tendency for the covariate in the population, e.g., the median weight, or a standard value (3), e.g., 70 kg. Other parameters in the typical parameter model relating age to $V_{\text{grp}}$ and $\text{CL}_{\text{grp}}$ may be theoretical constants such as the exponents in allometric models [Eqs. (3), (4)], or may be empirical parameters, such as $F_{\text{age},V}$, $F_{\text{age},CL}$, of a linear model [Eqs. (5), (6)]. An exponential model may be a more robust empirical model than the linear model for many models because the prediction is always positive [Eqs. (7), (8)]. $K_{age,v}$ and $K_{age,CL}$ are parameters of the exponential model that are approximately the fractional change in the parameter per unit change in the covariate value.

$$V_{grp} = V_{std} \cdot \left(\frac{Wt}{Wt_{std}}\right)^{1} \tag{3}$$

$$Cl_{grp} = CL_{std} \cdot \left(\frac{Wt}{Wt_{std}}\right)^{3/4} \tag{4}$$

$$V_{grp} = V_{std} \cdot (1 + F_{age,v} \cdot (Age - Age_{std})) \tag{5}$$

$$CL_{grp} = CL_{std} \cdot (1 + F_{age,CL} \cdot (Age - Age_{std})) \tag{6}$$

$$V_{grp} = V_{std} \cdot \exp(K_{age,v} \cdot (Age - Age_{std})) \tag{7}$$

$$CL_{grp} = CL_{std} \cdot \exp(K_{age,CL} \cdot (Age - Age_{std})) \tag{8}$$

*a. Additive and Proportional Fixed Effects Models for Group Parameters* When there is more than one covariate influencing the value of a group parameter, the effects may be combined in a variety of ways. If there is no mechanistic guidance for how to combine the covariate effects (the usual case), there are two empirical approaches that are widely used.

*b. Additive* The additive model requires a parameter $S_{wt,V}$ to scale the weight function predictions and a second parameter $S_{age,V}$ similar in function to the parameter, $F_{age,v}$ [Eq. (5)], but scaled in the units of $V$ rather than as a dimensionless fraction. Equation (9) illustrates the additive model using weight and age fixed effect models.

$$V_{grp} = S_{wt,v} \cdot \left(\frac{Wt}{Wt_{std}}\right)^{1} + S_{age,v} \cdot (Age - Age_{std}) \tag{9}$$

*c. Multiplicative* The multiplicative model combines Eqs. (3) and (5) so that $V_{std}$ retains a meaning similar to that in the population model [Eq. (2)]; i.e., the group value of volume of distribution when weight equals $Wt_{std}$ and age equals $Age_{std}$ will be the same as the population standard value and similar to the naïve population value $V_{pop}$ obtained when weight and age are not explicitly considered. It is usually more convenient to use multiplicative form of the model because it can be readily extended when new covariates are introduced without having to change the other components of the model or their parameter values. Equation (10) illustrates the multiplicative model using weight and age fixed effect models.

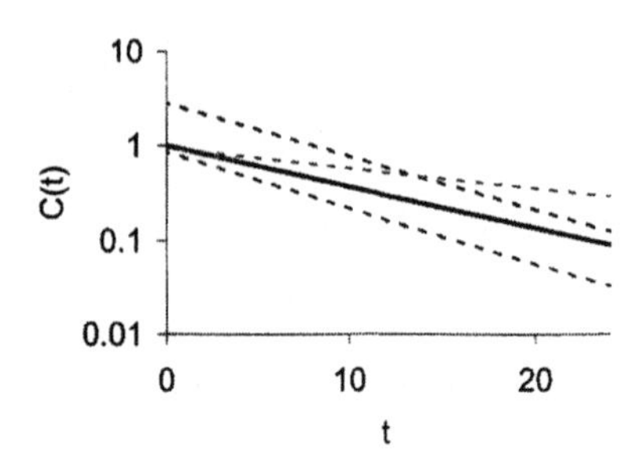


**FIGURE 2** Group IO simulation: systemic variability in two covariates (weight, age). Solid line is population IO model prediction. Dashed lines are group IO model predictions.

$$V_{\text{grp}} = V_{\text{std}} \cdot \left(\frac{\text{Wt}}{\text{Wt}_{\text{std}}}\right)^{1} \cdot (1 + F_{\text{age},v} \cdot (\text{Age} - \text{Age}_{\text{std}})) \tag{10}$$

#### 2.3.4.2 Group IO Model Simulation

Examples of group IO model simulations with systematic changes in both weight and age are shown in Figure 2. The group model [Eq. (11)] applies Eq. (10) for $V_{\text{grp}}$ and a similar expression for $\text{CL}_{\text{grp}}$ [based on Eqs. (4) and (8)].

$$C_{\text{grp}(t)} = \frac{\text{dose}}{V_{\text{grp}}} \cdot \exp\left(-\frac{\text{CL}_{\text{grp}}}{V_{\text{grp}}} \cdot t\right) \tag{11}$$

### 2.3.5 Individual IO Model

#### 2.3.5.1 Individual Parameter Model

Individual parameter values are simulated using a fixed effects model for the group parameter and a random effects model to account for stochastic variation in the group values. The random effects model samples a value $\eta_i$ (where the subscript $i$ refers to an individual) typically from a normal distribution with mean 0 and variability PPV (population parameter variability) [Eq. (12)].

$$\eta_i \sim N(0, \text{PPV}) \tag{12}$$

$\eta_i$ is then combined with the group parameter model to predict an individual value of the parameter, $\text{CL}_i$ [Eq. (13)].

$$\text{Cl}_i = \text{CL}_{\text{grp}} + \eta_i \tag{13}$$

The $\eta_i$ can come from a univariate or multivariate distribution. Multivariate distributions recognize the covariance between parameters and the importance of this is discussed in Section 2.5.2.

*a. Fixed Effect Models for Random Individual Parameters* There are two main sources of random variation in individual parameter values. The first is between subject variability (BSV) and the second is within subject variability (WSV) (4, 5). Within subject variability of an individual parameter may be estimated using a model involving an occasion variable as a covariate. The variability from occasion to occasion in a parameter is known as between occasion variability (BOV). BOV is an identifiable component WSV that relies on observing an individual on different occasions during which the parameter of interest can be estimated. Variability within an occasion, e.g., a dosing interval, is much harder to characterize, so from a practical viewpoint WSV is simulated using BOV. Other covariates may be used to distinguish fixed effect differences; e.g., WSV may be larger in the elderly compared with younger adults.

The total variability from both these sources may be predicted by adding the $\eta$ values from each source [Eq. (22)]. Representative values of BSV and WSV for clearance are 0.3 and 0.25, respectively (6).

$$\eta BSV_i \sim N(0, BSV) \tag{14}$$

$$\eta WSV_i \sim N(0, WSV) \tag{15}$$

$$\eta PPV_i = \eta BSV_i + \eta WSV_i \tag{16}$$

*b. Additive and Proportional Random Effects Models for Individual Parameters* Both additive [Eq. (13)] and proportional [Eq. (17)] models may be used with $\eta_i$. The proportional model is used more commonly because PPV approximates the coefficient of variation of the distribution of $\eta$. Because estimates of PPV are difficult to obtain precisely, it is often convenient to use a value based on an approximate coefficient of variation, e.g., a representative PPV might be 0.5 for clearance (approximately 50% CV).

$$CL_i = CL_{grp} \cdot \exp(\eta_i) \tag{17}$$

### 2.3.5.2 Individual IO Model Simulation

An example of individual IO model simulation is shown in Figure 3 based on Eq. (18). The figure illustrates the changes in concentration profile that might be expected using random variability from a covariate distribution model for weight and age (PPV = 0.3) and a parameter distribution model for $V$ and CL (PPV = 0.5) (Table 1).

$$C_i(t) = \frac{\text{dose}}{V_i} \cdot \exp\left(\frac{CL_i}{V_i} \cdot t\right) \tag{18}$$

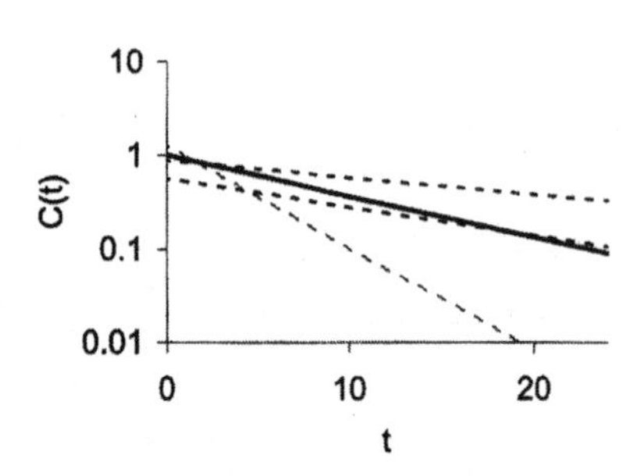


**FIGURE 3** Individual IO simulation: random variability in covariates (weight, age) and group parameters (*V*, CL). Solid line is population IO model prediction. Dashed lines are individual IO model predictions.

**TABLE 1** Simulation Model Parameters[a]

| Model | Level | Name | Value | Description |
|---|---|---|---|---|
| Covariate distribution | Population | $WT_{std}$ | 70 kg | Standard weight |
| | | $AGE_{std}$ | 40 y | Standard age |
| | Individual | $PPV_{wt}$ | 0.3 | Population parameter variability for weight |
| | | $PPV_{age}$ | 0.3 | Population parameter variability for age |
| Input Output | Population | Dose | 100 mg | Dose |
| | | $V_{std}$ | 100 L | Volume of distribution |
| | | $CL_{std}$ | 10 L/h | Clearance |
| | Group | $K_{age,V}$ | $0.01\ h^{-1}$ | Age and volume of distribution factor |
| | | $K_{age,CL}$ | $-0.01\ h^{-1}$ | Age and clearance factor |
| | Individual | $PPV_V$ | 0.5 | Population parameter variability for volume |
| | | $PPV_{CL}$ | 0.5 | Population parameter variability for clearance |
| | Observation | $RUV_{SD}$ | 0.05 mg/L | Residual unexplained variability, additive |
| | | $RUV_{CV}$ | 0.2 | Residual unexplained variability, proportional |
| Execution | Observation | LLQ | 0.05 mg/L | Lower limit of quantitation |

[a] Simulations illustrated in this chapter were performed using Microsoft Excel.

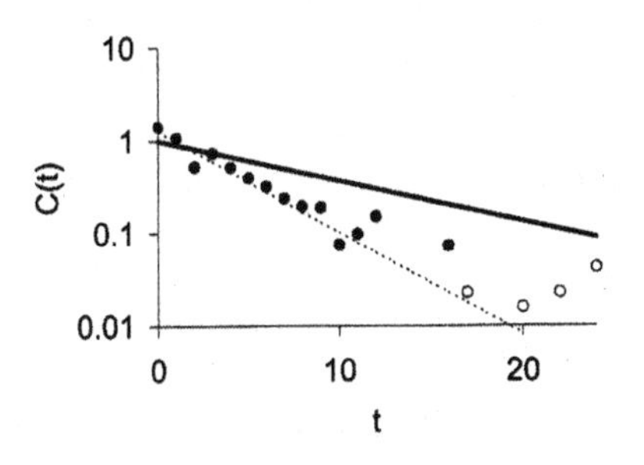


FIGURE 4 Observation IO simulation: random variability in covariates (weight, age), group parameters ($V$, CL), and residual unexplained variability (additive, proportional). Solid line is population IO model prediction. Dotted line is individual IO model prediction. Symbols are observation IO model predictions. Filled symbols are execution model predictions which will be used for data analysis.

### 2.3.6 Observation IO Model

The final level of the IO model hierarchy is used to predict observations. Observation values are simulated using individual IO model predictions and a random effects model to account for stochastic variation in the observation values.

#### 2.3.6.1 Observation Parameter Model

The random effects model samples a value $\varepsilon_j$ (the subscript $j$ is enumerated across all individuals and observations) typically from a normal distribution with mean 0 and variability RUV (random unidentified variability) [Eq. (19)].

$$\varepsilon_j \sim N(0, \mathrm{RUV}) \tag{19}$$

$\varepsilon_j$ is combined with the individual IO model to predict the observation. Common models include additive (Equation 20), proportional (Equation 21) and combined (Equation 22). The combined model most closely resembles the usual residual variability when pharmacokinetic models are used to describe concentration measurements.

a. *Additive*

$$C_{i,j}(t) = C_i(t) + \varepsilon_{\mathrm{add},j} \tag{20}$$

b. *Proportional*

$$C_{i,j}(t) = C_i(t) \cdot \exp(\varepsilon_{\mathrm{prop},j}) \tag{21}$$

c. *Combined*

$$C_{i,j}(t) = C_i(t) \cdot \exp(\varepsilon_{\mathrm{prop},j}) + \varepsilon_{\mathrm{add},j} \tag{22}$$

#### 2.3.6.2 Observation IO Model Simulation

An example of an observation IO model simulation is shown in Figure 4. Random variability in the observations was generated using a mixed additive ($RUV_{sd}$ = 0.05 mg/l) and proportional ($RUV_{cv}$ = 0.2) residual variability model.

Simulated observations less than the lower limit of quantitation (0.05 mg/l) are shown as open symbols in Figure 4. These observations would not be included in the analysis of this simulation. The removal of observations in this manner is an example of the application of an executive model. The IO model predicts the observation, but the execution model reflects local policy for removal of observations that are classified as unquantifiable.

## 2.4 SENSITIVITY ANALYSIS

A clinical trial simulation experiment should include an evaluation of how the conclusions of the simulation experiment vary with assumptions made about the models and their parameters (see Chapter 8 for more details). The nature of this sensitivity analysis will depend on the objectives of the simulation. If the objective is to determine the power of a confirming type trial, then the sensitivity of the predicted power of a trial design should be examined. Repeating the simulations with a different model, e.g., a linear instead of an Emax pharmacodynamic model, may do this. One may also examine the influence of the model parameters, e.g., changing the EC50 of an Emax model. The extent to which the power of the trial varies under these different scenarios of models and parameters is a key focus of a sensitivity analysis.

## 2.5 PARAMETERS

### 2.5.1 Source

There are three sources of model parameters for clinical trial simulation.

#### 2.5.1.1 Theory

Theoretical values are usually not controversial. However, although there is both empirical and theoretical support, there is still no widespread acceptance of the allometric exponent values for clearance and volume of distribution that are suggested by the work of West et al. (7, 8).

#### 2.5.1.2 Estimates from Data

The most common source will be estimates from prior analysis of data. Inevitably, it will be necessary to assume that parameter estimates obtained in a different population are suitable for the proposed clinical trial that is being simulated (see Chapter 5). It is particularly valuable to have standard, rather than naïve, population parameter estimates so that they can be coupled with a covariate distribution model in order to extrapolate to a population that has not yet been studied.

#### 2.5.1.3 Informed Guesses

Informed guesses are always a necessary part of a clinical trial simulation. For example, the size of a treatment effect will have to be assumed and the model performance modified by suitable adjustment of dosing and parameters in order to mimic an outcome of the expected magnitude.

### 2.5.2 Covariance

It is important to retain information about the covariance of individual IO model parameters in order to obtain plausible sets of parameters. While some covariance between parameters may be included in the simulation via the group IO model, e.g., if weight is used to predict $V_{grp}$ and $CL_{grp}$, there is usually further random covariance which cannot be explained by a model using a covariate such as weight to predict the group parameter value.

The need to include parameter covariance in the model is especially important for simulation. It can often be ignored when models are applied to estimate parameters for descriptive purposes, but if it exists and it is not included in a simulation then the simulated observations may have properties very different from the underlying reality. For example, if clearance and volume are highly correlated, then the variability of half-life will be much smaller than if the clearance and volume were independent.

The methods for obtaining samples of parameters from multivariate distributions are the same as those used for obtaining covariates (see Chapter 3). They may be drawn from parametric distributions, e.g., normal or log normal, or from an empirical distribution if there is a sufficiently large prior population with adequate parameter estimates.

### 2.5.3 Posterior Distribution of Parameters

It is worth remembering that point estimates of parameters will have some associated uncertainty. It is possible to incorporate this uncertainty by using samples from the posterior distribution of the model parameter estimates rather than the point estimate. For instance, if clearance has been estimated and a standard error of the estimate is known, then the population clearance used to predict the group clearance could be sampled from a distribution using the point estimate and its standard error.

### 2.5.4 Parameterization

The choice of parameterization of a model is often a matter of convenience. A one-compartment disposition model with bolus input may be described using Eq. (23) or (24). The predictions of these models, with appropriate parameters, will be identical.

$$C(t) = \frac{\text{dose}}{V} \cdot \exp\left(\frac{\text{CL}}{V} \cdot t\right) \tag{23}$$

$$C(t) = A \cdot \exp(-\alpha \cdot t) \tag{24}$$

The apparent simplicity of Eq. (24) may be appealing, but it hides important features when applied to clinical trial simulation. An explicit value for the dose is not visible and doses are essential for clinical trials of drugs. The rate constant $\alpha$ appears to be independent of the parameter $A$, but when it is understood that both $A$ and $\alpha$ are functions of volume of distribution it is clear that this population level interpretation of independence is mistaken. Finally, because clearance and volume may vary differently as a function of some covariate such as weight [see Eqs. (3) and (4)], the value of $\alpha$ will vary differently at the group and individual level from the way that $A$ differs. If the model parameterization corresponds as closely as practical to biological structure and function, then the interaction between different components of the model is more likely to resemble reality.

## 2.6 Conclusion

The input-output model brings together the warp of scientific knowledge and weaves it with the weft of scientific ignorance. The art of combining signal with noise is the key to successfully simulating the outcome of a clinical trial and to honestly appreciating that the future cannot be fully predicted.

## REFERENCES

1. NHG Holford, M Hale, HC Ko, J-L Steimer, LB Sheiner, CC Peck. Simulation in drug development: good practices, 1999. http://cdds.georgetown.edu/sddgp723.html
2. GEP Box. Robustness in the strategy of scientific model building. In: RL Launer, GN Wilkinson, eds. Robustness in Statistics. New York: Academic Press, 1979, p 202.
3. NHG Holford. A size standard for pharmacokinetics. Clin Pharmacokinet 30:329–332, 1996.
4. MO Karlsson, LB Sheiner. The importance of modeling interoccasion variability in population pharmacokinetic analyses. J Pharmacokinet Biopharm 21(6):735–750, 1993.

5. NHG Holford. Target concentration intervention: beyond Y2K. British J Clin Pharmacol 48:9–13, 1999.
6. NHG Holford. Concentration controlled therapy. In: A Breckenridge, ed. Esteve Foundation Workshop. Amsterdam: Elsevier Science, 2001.
7. GB West, JH Brown, BJ Enquist. A general model for the origin of allometric scaling laws in biology. Science 276:122–126, 1997.
8. GB West, JH Brown, BJ Enquist. The fourth dimension of life: fractal geometry and allometric scaling of organisms. Science 284(5420):1677–1679, 1999.

# 3

# Defining Covariate Distribution Models for Clinical Trial Simulation

**Diane R. Mould***
Georgetown University, Washington, D.C., U.S.A.

## 3.1 BACKGROUND

### 3.1.1 Clinical Trial Simulation

In modeling and simulation analysis, it is important to distinguish between a system's structure or the inner constitution, and its behavior, which is the observable manifestation of that system. External behavior can also be described in terms of the relationship that the system imposes between its input and output time histories. The internal structure of the system defines the system states and transition mechanisms, and maps its states to output transitions. Knowing the system structure allows the modeler to analyze and simulate its behavior under specific input conditions. More commonly, however, the system structure is not known, and therefore a valid representation of the structure must be inferred from its behavior. These inferences are the basis of the assumptions that define the representative system.

Consequently, simulation analyses are conducted on the basis of numerous assumptions (1), which include not only the choice of structural and stochastic models used to describe the pharmacokinetics and pharmacodynamics of a drug, but also models that deal with study conduct including compliance and dropout,

* *Current affiliation*: Projections Research Inc., Phoenixville, Pennsylvania, U.S.A.

and those that describe patient characteristics such as demographics, co-morbid diseases, and concomitant medications. Because these assumptions are critical to the results obtained from a simulation analysis, it is important to understand the consequences of each and to describe them based upon the underlying information associated with them. Therefore, the assumptions made for any simulation work should be described in the simulation plan, as well as in any documentation describing the results of that simulation analysis. To date, an adequate description of the models and assumptions used to generate the database of virtual patients has not typically been included in published works.

Assumptions can be broadly classified in one of three categories: those that are based on data and experimentation; those that are not directly supported by observations, but are theoretically justified; and assumptions that are conjectural, but necessary for the simulation. The risks inherent with the use of an assumption generally increase as assumptions become more conjectural. Therefore, in order to conduct a simulation analysis that mimics reality, all assumptions should be based on observed data to the greatest extent possible. Wherever possible, input from experts should be solicited and used.

The fact that bias in patient selection may influence the results of clinical studies was one of the major reasons for the development of randomized controlled clinical trials in medical research. However, bias can also be important at other stages of a trial, especially during analysis. Withdrawing patients from consideration because of ineligibility based on study entry criteria, protocol compliance, or because of associated poor quality (i.e., missing) data may be a source of systematic error (2). Consequently, these factors must be considered and described in any trial simulation experiment.

An important concept of simulation analysis is the construction of the model system. Model systems are typically constructed through decomposition, based on how the system can be broken down into distinct component parts. These components are then coupled together (via composition) to form a larger system. The ability to create larger and larger systems is referred to as hierarchical construction. In some model systems, the components are coupled together using specific inputs and outputs. This is referred to as a modular construction. System construction is particularly relevant to clinical trial simulation, which is generally a complex, modular system.

A clinical trial simulation model typically consists of three components (1): a covariate distribution model, an input-output (IO) model, and a clinical trial execution model. These submodels and their interrelationships are depicted graphically in Figure 1. The IO model describes information about both the clinical indication being investigated and drug that is being tested. It generally consists of a series of structural models describing the pharmacokinetics and pharmacodynamics of the drugs being tested in the clinical trial, and perhaps a pharmacoeconomic model. The covariate distribution model describes patient-specific

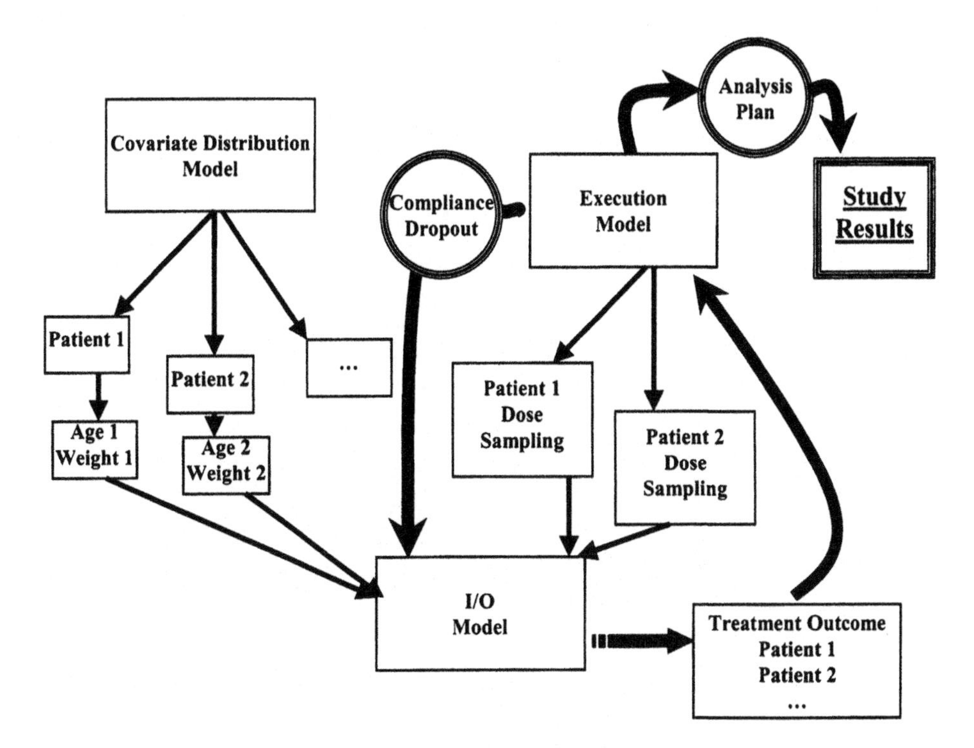


**FIGURE 1** The covariate distribution model, IO model, and execution model and their interrelationships for a clinical trial simulation.

aspects, such as the vector of patient demographics and characteristics (covariates). It may also include aspects of compliance, if the patterns of compliance are thought to be associated with specific patient characteristics. This information vector is usually associated with the systematic differences between patients in terms of observed pharmacokinetics and pharmacodynamics, and is one source of variability in individual parameter values (Figure 2). Covariate information is used, in turn, to predict IO model parameters for a virtual patient with a particular combination of demographic and patient characteristics. Lastly, pharmacoeconomic models may be linked to the clinical trial simulation.

For the purposes of simulating the pharmacokinetic and pharmacodynamic behavior of a compound, this usually involves a rather straightforward application of IO models that have been previously constructed in the course of drug development. If the models used in the simulation analysis were developed using population-based methods, then it is likely that these models include terms for covariate effects such as demographic or patient characteristics, concomitant

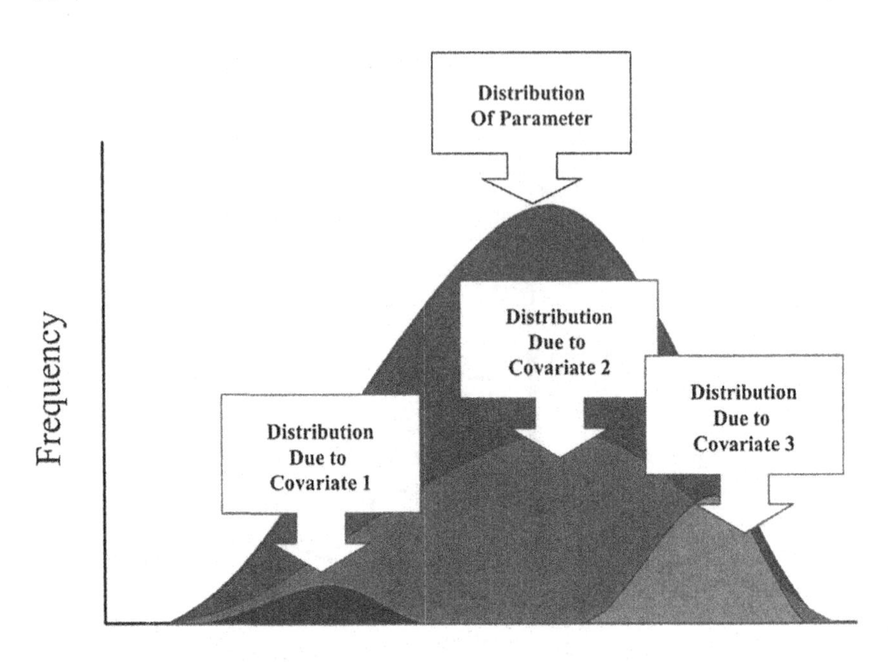


**FIGURE 2** The variability of a parameter value is often influenced by covariate effects. In this figure, the overall distribution of parameter values is shown in dark gray. The distribution of parameter values for several covariates is shown in the foreground. It should be noted that if the total parameter distribution includes random variability, then the area of the total distribution should be larger than the sum of the areas of the covariate specific distributions.

medications, and compliance patterns. Consequently, simulation of the matrix of covariates for the virtual patient population in the simulation trial assumes a greater importance, since the combinations of covariates, or the covariate vector that is associated with each virtual patient becomes important in determining the outcome of the simulated study. The need to be able to realistically re-create patient vectors becomes even more important when relevant patient demographics or characteristics are expected to change over the course of the study, e.g., renal function and age.

A model system is referred to as being closed under composition if the structure and behavior of the model system can be exactly expressed in terms of the original system's theoretical performance (3). This would be referred to as a fully decomposed system, which generally has a high level of specification within the system. Closure under composition guarantees that the model will exhibit well-defined behavior. However, as mentioned previously, the structure of most systems being modeled is based on inference and therefore the component model systems are not always completely specified. Therefore, it is usually necessary to test the behavior of the model system under a series of different inputs in order to evaluate its behavior. Furthermore, given the fact that highly specified systems generally perform slowly, it is often beneficial to make simplifying assumptions to enhance performance. In this aspect, the analyst must make an informed decision to balance model accuracy with speed, and to understand the inherent limitations imposed on the model system when simplifying assumptions are made. This process is referred to as model qualification, and is an important part of modeling and simulation analysis.

The aim of this chapter is to describe the different aspects and requirements of the covariate distribution models that are used to develop covariate vectors for virtual patients in a simulated trial. Several methods that may be useful for generating the covariate vectors will be covered. Model qualification is also described briefly.

### 3.1.2 Definition of Terminology

Before entering into a discussion on the available methods for creating a virtual patient population, it is imperative to define terminology that will be used in this chapter.

- Case deletion: Removal of subjects with missing covariate values.
- Covariates: Attributes such as demographic or patient characteristics that have been found to be influential for IO model parameters.
- Exchangeable quantities: In nonparametric resampling, an exchangeable quantity is a vector that is independent and can be employed as a basis for sampling. For instance, disease severity or age status (i.e., elderly versus young), which are covariates that are often critical for enrollment, may be a basis for patient selection in resampling algorithms.
- Hierarchical random effects model: Equivalent to population-based IO models, these models are comprised of a series of individual level models, which describe the behavior of each patient in the trial. The population model is composed of the individual models with terms describing the parameter variability seen across the individual model parameters.
- Imputation: Filling in missing data with plausible values. The plausible values used for missing values can be based on a model or have a theoretical basis.
- Joint function or joint distribution: See *Multivariate distribution.*
- Multivariate distribution: A distribution consisting of several correlated

components. Multivariate distributions are typically based on normal distributions, with the correlations being specified. For instance, creatinine clearance and age may be considered to be correlated. The range of possible values of creatinine clearance for a patient will be determined by age based on the underlying correlation and the characteristics of the distributions for each covariate.

- Nonstochastic simulation: A simulation that does not include effects of random variability.
- Patient covariate vector: The patient demographics and characteristics associated with a particular patient.
- Patient characteristics, e.g., smoking status, creatinine clearance.
- Patient demographics, e.g., age, weight, sex.
- Sensitivity analysis: A simulation study that examines the consequences of particular assumptions, such as the magnitude of drug effect, on the anticipated outcome of a trial.
- Simulation team: A group of experts, including a biometrician, modeler, clinical and pharmacological experts that have the necessary information to select assumptions used in a clinical trial simulation.
- Stochastic simulation. A simulation that includes random variability.
- Virtual patient: A simulated patient.

### 3.1.3 Software

The choice of methodology used for generating a virtual patient population is dependent to some extent on the software used for the simulation. At present, there are several software packages available for simulation of clinical trials, with the predominant packages being NONMEM (4) and the Trial Simulator (5). Other packages and proprietary software are also available for use in trial simulation. Both NONMEM and the Trial Simulator provide a means of either using a table of constructed patient vectors of covariates, or simulating virtual patient populations. Both packages can also be used for stochastic and nonstochastic simulation. In addition to these, it should be noted that many other statistical and mathematical software could be used to perform clinical trial simulation.

## 3.2 OVERVIEW OF AVAILABLE METHODS

As a consequence of the growing popularity of population-based modeling, sponsors involved in drug development have a new awareness of the importance of patient covariates on pharmacokinetics and pharmacodynamics. The use of nonlinear hierarchical random effects models is becoming a standard analysis method for pharmacokinetic and pharmacodynamic data (6). Hierarchical random effects models treat the parameters of the individual-level models as being random, with the distributions of individual parameters being affected by the influence of covariates (Figure 2), or as a consequence of study conduct. In other words, the distributions of the individual parameters are predicted by the covariates, e.g., mean and (usually) variance. The relevance of these distributions is that IO models used for simulation studies must describe the variability between patients as a function of covariates provided by the covariate distribution model, as well as account for within patient variability over the course of the study (Figure 3). In addition to the systematic differences between patients due to covariate influences, IO models also include terms for random differences between patients as well.

Covariate distribution models define the distribution and interrelationships of covariates in the population to be studied in the trial, e.g., ranges of age, or the proportion of patients with a specific stage of disease expected to enroll in

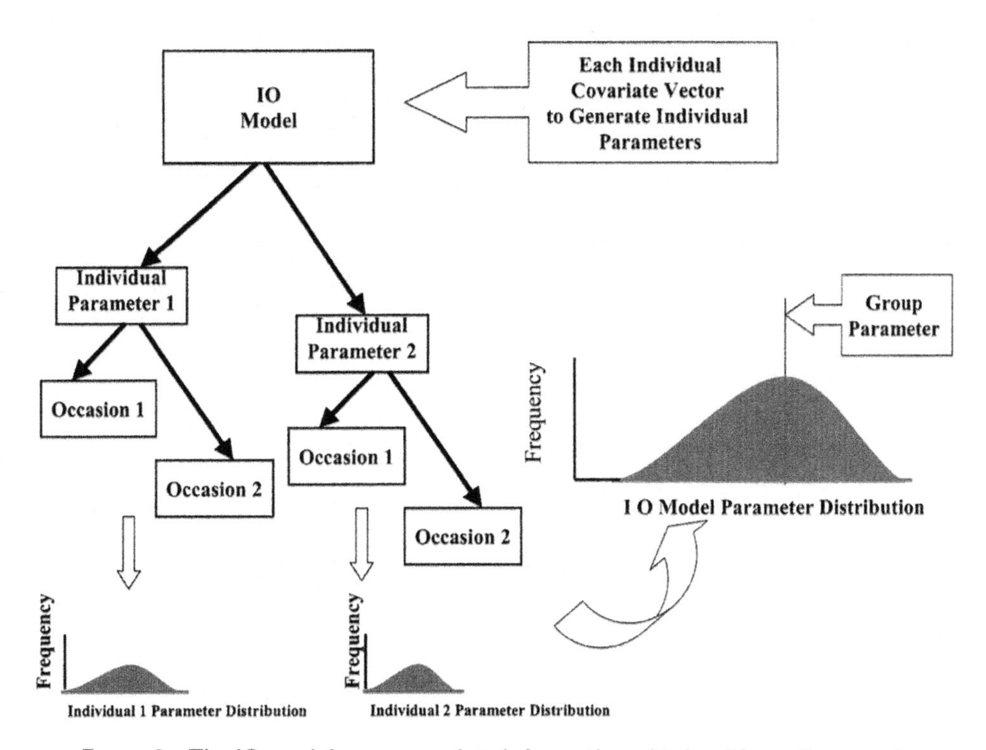


**FIGURE 3** The IO model uses covariate information obtained from the covariate distribution model and accounts for between subject variability and within subject variablity, thus generating a realistic distribution of parameter values for each individual model. These distributions, in turn, describe the overall variability in the population model.

a study. Because of the influence of the covariates on the IO model parameters, it is important to be able to simulate patients, with associated covariates that accurately reflect those patients expected to enroll in the study. There are several methods used to generate the vectors of covariates associated with virtual subjects. These methods include the following: use of specific patient inclusion and exclusion criteria, resampling techniques where patient vectors obtained from similar trials are resampled to create a new virtual patient population, use of univariate distributions, use of conjectural joint (multivariate) distribution models based on the expected distributions of covariates, and use of a joint function which describes the interrelationships between covariates based on observed data. Simulation study designs can make use of one or more of these approaches to ensure that the virtual patients describe a realistic distribution of patients likely to be enrolled in the trial being simulated.

### 3.2.1 Patient Inclusion and Exclusion Criteria

Defining a series of patient inclusion and exclusion criteria is a necessary step prior to beginning a clinical trial simulation. In a clinical trial, inclusion and exclusion criteria define, on a broad level, individual patient demographics and characteristics that must be met before being enrolled in a study. For a simulation analysis, inclusion and exclusion criteria serve as a means for removing virtual patients whose associated demographics or characteristics do not match what is expected in the actual study. Therefore, such criteria serve to restrict variability in the simulated trial to the level expected in the actual trial. Regardless of other methods employed to generate a virtual patient population, inclusion and exclusion criteria should always be used as the final determination of the acceptability of any virtual patient.

If a protocol has been developed prior to the initiation of the simulation work, then the inclusion and exclusion criteria used in the simulation analysis should match those defined in the protocol. However, a simulation study may be conducted prior to protocol development. In these situations, the clinical specialist member of the simulation team must define inclusion and exclusion criteria for the simulation study that will reflect the criteria used in the actual trial.

### 3.2.2 Resampling Methods

Resampling, or nonparametric bootstrapping, is one of the most straightforward methods available for constructing a virtual patient population (7, 8). Using this approach, a data set consisting of patient data (i.e., vectors of demographic information, patient characteristics, co-morbid disease and concomitant medication information) can be created by randomly selecting patient vectors for inclusion into the virtual patient population. Such a data set can be constructed using compiled data obtained from other studies in the same or related patient populations.

In general, resampling is done with replacement of individuals, so that the new data sets will have different distributions of covariates from the original data set.

In clinical trials, there are usually several prognostic factors known or believed to influence the patient's ability to respond to treatment. Therefore, the clinical trial must be designed so that treatment balance is simultaneously achieved across all such patient factors (9). In generating a virtual patient database using resampling, the criteria for selection of patients with such prognostic factors, such as a particular disease severity, may be governed either by chance or by expectations of enrollment in the actual trial. For example, an early phase study in patients with mild to moderate disease might be expected to enroll a greater proportion of patients with moderate disease than mild disease. In this case, resampling would be conducted using patient and prognostic factor information as the exchangeable quantities, with the proportion of patients in each class of prognostic factor being held to a predefined level. Prior to using a covariate as an exchangeable quantity, however, it is important to determine that this assumption is valid. The exchangeable values do not need to be independent, but their joint density under the null hypothesis must be invariant under permutation, which means that if you split the data set based on one particular parameter such as a prognostic factor, the other distributions remain unaffected; e.g., the age distribution remains appropriately partitioned between groups such that there is no evidence for a propensity of older patients in one group.

In general, many covariates are correlated, and these correlations need to be maintained during the creation of virtual patients. For example, in pediatric patient studies, age and weight are generally related, and both may be relevant to the pharmacokinetics of the agent being simulated (10). One of the issues in resampling is the identification of exchangeable quantities from which to resample. Covariates obtained from different individuals are generally assumed to be independent, whereas those obtained from the same individual are not independent. Therefore, the vector of correlated covariates ($\mathbf{X} = X_1, X_2, \ldots, X_n$; Figure 4) must be maintained for each individual in the resampling algorithm. This approach then generally involves resampling of individuals within specified groups, thereby retaining blocks of observations and preserving correlations, but providing a set of virtual patients having the overall characteristics expected in the actual trial.

In some cases, the data set used for the generation of the virtual patient population may be incomplete due to missing data or censoring. In this instance, missing data can be dealt with in a number of ways, including case deletion (removal of subjects with missing covariate values), imputation (filling in missing data with plausible values, semiparametric resampling). Case deletion is generally considered to be inferior to the imputation methods because of the inherent loss of data. Imputation can be limited to substituting a mean value for the missing data or it can involve defining a joint function for the covariates and generating

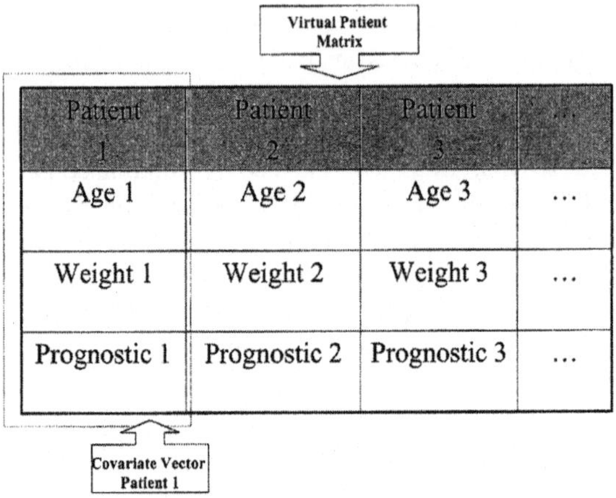

Virtual Patient Matrix

| Patient 1 | Patient 2 | Patient 3 | ... |
|---|---|---|---|
| Age 1 | Age 2 | Age 3 | ... |
| Weight 1 | Weight 2 | Weight 3 | ... |
| Prognostic 1 | Prognostic 2 | Prognostic 3 | ... |

Covariate Vector Patient 1

**FIGURE 4** Simulated patients with their associated covariates. The associated covariates are referred to as a covariate vector, with the series of patient vectors comprising the patient matrix.

a reasonable value for replacement of missing data based on that function. The application of joint functions is discussed in the next section.

#### 3.2.2.1 Advantages and Disadvantages of Resampling

Resampling offers a very straightforward method for creating a virtual patient population with associated covariate vectors that have a good probability of being realistic. However, because the virtual patient population is drawn from a pool of existing patient information, the patient matrix will not include patients with novel combinations of covariates. Thus, if the database from which patient covariate vectors are drawn is restricted, the virtual patient database will be similarly restricted and may not reflect the wider variety of patients that could be expected to enroll in an upcoming trial. Therefore, this approach is usually best when applied to a relatively large historical database of patient information.

### 3.2.3 Creating Multivariate Distributions of Covariates

Simulation of virtual patients with entirely new covariants provides an alternative to resampling strategies. Therefore, if no data set containing appropriate patient demographic information exists, or if the simulation team decides not to use a resampling approach to generate virtual patients, then a virtual patient data set must be created entirely through simulation. There are two methods for generating covariate vectors: through a series of univariate distributions or by using a multivariate joint function. Both methods are examples of parametric resampling approaches.

#### 3.2.3.1 Univariate Distributions

Constructing covariate vectors using a series of univariate distributions is typically accomplished using a normal distribution that empirically describes the distribution that the analyst wishes to create. Normal distributions are commonly employed for this, since normal random variates can be transformed directly into random variates from other distributions, such as the lognormal distribution (11). Given that a covariate $X$ is assumed to be described by a normal distribution N(0, 1) with a mean value $\mu$ and standard deviation $\sigma$, then the simulated covariate $X'$ can be sampled using the same distribution:

$$X' \approx N(\mu, \sigma) \qquad \text{if } X' = \mu + \sigma X$$

One of the early methods used to generate normally distributed variables was that of Box and Muller (12). This method requires the generation of two uniformly distributed random numbers $U_1$ and $U_2$ with independent identical distributions (IID) $U(0, 1)$. Covariates $X_1'$ and $X_2'$ are simulated using

$$X_1' = \left[\sqrt{-2\ln(U_1)}\right]\cos(2\pi U_2)$$

$$X_2' = \left[\sqrt{-2\ln(U_1)}\right]\sin(2\pi U_2)$$

The method of Box and Muller was improved by Marsaglia and Bray (13) to remove the requirement for the evaluation of trigonometric functions using the following algorithm:

$$V_i = 2U_i - 1 \qquad i = 1, 2$$
$$W = V_1^2 + V_2^2$$

If $W > 1$ then $V_i$ must be regenerated, if $W \leq 1$ then

$$Y = \sqrt{\frac{-2\ln W}{W}} \qquad X_1' = V_1 Y \qquad X_2' = V_2 Y$$

Distributions of covariates may also be empirically described using a log normal distribution. This can be accomplished by generating $Y \approx N(\mu, \sigma^2)$ and returning a simulated covariate of $X' = e^Y$ using the approach described previously. However, it should be noted that $\mu$ and $\sigma^2$ are not the mean and variance of the LN($\mu$, $\sigma^2$) distribution.

If $X' \approx \text{LN}(\mu, \sigma^2)$, $\mu_1 = E(X')$, and $\sigma_1^2 = \text{VAR}X')$,

$\mu_1 = e^{\mu+\sigma 2/2}$ and $\sigma_1^2 = e^{\mu+\sigma 2}\,(e^{\sigma 2} - 1)$

If a log normal distribution with a specific mean $\mu_1$ and variance $\sigma_1^2$ is required in the simulation, then $\mu$ and $\sigma^2$ must first be solved for in terms of $\mu_1$ and $\sigma_1^2$. Application of the following formulas gives the relationships.

$$\mu = \ln\left(\frac{\mu_1^2}{\sqrt{\mu_1^2 + \sigma_1^2}}\right)$$

$$\sigma^2 = \ln\left(\frac{\sigma_1^2 + \mu_1^2}{\mu_1^2}\right)$$

This is an important distinction as some of the available commercial packages for clinical trial simulation offer the analyst the option of generating a log normal distribution but do not make the transformation of $\mu$ and $\sigma^2$ to $\mu_1$ and $\sigma_1^2$ readily available.

Relating the example above to more standard population-based modeling terminology, if one estimates individual clearance (CL) using the function $CL = CL_{pop} \cdot e^{\eta CL}$, then we can also say $\ln(CL) = \ln(CL_{pop}) + \eta$. Furthermore, we can say that ln(CL) has a normal distribution. Consequently, because of the properties of normality, $\ln(CL_{pop})$ is also equal to the mean, median, and mode of ln(CL). When the distribution of CL is exponentiated, the distribution extrema are preserved, making $CL_{pop}$ the mode of CL. Exponentiation is a monotone transform, such that every value that is lower following log transformation will also be lower following exponentiation; the same is true for higher values. It follows, therefore, that $CL_{pop}$ must also be the median of the distribution of CL. However, this relationship doesn't hold for the mean of CL, because the mean value of a distribution depends on distance of points from each other, and the exponential transform modifies these by compressing values that are less than 0 and spreading out values that are greater than 0. Finally, for a random variable $x$, the geometric mean $GM(x) = \exp(\text{mean}(\log(x)))$. As above, $\ln(CL_{pop})$ is the mean of ln(CL); hence $GM(CL) = CL_{pop}$. In summary, $CL_{pop}$ is equal to the GM(CL), the median(CL), and the mode(CL), but $CL_{pop}$ is not equal to the mean(CL). To determine mean(CL), one must employ the formulas used above.

A complete vector of covariates $\mathbf{X}' = (X_1', X_2', \ldots, X_n')$ can be created for each virtual patient by randomly sampling independently for covariate information from each univariate distribution. Random sampling in this fashion will result in a vector of covariates, which meets the specifications of the original distributions. However, patient demographics are frequently correlated, and the consequences of ignoring these correlations may be that a proportion of the virtual patient population generated in this fashion has combinations of covariates that are unlikely to be seen in the actual trial. Ignoring these correlations may result in serious inaccuracies in the simulation (14). It should also be noted that if significant covariance does exist and the characteristics are generated from univariate distributions, then the variance would be artificially inflated. Therefore, if the individual components of the covariate vector cannot be considered to be independent, it becomes necessary to specify that the covariate vector elements be sampled from a specified multivariate joint distribution.

### 3.2.3.2 Joint Multivariate Distributions

The intersection of two or more observations is of interest in a clinical trial simulation. In the framework of covariate models, this applies to the observation of covariates such as height and weight in a specific individual. In this example, there are a range of weights that may be commonly seen in individuals having a specific height, and other weights that may be unlikely to occur. Specifically, a patient who is 6 ft tall is unlikely to weight 50 lb; a more reasonable intersection of height and weight might be 150–250 lb for a 6-ft-tall individual. A specific set or vector of covariates can be expressed in terms of the intersection of covariate values using a joint distribution. A joint distribution function describes a probability function for multivariate distributions that allows selection of a vector of covariates based in part on the correlations between the covariates. An example of a multivariate distribution between two covariates described by a frequency plot is given in Figure 5.

Therefore, in order to assure that a simulated vector of patient demographics is realistic, the expected correlation between these covariates should be defined. This can be accomplished in one of two ways, either by creating a multivariate joint distribution using assumed correlations, or by using a joint function model where the correlations are defined by fitting patient demographic and char-

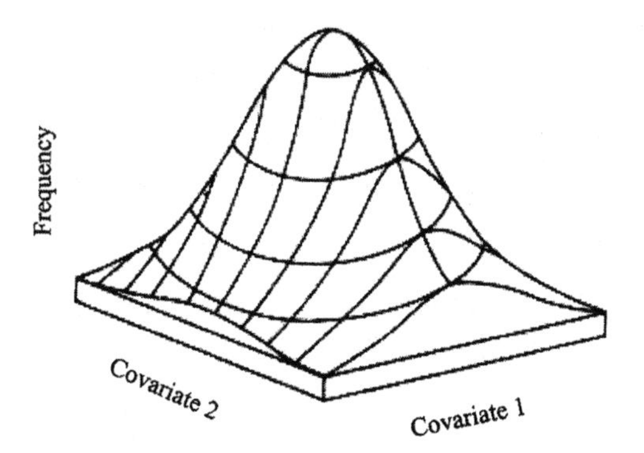


**FIGURE 5** An example of a joint function distribution of two covariates. The areas where the column heights are highest are areas of intersection for these two covariates, or areas where the probability of having both covariates in the same virtual patient are high.

acteristic information obtained in other trials with similar patient demographics. Even if it is not possible to describe the full joint distribution of all relevant covariates ($X_1, X_2, \ldots, X_n$), it may be useful to generate the covariate vector so that the individual covariates have specified univariate distributions (marginal distributions) so that the correlations $\rho_{ij}$ between two covariates $X_i$ and $X_j$ can be specified by the analyst. This would be analogous to having only the variance terms for two covariates but not the covariance terms. As with the univariate approach, normal or lognormal distributions are typically employed.

*a. Conditional Distributions* If a fully specified joint distribution function describing a series of covariates ($F_{x1,x2,\ldots,xn}(x_1, x_2, \ldots, x_n)$) exists from which a random vector of covariates $\mathbf{X}' = (X'_1, X'_2, \ldots, X'_n)^T$ (here $A^T$ is denoted as the transpose of vector $A$) can be generated, and if it is possible to define a conditional distribution for each element $X'_k$ given $X_i = x_i$ for $i = 1, 2, \ldots, k-1$, such that the conditional distribution is denoted as $F_k = (x_k|x_1, x_2, \ldots, x_{k-1})$, then a general algorithm for generating a random covariate vector of covariates from the joint distribution function $F_{x1,x2,\ldots,xn}$ is

1. Generate $X'_1$ with a distribution function $F_{x1}$.
2. Generate $X'_2$ with the distribution function $F_2(.|X'_1)$.
3. Generate $X'_3$ with the distribution function $F_3(.|X'_1, X'_2)$.
4. . . .
5. Generate $X'_n$ with the distribution function $F_n(.|X'_1, X'_2, \ldots, X'_{n-1})$.

A limitation of this algorithm is that in step 2 the conditional distribution used for the selection $X'_2$ is based only on $X'_1$, therefore requiring an inherent selection order for the simulation process. This is also true for the selection of the ensuing covariates in the simulated vector, with each additional covariate becoming more and more constrained by the conditional probabilities. In this situation, if the analyst inappropriately specified the selection order, the combinations of covariates may be inappropriately generated. For example, if creatinine clearance were selected first, followed by age, then the distribution of weights will be very restricted. It might be more appropriate to select age first, then weight, then creatinine clearance since creatinine clearance is dependent on the age and weight of a patient. In any case, such a completely specified distribution is generally not practical in the clinical trial simulation setting because such systems often exhibit slow performance. Therefore, the simulation framework is more commonly described using a series of joint functions, with the selection order being determined empirically.

*b. Multivariate Normal Distributions* An $n$-dimensional miltivariate normal distribution with a mean vector $\boldsymbol{\mu} = (\mu_1, \mu_2, \ldots, \mu_n)^T$ and a covariance matrix $\Sigma$, where the $(i, j)$th element is $\sigma_{ij}$ having the joint distribution function $[N_n(\boldsymbol{\mu}, \Sigma)]$, is given below:

$$f(x) = (2\pi)^{-n/2}|\Sigma|^{-1/2} \exp\left[\frac{-(\mathbf{x} - \boldsymbol{\mu})^T \Sigma^{-1} (\mathbf{x} - \boldsymbol{\mu})}{2}\right]$$

In this equation, $\mathbf{x} = (x_1, x_2, \ldots, x_n)^T$ is any point in the $n$-dimensional real space and $|\Sigma|$ is the determinant of $\Sigma$. If the covariate vector $\mathbf{X} = (X_1, X_2, \ldots, X_n)^T \approx N_n(\boldsymbol{\mu}, \Sigma)$, then $E(X_i) = \mu_i$ and $\text{Cov}(X_i, X_j) = \sigma_{ij} = \sigma_{ji}$ such that $\Sigma$ is symmetrical and positive definite.

Working from this joint function distribution, the conditional distribution algorithm described previously can be applied. However, since $\Sigma$ is symmetric and positive definite, $\Sigma$ can be factored uniquely as $\Sigma = CC^T$ by decomposition, where $C$ represents an $n \times n$ matrix that is lower triangular. The IMSL function CHFAC (15) can be used to factorize $\Sigma$ into the component $C$ matrix, and there are other factorizing algorithms available in statistical software packages. If $c_{ij}$ is the $(ij)$th element of $C$, then one approach for generating a covariate vector $\mathbf{X}'$ with the requisite multivariate normal distribution is

1. Generate $Z_1, Z_2, \ldots, Z_n$ as IID $N(0, 1)$ random variates.
2. For $i = 1, 2, \ldots, n$, let $X_i = \mu_i = \sum_{j=1}^{i} c_{ij} Z_j$.
3. Return $\mathbf{X}' = (X_1', X_2', \ldots, X_n')^T$.

The use of a multivariate log normal distribution for generating a covariate vector requires special consideration (10, 16, 17) that is beyond the scope of this text.

c. *Joint Multivariate Distributions Based on Modeled Correlations* For most simulation work, the correlation terms are largely conjectural. However, covariate information, like any other data, can be described using a model. The use of joint functions for the purpose of estimating missing covariate information has been described previously (18, 19).

In the second example (19), a relatively large fraction (19.6%) of patients were missing weight observations. A joint modeling method was used to predict the missing weight information based on body surface area, creatinine clearance, and sex. This approach is somewhat different than the traditional method of dealing with missing data such as case deletion (removing incomplete data records), or replacing missing covariate values with median values. Removing incomplete records would have substantially reduced the size of the data set, and if the missing data are not missing completely at random, this can introduce bias. Imputing missing data by replacing with median values introduces another type of bias in that the distributional characteristics of the covariates are altered, which can potentially alter the covariate relationship to the pharmacokinetic behavior. A joint function approach provided a means of replacing the missing data using reasonable values based on a multivariate function describing the correlations between covariates. The covariance matrix for these covariates provided additional information that was utilized in the estimation of missing weight. Where weight data were available, the agreement between observed and predicted weights was good (19).

The use of such joint functions have therefore shown utility in population-based modeling, being more approapriate than the more traditional approach of replacing missing covariate information with a median value, and may also be used to define the joint multivariate functions used in a simulation study. This approach has the advantage of being useful during both the modeling and simulation processes and, depending on the choice of software, can have the additional benefit of providing estimates of $\Sigma$.

#### 3.2.3.3 Advantages and Disadvantages of Using Univariate or Multivariate Distributions

Unlike resampling, generating virtual patient populations using univariate or multivariate distributions are not restricted to previously assessed covariate combinations. However, the distributions of covariates and their interrelationships should be well understood prior to using these methods. This may involve defining a joint function for relevant covariates from a database of related patients or it may require making assumptions about the behavior of relevant covariates.

Although simple to implement, the use of univariate distributions is generally not recommended unless the covariates are known to be independent, or if there is only one covariate that needs to be generated for the simulation analysis. The use of multivariate distributions is a better way to generate a matrix of new virtual patients. The complete covariance matrix defining correlations may require some conjectural assumptions, however.

### 3.2.4 Discrete Functions

Although continuous functions are more common, there are also cases where the use of a discrete function may be required. One example of such a case might be in simulating patients with rheumatoid arthritis (RA). RA patients occasionally "flare" in their disease status, causing temporary elevations in cytokines, which may result in altered pharmacokinetics or pharmacodynamics of some therapeutic agents. This transient flaring of disease state could be associated with inherent changes in the covariate vector during the period that the disease is flaring. Furthermore, the patient status of flare is an example of a discrete covariate, and can be simulated using discrete probability functions. This example will be extended in Section 3.2.4.3.

Only three basic discrete functions are described here. The reader should note that there are numerous other discrete functions that can be employed to describe discrete events. Care should be used to select the distribution function that most closely replicates the stochastic nature of the process that is being simulated.

#### 3.2.4.1 Bernoulli Distribution

The most commonly used function for simulation of discrete events is the Bernoulli distribution (11). This distribution describes a random occurrence of two possible outcomes, and it is often employed in the generation of other discrete random variates, such as binomial and geometric functions. The Bernoulli probability parameter $p$ is defined as $p \in (0, 1)$ where $p$ is the probability of an event occurring. The Bernoulli distribution has the following distribution:

$$F(x) = \left\{ \begin{matrix} 0 & \text{if} & x < 0 \\ 1 - p & \text{if} & 0 \leq x \leq 1 \\ p & \text{if} & 1 \leq x \end{matrix} \right\}$$

such that the distribution returns a value of 0 if $F(x) < p$ and 1 otherwise. The generation of a Bernoulli function is straightforward, and generally uses the following algorithm:

1. Generate $U \sim U(0, 1)$.
2. If $U \leq p$, return $X' = 1$; otherwise return $X' = 0$.

#### 3.2.4.2 Binomial Distribution

The binomial distribution is a generalization of the Bernoulli distribution that includes more than one deviate. It is helpful to think of the binomial distribution as describing the probability $p$ of having a successful outcome in a given number of trials $t$, or the number of patients having a particular covariate in a group. As with the Bernoulli distribution, the binomial distribution returns only a 0 or 1. To generate random variates that follow a binomial distribution, bin($t$, $p$), the following algorithm can be implemented:

1. Generate $\mathbf{Y}_1, Y_2, Y_3, \ldots, Y_t$: IID Bernoulli random variates.
2. Return $X = \sum_{i=1}^{t} Y_i$.

As can be seen, the Bernoulli distribution is used to generate the random variate used in the binomial distribution.

#### 3.2.4.3 Poisson Distribution

In the example of the generation of a flare situation for a patient with RA, or in the simulation of patients with regularly occurring events, a Poisson distribution may be more appropriate. The Poisson distribution is commonly used to define the number of events occurring in a specified interval of time when these events are assumed to occur at a constant rate per unit interval. The distribution is given below:

$$F(x) = \begin{cases} 0 & \text{if } x < 0 \\ e^{-\lambda} \sum_{i=0}^{\lfloor x \rfloor} \frac{\lambda^i}{i!} & \text{if } 0 \le x \end{cases}$$

The generation of Poisson variates is as follows:

1. Let $a = e^{-\lambda}$, $b = 1$, and $i = 0$.
2. Generate $U_{i+1} \approx U(0, 1)$ and replace $b$ by $bU_{i+1}$.
3. If $b < a$, then return $X = i$; otherwise go to step 4.
4. Replace $i$ by $i + 1$ and go back to step 2.

The algorithm is supported by noting that $X = i$ if and only if $\sum_{j=1}^{i} Y_j \le 1 < \sum_{j=1}^{i+1} Y_j$, where $Y_j = -1/\lambda \ln(U_j) \approx \exp 1/\lambda$ and the $Y_j$ are independent.

The Poisson distribution can be combined with the Bernoulli distribution to create a function that describes erratic events occurring with a given probability over a period of time, such as a patient experiencing epileptic seizures. There are other empirical discrete functions that can also be used to describe this kind of covariate as well.

## 3.3 IMPLEMENTATION OF COVARIATE DISTRIBUTION MODELS IN A SIMULATION

Once the covariate distribution models have been established for the trial simulation, they need to be implemented in terms of protocol specifications. This generally implies ensuring that the virtual patient meets the specified patient inclusion and exclusion criteria. This would require that the virtual patient's covariate vector is checked against the protocol criteria and if the vector is not acceptable, then the virtual patient must be regenerated. An example of this implementation is shown in Figure 6. It is possible that the analyst may accept a random fraction of the inappropriate virtual patients in order to represent protocol violations—although this is part of the "execution model" concept. This is often referred to as a two-stage implementation, where virtual patients are generated using nonparametric or parametric methods, and the resulting covariate vector is then checked against the protocol-defined criteria for acceptability.

## 3.4 QUALIFICATION OF COVARIATE DISTRIBUTION MODELS

As with other aspects of clinical trial simulation, the analyst should attempt to show that the model used for creating the virtual patient population matrix reflects the distribution of covariates in the expected patient enrollment. Models used to generate covariate vectors should be tested and qualified with the same scientific

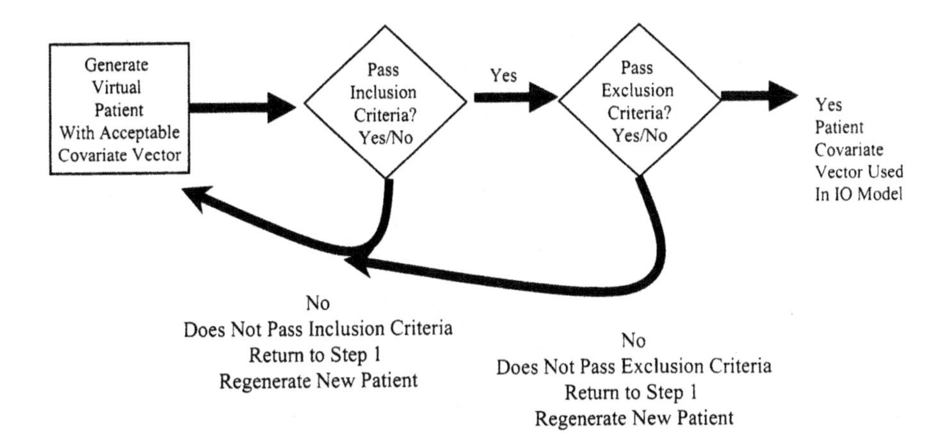


**FIGURE 6** Implementation of a two-stage generation of suitable covariate vectors for a simulation study. Virtual patients are generated according to either a parametric or nonparametric resampling approach. The covariate vector for the simulated patient is then compared against the inclusion and exclusion criteria. If patients are acceptable, then they are used for the simulation study.

rigor that is given to any other IO model employed in the analysis. Model verification or qualification is the attempt to establish that the behavior of the model system reflects the original system. Qualification can be accomplished using formal proofs of correctness, but are more commonly done through extensive testing. Comparisons between the model and the system are usually done at the final output, or model behavior, level, although a thorough qualification should test the system behavior at the model component level as well. Model qualification follows a hierarchy: replicative, predictive, and structural validity (20, 21). Replicative qualifications are those where the model can successfully reproduce known behavior from a system. Predictive qualification is when a model can accurately predict system behavior outside the limits of its initial specification. Structural validity is when the model is defined as closed. Most model qualifications evaluate only replicative behavior, and predictive qualification is generally only assessed during repeated use and development of the model system.

There are several aspects that should always be considered before developing a simulation model that will make qualification easier, which can be applied to any of the models employed in a simulation model. These steps are:

1. Develop a simulation team comprised of experts in relevant areas.
2. Use any expertise and data gained from similar analyses to develop the simulation model. Make use of intuition and related experience.

3. Develop a model that has prima facie validity.
4. If a data set exists that is similar to that which you are trying to simulate, verify the simulation methodology by comparing model output with existing data. This is the most definitive test of a simulation model.
5. Generate a sufficient number of replicate patient matrices to assess the impact of the covariates. This may also help to reduce the likelihood of making decisions based on spurious covariate vectors appearing in the data set.
6. Assess the distribution of covariate vectors graphically to manually check for potential outliers.

The selection of the method of qualification for the covariate distribution model is dependent on the method used for generation of the virtual patient database. If resampling is used, then a visual check of the virtual patient covariates using frequency histograms may be sufficient. However, if the resampling algorithm was based on specific prognostic factors, a sensitivity analysis, where the expected distributions of patients having specified prognostic factors is varied, may be necessary to demonstrate the consequences of patient enrollment that does not match expectations.

When sampling from univariate or multivariate distributions has generated the virtual patient database, then more elaborate testing may be required. One of the first tests that can be employed is a visual check of covariate factors to assure that the covariate combinations are reasonable. Approaches such as a posterior predictive check (22) may be appropriate to demonstrate that the covariate models can be used reliably. The application of other software tools such as those used to scan actual data sets for reasonable covariate values (e.g., rule-based systems) may be beneficial in assuring that the virtual patient population that has been created is reasonable.

The ultimate test of a simulation experiment is to compare predictions against the outcome of the actual study. Consequently, there are some good examples where these comparisons have been made (23). During drug development, simulations based on data obtained in early phase studies may not result in accurate predictions. However, data obtained in subsequent studies can be used to alter or correct the models used for simulation, thereby improving the quality of the simulation analysis for later studies. In a learning versus confirming paradigm (24) the quality of the model would be expected to improve with more use. This is one example of predictive qualification.

In Section 3.1, the concept of a closed system was introduced. Theoretically, a closed system would have the highest resolution of output, because it is completely specified (3, 21). However, as a consequence of being completely specified, the model system is also generally complex. Typically, the performance of a simulation system will decrease as the model system increases in complexity (3, 21). Furthermore, a more complex system may be unsolvable numerically. Therefore, reducing the complexity of the model often results in improved performance but also entails a loss of validity of the model system. In clinical trial simulation, it is often necessary to adopt a simplified system because of the difficulties inherent in developing a fully specified system. In cases where it is necessary or desirable to simplify a model system, a level of error tolerance must also be specified a priori for testing purposes. The tolerance limits should be determined based on the purpose for which model will be used. For instance, tolerance limits for a pharmacodynamic model should alllow at least the level of error seen in an experimental situation. When multiple components of a model system are simplified, particularly in the case of any modular system, error propagation should be examined. Small deviations in behavior in one component may be acceptable at the component level, but when added to other deviations in other components, the resulting model system may not adequately capture the system being modeled. A model component is "error critical" if an error in its behavior propagates in successive transitions to other components and ultimately to the system behavior or output of interest (21). Sensitivity analysis is a key method for identification of error critical components.

## 3.5 SUMMARY

Clinical trial simulation has become an important tool in the pharmaceutical sciences. Models used in clinical trial simulation have progressed from simple decision trees and nonstochastic models to extremely complex systems, yet all are built using a logical process based on objective evaluation of available data. In general, modelers justify both the structure and content of the model and then test their assumptions using a comprehensive process of sensitivity analysis and model validation (20, 24). As with clinical trials, simulation analyses sometimes produce results that are later found to be invalid as new data become available. In the absence of data, assumptions are often based either on the recommendations of experts or on inadequately defined notions of the standard of care or medical necessity. Because such assumptions do not require the rigorous process of data collection, synthesis, and testing, it is usually not possible to examine these assumptions fully, although the consequences of using these assumptions can be tested during the simulation experiment.

When covariates affecting the pharmacokinetics or pharmacodynamics of a new chemical entity have been identified, or if prognostic factors are thought to be relevant to the expected outcome of a specific course of treatment, then the assumptions inherent in the selection and use of a covariate model become important in understanding the results of a clinical trial simulaton. In such cases, the development of a covariate distribution model requires more attention than when covariates are not identified or when only one primary covariate has been identified.

For the most simple case, when only one covariate has been found, covariate models can be based on simple univariate distributions, or a virtual patient data base can be readily generated using resampling techniques. Covariate model qualification is of less relevance and the process is fairly straightforward. However, when multiple covariates have been identified, then either resampling from similar patient population or a joint function or multivariate model must be constructed. In these more complex situations, the covariate distribution model should be tested to reliability to assure that the covariate vectors contain covariate combinations that are reasonable in the target patient population.

When the relationships between covariates are not known explicitly, these relationships must necessarily be based on conjecture if no relevant data base is available for resampling. In such cases, the covariate model will contain elements that are conjectural. Whenever possible, it is generally best to base assumptions on observed data rather than on conjecture. Sensitivity analysis of conjectural models together with other means of testing the models and assumptions used in the generation of a virtual patient population are necessary parts of any well-conducted simulation analysis.

The development of a covariate distribution model should therefore be based on necessity. If covariates are not identified as being clinically relevant, then a covariate distribution model can be excluded from the simulation model. Common sense should always be used when developing and adding models to a simulation analysis, and the level of complexity of the simulation model system should be constrained by necessity.

## REFERENCES

1. Editors: NHG Holford, M Hale, HC Ko, J-L Steimer, CC Peck. Contributors: P Bonate, WR Gillespie, T Ludden, DB Rubin, LB Sheiner, D Stanski. Simulation in Drugs Development: Good Practices, 1999. http://cdds.georgetown.edu
2. GS May, DL DeMets, LM Friedman, C Furberg, E Passamani. The randomized clinical trial: bias in analysis. Circulation 64(4):669–673, 1981.
3. B Ziegler, H Praehofer, TG Kim. Theory of Modeling and Simulation, 2nd ed. San Diego, CA: Academic Press, 2000, pp 4, 367–389.
4. SL Beal, AJ Boeckmann, LB Sheiner. NONMEM Users Guides, Version V. San Francisco, CA: University of California at San Francisco, NONMEM Project Group, 1999.
5. Pharsight Corporation Trial Simulator, Version 2.0 Users Guide, Mountain View, CA: Pharsight Corporation, 2000.
6. FDA Guidance for Industry Population Pharmacokinetics, 1999. http://www.fda.gov/cber/gdlns/popharm.pdf
7. AC Davison, DV Hinkley. Bootstrap Methods and Their Application (Cambridge

Series in Statistical and Probabilistic Mathematics, No. 1). Cambridge, UK: Cambridge University Press, 1997, pp 14–30, 143, 145.
8. A Yafune, M Ishiguro. Bootstrap approach for constructing confidence intervals for population pharmacokinetic parameters I: use of a bootstrap standard error. Statistics Med 18:581–599, 1999.
9. SJ Pocock, R Simon. Sequential treatment assignment with balancing for prognostic factors in the controlled clinical trial. Biometrics 31(1):103–115, 1975.
10. BJ Anderson, GA Woollard, NH Holford. A model size and age changes in the pharmacokinetics of paracetamol in neonates, infants and children. Br J Clin Pharmacol 50(2):125–134, 2000.
11. AM Law, WD Kelton. Simulation Modeling and Analysis, 2nd ed. New York: McGraw-Hill, 1991, pp 335–348, 490–503.
12. GEP Box, ME Muller. A note on the generation of random normal deviates. Ann Math Statist 29:610–611, 1958.
13. G Marsaglia, TA Bray. A convenient method for generating normal variables. SIAM Rev. 6:260–264, 1964.
14. CR Mitchell, As Paulson, CA Beswick. The effect of correlate exponential service times on single tandem queues. Nav Res Logist Quart 24:95–112, 1977.
15. IMSL Users Manual: Stat/Library vol. 3, Houston, TX: IMSL, 1987.
16. ME Johnson. Multivariate Statistical Simulation. New York: John Wiley, 1987.
17. ME Johnson, C Wang, JS Ramberg. Generation of continuous multivariate distributions for statistical applications. Am J Math Management Sci 4:96–119, 1984.
18. MO Karlsson, EN Jonsson, CG Wiltse, JR Wade. Assumption testing in population pharmacokinetic models: illustrated with an analysis of moxonidine data from congestive heart failure patients. J Pharmacokinet Biopharm 26(2):207–246, 1998.
19. DR Mould, NHG Holford. The pharmacokinetics and pharmacodynamics of topotecan. Proceedings of EORTC/PAMM, Verona, Italy, 2001.
20. SD Ramsey, M McIntoch, R Etzioni, N Urban. Simulation modeling of outcomes and cost effectiveness. Hematol Oncol Clin North Am 14(4):925–938, 2000.
21. D Cloud, L Rainey. Applied Modeling and Simulation: An Integrated Approach to Development and Operation. New York: McGraw-Hill, 1998.
22. Y Yano, SL Beal, LB Sheiner. Evaluating pharmacokinetic/pharmacodynamic models using the posterior predictive check. J Pharmacokinet Biopharm 28(2):171–192, 2001.
23. JC Kimko, SS Reele, NH Holford, CC Peck. Prediction of the outcome of a phase 3 clinical trial of an antischizophrenic agent (quetiapine fumarate) by simulation with a population pharmacokinetic and pharmacodynamic model. Clin Pharmacol Ther 68(5):568–577, 2000.
24. LB Sheiner. Learning versus confirming in clinical drug development. Clin Pharmacol Ther 61(3):275–291, 1997.

# 4

# Protocol Deviations and Execution Models

**Helen Kastrissios**
University of Queensland, Brisbane, Australia

**Pascal Girard**
Pharsight, Lyon, France

## 4.1 GOALS OF CLINICAL TRIAL SIMULATION

An important goal of clinical trial simulation (CTS) is to develop well-designed protocols that will maximize the ability to address the stated aim(s) of a proposed clinical trial. The first step in this process is to identify a useful input-output model (IO), including the model structure and its parameters, which will adequately reproduce salient characteristics that clinicians wish to observe in a future clinical study (see Chapter 2). Examples of such characteristics include drug (and metabolite) concentrations, biomarkers of therapeutic or toxicological response (e.g., changes in serum cholesterol, blood pressure, $CD_4$ cell counts, coagulation time, neutrophil counts, hepatic and renal markers, QT prolongation or the incidence of an event, such as drug-induced rash) or clinical outcomes (e.g., time to AIDS conversion, survival time, recovery from stroke, improvement in cognitive scales).

If the process of IO model identification is successful, a subsequent consideration is to evaluate the influences of protocol deviations on outcomes of interest. CTS provides an invaluable tool to "push" experimental study designs to the point of failure. In the same way, aeronautical engineers use flight simulators to evaluate aircraft under adverse conditions to identify conditions that could cause the aircraft to crash so that the design can be improved, long before the takeoff of the prototype. Herein lies one of the most powerful uses of CTS: the ability to perform many virtual studies allowing for the occurrence of variations in carrying out the clinical protocol so as to identify weaknesses or limitations in the proposed study design. Using this approach, the type, extent, and combination of protocol deviations that may be evaluated are limited by the ability to imagine the possibility of these events occurring.

## 4.2 DEFINITION OF EXECUTION MODELS

Execution models describe protocol deviations from a specified study design. When a clinical trial is planned, it is generally supposed that it will be executed according to a specific protocol that defines all aspects of the experimental design, from its beginning to its completion. For example, the following characteristics must be precisely defined in any clinical protocol:

- Characteristics (inclusion/exclusion criteria) of patients or healthy volunteers (that we will name indifferently as subjects)
- Number of subjects to be accrued
- Treatments and allocation mechanism
- Blinding of investigators and/or subjects to the treatment allocation
- Dosage regimen (dose and timing of doses)
- Measurements to be performed (type, date, and time)
- Frequency of follow-up evaluations
- Study length

Adherence to the protocol will allow estimation of the treatment outcome (safety and efficacy) with sufficient statistical power, or at least that is what is assumed. In reality, however, deviations from the protocol may lead to failure of the study to achieve its stated aims.

In anticipation of protocol deviations that contribute to inflated residual variability and decreased study statistical power, trial designers tend to overpower studies in a rather arbitrary way. It should be emphasized that in certain cases, the increase of number of patients is not sufficient to compensate for this decrease in power and in other cases this overpowering may result in unneeded larger studies that have financial consequences on the overall drug development program. It is difficult to estimate quantitatively the consequences of one protocol deviation on statistical study power and, a fortiori, it is almost impossible to do it for a combination of protocol deviations. The only way to study the consequences of model deviations is by using modeling and simulation techniques, and more specifically longitudinal stochastic models, which are the only ones that can describe individual behaviors.

## 4.3 SOURCES OF PROTOCOL DEVIATIONS AND MODELS TO SIMULATE THEM

Table 1 lists several sources of protocol deviations. They may be investigator related (e.g., wrong inclusion, wrong treatment allocation), treatment related (e.g., switch to another treatment), or purely patient related (e.g., missed visit, definitive dropout). Figure 1 exemplifies some of the complex interactions that can be found between drug-disease and execution models. The ones with compliance can be among the more complex: We usually expect that changes in treatment compliance will change the outcomes; but it is also highly likely that changes in non-silent biomarkers such as pain and subjective self-measurements on various scales may in turn change patients' compliance, because patients tend to forget their treatment more frequently when they feel subjectively better, because they may increase their doses when they feel worse or skip them altogether if they believe the treatment has little effect or no effect at all.

In general, in comparison to IO model identification, implementing an execution model should be relatively straightforward. The execution model has, of course, to be structurally close to reality, but its parameterization will essentially depend on the ability of the clinical team to identify the magnitude and scope of possible protocol deviations.

All protocol deviations are unexpected by nature, and so only probabilistic models can be used to simulate them. There are essentially two classes of model that can be used: time-independent models and time-dependent models. Time-dependent distributions describe discrete events that are conditional on the occurrence of a previous event; that is, they have an order or memory. On the other hand, a time-independent event is discrete and memoryless. Models associated with each source of protocol deviations are listed in Table 1, and their properties are briefly reported in the next sections.

### 4.3.1 Time-Independent Distributions

#### 4.3.1.1 Binomial and Multinomial Distributions

These distributions are used when it is assumed that the protocol deviation is a discrete event (binomial for a binary event and multinomial for more than two categories) that may or may not occur. Typical cases are probability of wrongly including a patient who does not fulfill inclusion criteria; probability of being allocated to the wrong arm; probability of not taking the dose at each time; probability of not having the measurement. If, in a sample of $n$ subjects, the event

**Table 1** Types of Protocol Deviations and Possible Execution Models

| Nature of protocol deviation | Type | Candidate models to simulate protocol deviation |
|---|---|---|
| 1. Patient is wrongly included (does not meet inclusion criteria) | D | Binomial/multinomial |
| 2. Less subjects than expected are included | C | Logistic |
| 3. Patient receives the wrong treatment (e.g., placebo instead of active) | D | Binomial/multinomial |
| 4. Patient receives the wrong dose | D | Binomial/multinomial |
| 5. Patient crosses over to the alternate treatment | D | Binomial/multinomial/time-to-event hazard model |
| 6. Patient takes a forbidden co-medication | D | Binomial/multinomial |
| 7. Patient takes fewer or extra dose(s) of treatment than prescribed, but the remaining doses are taken on time | D | Binomial/multinomial or Markov |
| 8. Patient takes all doses but does not take them on time | C | Normal or uniform |
| 9. Patient stops taking the treatment but remains on the study | D | Time-to-event hazard model |
| 10. Patient or clinical team does not comply with measurement times, but all measurements are recorded | C | Normal or uniform |
| 11. Patient or clinical team misses some measurements but completes the study | D | Binomial/multinomial or Markov |
| 12. Measurements are incorrect and thought missing (deficient measurement technique) | D/C | Binomial/multinomial + normal or uniform distributions |
| 13. Measurement times are switched | D | Binomial/multinomial |
| 14. Patient drops out before the end of the study | D | Time-to-event hazard model |

C: Continuous deviation from protocol
D: Discrete deviation from protocol

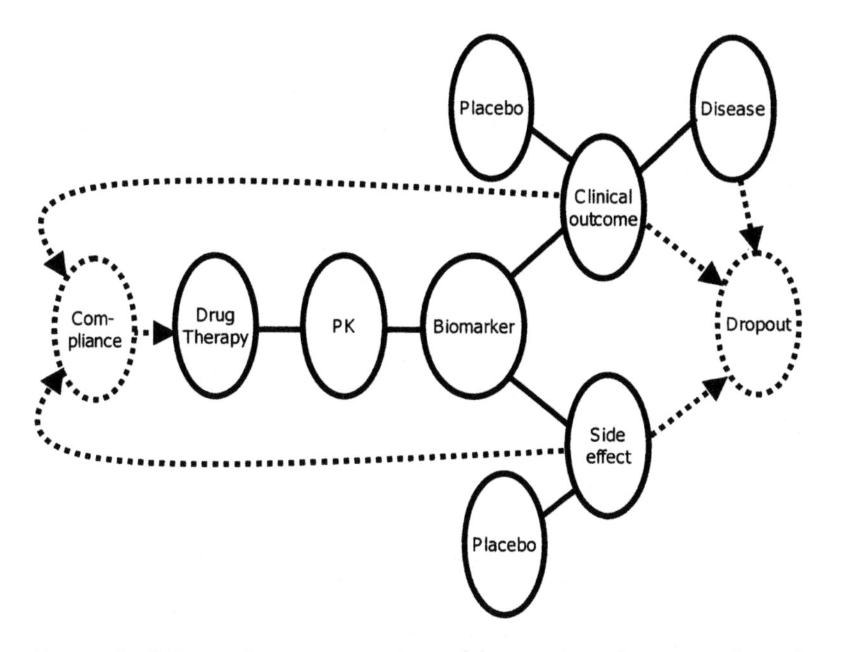


**FIGURE 1** Schematic representation of interactions between drug-disease model and two aspects of execution model: the compliance model and the dropout model. Drug-disease models interactions are shown with continuous lines, while execution models are presented with dotted line arrows. Notice that compliance influences PK/PD models, which in turn, by feedback mechanisms, may influence compliance.

occurs with a frequency of $\theta$, such that $0 \leq \theta \leq n$, and with a relative frequency of $\pi$, such that $0 \leq \pi \leq 1$, then $p(\theta) \approx \text{Bin}(\theta \mid n, \pi)$. The density function is

$$p(\theta|n) = \binom{n}{\theta} \cdot \pi^{\theta} \cdot (1 - \pi)^{n-\theta} \tag{1}$$

and the expected value of $\theta$, $E(\theta) = n \cdot \pi$ and the variance, $\text{var}(\theta) = n \cdot \pi \cdot (1 - \pi)$.

If the protocol deviation is a discrete event that may be grouped into $k$ categories and the multinomial distribution model is defined with parameters $n$ and $\pi_1, \ldots, \pi_k$, then $p(\theta) \approx \text{Multin}(\theta \mid n; p_1, \ldots, p_k)$. Thus,

$$p(\theta_j \mid n) = \frac{n!}{\Pi_{j=1}^{k} \theta_j!} \cdot \sum_{j=1}^{k} \pi_j^{\theta} \tag{2}$$

and $E(\theta_j) = n \cdot \pi_j$ and $\text{var}(\theta_j) = n \cdot \pi_j \cdot (1 - \pi_j)$.

#### 4.3.1.2 Logistic Distributions

This class of distribution can be useful to simulate a proportion. For example, imagine that for any reason the proportion of included patients is supposed to be 100* ρ% of the planned number, with $0 \leq \rho \leq 1$, and that this proportion is changing from one replicated study to another. This can be achieved by using the logistic model:

$$\rho = \frac{\exp(\alpha + \psi)}{1 + \exp(\alpha + \psi)} \tag{3}$$

where α is a fixed parameter, and ψ a random (normal for example) effect representing uncertainty on the proportion of patients that will be accrued. Within each replicated study, a patient, with correct inclusion criteria, will have the probability ρ to be accrued.

#### 4.3.1.3 Univariate and Multivariate (Log)Normal Distributions

These distributions are used when it is assumed that the protocol deviation event may be described by a continuous random variable such that $p(\theta) \approx N(\theta \mid \mu, \sigma^2)$ for a univariate model and $p(\theta) \approx N(\theta \mid \mu, \Sigma)$ for a multivariate normal model. Typically, a departure of time from a nominal designed time maybe either normally (or uniformly; see below) distributed (in an interval) around this nominal time.

#### 4.3.1.4 Uniform Distributions

Assuming that the protocol deviation event may be described by a continuous random variable defined within a finite interval $[a, b]$, in which all possible events are equally likely to occur, then $p(\theta) = U(\theta \mid a, b)$. The probability distribution for the event occurring is

$$p(\theta) = \frac{1}{b - a} \tag{4}$$

where $E(\theta) = (a + b)/2$ and $\text{var}(\theta) = (b - a)^2/12$.

### 4.3.2 Time-Dependent Distributions

#### 4.3.2.1 Conditional Models as Discrete Markov Chains

A first-order discrete Markov chain is characterized by the basic following property:

$$P(x_t \mid x_{t-1}, x_{t-2}, \ldots, x_1) = P(x_t \mid x_{t-1}) \tag{5}$$

where $x_t$ is a random discrete variable measured at time $t$.

This property simply expresses that the probability of observing a random discrete event $x_t$ at time $t$ is dependent on the observation of this event at time $t - 1$ and is independent of all past events. A second-order discrete Markov chain would extend this property to time $t - 2$, and so on for higher orders. More details on discrete Markov chain are given in Section 4.4 and exemplified for a compliance model.

#### 4.3.2.2 Hazard Models

These distributions are invaluable tools to simulate dropouts, time to switch to another medication (e.g., medication rescue), time to definitively stop taking medication, etc. Assuming that the time to a specified protocol deviation event $u$ (or a transformation of $u$) follows a distribution function $F$, then the probability of the event occurring may be expressed:

$$p(\theta) = \int_0^T f(u) \cdot e^{-\lambda \cdot u} \cdot du \tag{6}$$

where the hazard $\lambda$ is defined as the probability that the event happens given that it has not happened at a specified time $t$ during the period of the study $T$.

## 4.4 MODELS FOR VARIABLE COMPLIANCE WITH ORAL DRUG THERAPY

Noncompliance with oral medications is common in ambulatory patients, both in clinical practice and in clinical trial settings. Subjects' dosing patterns may differ relative to the prescribed regimen in terms of the amount of the dose, the timing of doses, and the duration of treatment (1–3). Consequences of variable dose timing on treatment outcomes are determined by the magnitude of erratic dosing about the prescribed dosing times, the number and frequency of sequentially missed doses, or "drug holidays," and the pharmacological properties of the drug (4).

### 4.4.1 Models of Dose Taking and Dose Timing

Individual deviations from prescribed oral drug regimens have been quantified using electronic monitoring devices and show large between-subject variability in the timing of doses relative to the prescribed interdose interval. In comparison, indices of dose-taking compliance (the quantity of the dose) are usually less variable (4). Therefore, several investigators have simulated dosing histories based on the assumptions that the prescribed number and quantity of the dose were taken as prescribed, but at variable dose times (5, 6). In these studies, the $i$th dosing interval for the $j$th patient, $\tau_{ij}$, was drawn from a normal distribution with mean $\tau$, the prescribed dosing interval, and variance $\sigma^2$. Negative values thus generated were truncated to zero, resulting in a double dose event at that time [7]. Earlier attempts proposed a model where the number of doses taken at each dose time was simulated using a multinominal distribution allowing for 0, 1, 2, . . . , $m$ doses taken at each dosing time (8, 9). In the latter paper, the number of doses taken within each nominal dosing time interval was modeled using a Markov model (see Section 4.4.2) giving the probability of taking no dose, one dose, or two doses and more (9). The model was fit to MEMS data (4) collected in AIDS Clinical Trials Group protocol 175 (ACTG175). In order to fully mimic the noncompliance stochastic phenomenon in the simulated series, random departures from nominal dose timing were added to nominal times and were obtained by resampling actual differences between nominal dose timing and actual measured ones (9).

Another summary statistic of dose-taking compliance describes the fraction of study days on which the patient took the prescribed number of doses. Based on data collected in a lipid-lowering study, it was shown that this compliance parameter may be bimodal and its frequency distribution may be described using a mixture of beta density functions (10).

### 4.4.2 Markov Model of Medication Compliance

A hierarchical Markov model for patient compliance with oral medications was developed conditional upon a set of individual-specific nominal daily dose times and individual random effects that are assumed to be multivariate normally distributed (11). The first-order Markov hypothesis supposes that the subject-specific probability of not taking a dose or taking one or more doses at any given dose time depends on the number of doses taken at the dose time immediately previous to the one in question and is independent of any previous dosing events. More formally, let us define $Y = (y_1, y_2, \ldots, y_n)$ a random vector indicating whether the patient has not taken his treatment ($y_i$ = NT) or taken it ($y_i$ = T) at $i$th time. The 2-state Markov chain model is fully defined by giving the two conditional probabilities of not taking the treatment given it was not taken the time before, and taking it given it was taken the time before:

$$\begin{aligned} p(y_i = \mathrm{NT} \mid y_{i-1} = \mathrm{NT}) &= P_{00} \\ p(y_i = \mathrm{T} \mid y_{i-1} = \mathrm{T}) &= P_{11} \end{aligned} \tag{7}$$

from which we derive:

$$\begin{aligned} p(y_i = \mathrm{T} \mid y_{i-1} = \mathrm{NT}) &= P_{01} = 1 - P_{00} \\ p(y_i = \mathrm{NT} \mid y_{i-1} = \mathrm{T}) &= P_{10} = 1 - P_{11} \end{aligned} \tag{8}$$

Those four probabilities can be arranged into the Markov transition matrix:

$$P = \begin{bmatrix} P_{00} & P_{01} \\ P_{10} & P_{11} \end{bmatrix} \tag{9}$$

Simulation can be easily produced with a pseudo-random uniform number generator. The following simple Splus function shows how to simulate one series of dose:

```
Markov <-function (L, p00, p11)
{
   p01 <- 1-p00
   p10 <- 1-p11
   NT <- sample(c(−1,1),size=L,replace=T,prob=c(p00,
p01))
   T <- sample(c(−1,1),size=L,replace=T,prob=c(p10,
p11))
   Y <- rep(1,L)
   if (p==000) N <- T
     else {
        for (i in 2:L){
           if (Y [i−1]==−1) Y [i]<-NT[i]
              else Y [i]<-T[i]
        }
   }
   X
}
# Series of 70 doses, with P(Yi=NT | Yi−1=NT)=30% and
P(Yi=T | Yi−1=T)=90%
Y <- markov (70, 0.3, 0.9)
```

In the field of modeling compliance, these Markov chains are sometimes and improperly named "two-coins" models. This corresponds to a patient that has 2 "virtual coins" that will be tossed alternately depending on whether the previous dose was or was not taken. Usually coins are equilibrated and have a 50/50 probability of getting heads/tails. Similarly, the one-coin model is a simple probability model where the unconditional probability of taking the dose is 50% (see Figure 2).

In reality these probabilities vary from one individual to another and can take values between 0 and 1, which suggests that we need a "biased-coin" model. This interindividual probability can easily be implemented in the model by using mixed effect logistic regression (11):

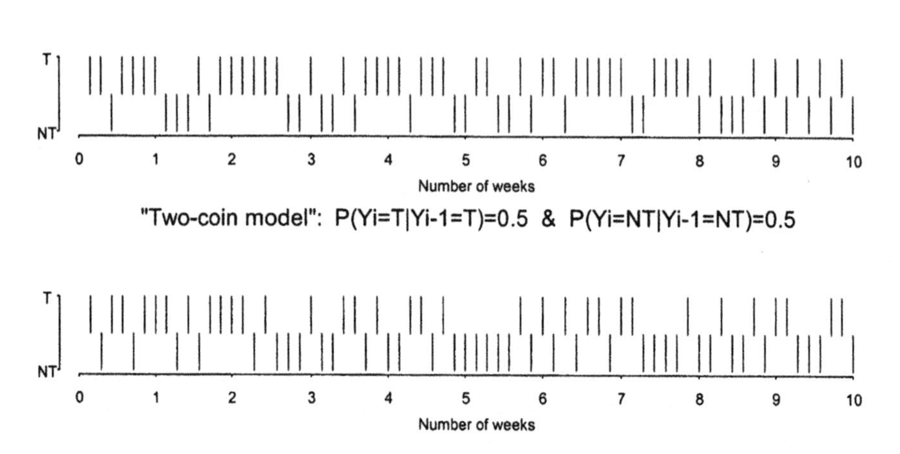


**FIGURE 2** Compliance simulation profiles using the so-called one-coin and two-coin models. Each bar represents one dosing time where the patient takes (T) or does not take (NT) the treatment. Notice that the notion of "coin model" restricts the probability of taking (T) or not taking (NT) to 0.5.

$$
\begin{aligned}
P_{00} &= \frac{\exp(\alpha_1 + \beta_1 X + \eta_1)}{1 + \exp(\alpha_1 + \beta_1 X + \eta_1)} \\
P_{11} &= \frac{\exp(\alpha_2 + \beta_2 X + \eta_2)}{1 + \exp(\alpha_2 + \beta_2 X + \eta_2)}
\end{aligned}
\tag{10}
$$

where

$\alpha_1, \alpha_2$ = intercept parameters
$\beta_1, \beta_2$ = vectors of covariate parameters
$X$ = matrix of patient individual covariates
$\eta_1, \eta_2$ = two random effect parameters, with mean 0 and variance $\Omega$, that model random interindividual effects

This model has great flexibility and allows description of almost all different compliance profiles. $P_{11}$ controls the fact that the patient may stop taking the treatment, while $P_{00}$ controls the fact that the patient may go on not taking the treatment. The use of covariates allows to control, for example, the date at which the patient will have a drug holiday. Figure 3 shows three very different profiles: top patient has on average 2 missing doses per week; middle patient shows 2 drug holidays per month on average; and bottom patient shows a "treatment drop-out," which means that this patient definitively stops taking the treatment but remains on the study (12).

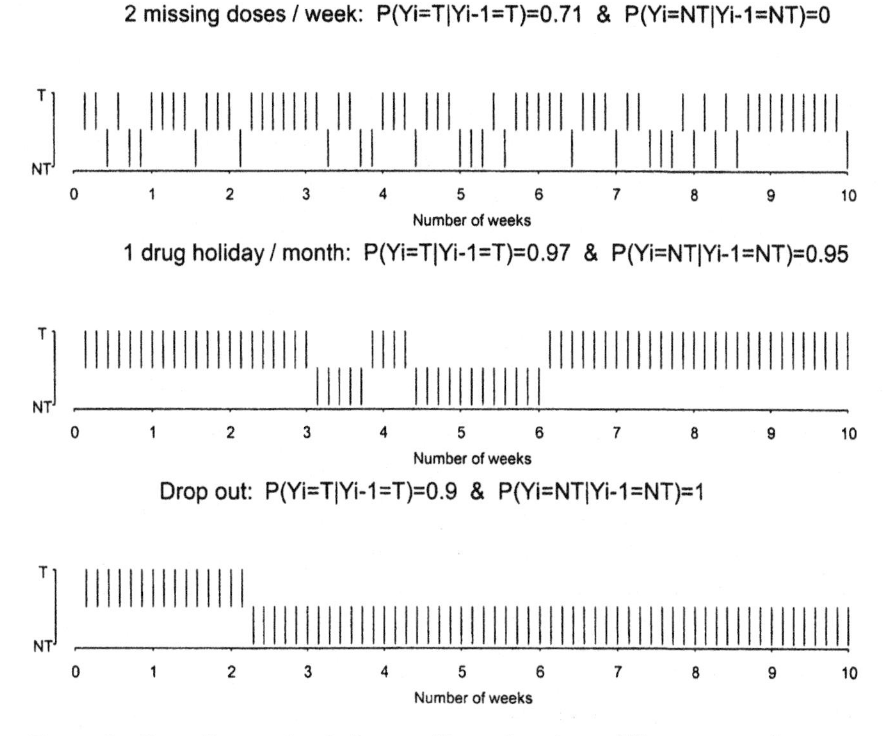


**FIGURE 3** Compliance simulation profiles using three different sets of parameters.

If dose timing might be of importance, a set of errors between actual dose times associated with each subject-specific nominal time is assumed to be multivariate normally distributed conditional on patient covariates and the number of actual dose times.

## 4.5 CONSEQUENCES OF PROTOCOL DEVIATIONS

Consequences of protocol deviations on clinical trial outcomes depend on their qualitative and quantitative natures. Thus, while the consequence of one type of protocol deviation can be easily evaluated (e.g., deviation 2 in Table 1 and study power), some are more difficult to evaluate than others (e.g., noncompliance to treatment). It follows that the combination of several deviations of varying degrees may lead to unexpected consequences on study outcomes. Accordingly, the study of multiple deviations on clinical trial outcomes can usually only practically be studied using in silico or in numero technologies such as those utilized to implement CTS. This allows:

1. Identification of the type, extent, and combination of deviations that may have an impact on study outcomes, and in particular, those that can lead to failure of the study to achieve its aims
2. Estimation of acceptable tolerance limits within the study design that will improve the likelihood of success of the study (i.e., a sensitivity analysis)
3. Identification of changes that might be made to the study design in order to safeguard against protocol violations and increase robustness of the study design

## 4.6 MODELS TO ADJUST FOR PROTOCOL DEVIATIONS

Models to adjust for protocol deviations, when those deviations are unknown, essentially consist in sophistication of random inter- and intraindividual variability models. We give below some examples of such models. Nested random effects models allow multiple sources of random effects to be modeled simultaneously. For example, in traditional population pharmacokinetic models, random residual variability is nested within between-subject variability. However, there are other sources of random variability that may be modeled, including between-occasion variability (13) and between-study variability (14), although others such as between-batch variability are also possible (15). It may also be important to consider an interaction between the residual variance and the between-subject random effects. This allows for estimation of individual residual variances, which increases considerably the flexibility of model development. Some authors have also tried to develop methods to adjust for noncompliance, but these models require highly complex techniques (see, for example, Ref. 16).

## 4.7 AN ILLUSTRATIVE EXAMPLE

Suppose a randomized, double-blind, placebo-controlled phase II clinical trial will be conducted to evaluate the effectiveness of a new oral formulation of drug X in 200 subjects. A single daily oral dose of drug X, or placebo, will be administered for 100 days. Patient selection assumes that 95% of subjects are responders (R) and 5% are nonresponders (NR). The primary endpoint is the occurrence of an unwanted clinical event ($Y = 1$) during the period of treatment. A binary model is used, such that $\text{RISK}(Y = 1 \mid \text{placebo}) = 30\%$.

We assume that RISK is reduced by adequate exposure to the drug or "therapeutic coverage" (5). For simplicity, exposure to the drug is calculated as the percentage of doses taken relative to the prescribed number of doses, and it is assumed that a threshold of 50% of doses must be taken to achieve a therapeutic response. Thus,

$$TC = \begin{cases} 0, & \text{if (exposure} - \text{threshold)} < 50, \text{ else} \\ (\text{exposure} - \text{threshold}) \end{cases}$$

More than 80% of doses were assumed to have been taken as prescribed by 60% of subjects. For active treatment, RISK is decreased by 20% in R and 5% in NR for a 10% increase in therapeutic coverage over the threshold.

One hundred simulations were performed to determine the difference in outcome expected between the placebo and treatment groups under the stated conditions and study power. Various scenarios (A2–A5) were simulated and the results were compared with the pivotal simulation (A1), described above:

A2: Increased number of NR to 30% due to wrongly included subjects
A3: Dropout rate of 20%
A4: Increased noncompliance with the assigned treatment so that only 30% of subjects are more than 80% compliant
A5: Combination of A2, A3, and A4

Assuming no deviations from the specified protocol (pivotal simulation, A1), the study power is 0.93 to detect a treatment effect resulting in a 10% decrease in the occurrence of the endpoint of interest. In comparison, each of the protocol deviations detailed in simulations A2–A5 resulted in both a significantly ($p < 0.0125$) reduced study power and a reduced expected treatment effect (Table 2 and Figure 4).

## 4.8 SOFTWARE TOOLS

Execution models may be implemented using any statistical modeling software, such as Splus (Insightful Corporation, Seattle, WA) and SAS (SAS Institute, Cary, NC), or using specialized clinical trial simulation software, such as Clinical Trial Simulator (Pharsight Corporation, Mountain View, CA). In addition, execu-

**Table 2** Results of CTS Example: Effect of Protocol Deviations

| | A1 | A2 | A3 | A4 | A5 |
|---|---|---|---|---|---|
| Difference in effect between treatment and placebo groups[a] | 10 | 7.5 | 7 | 7 | 5 |
| Study power[a] | 0.93 | 0.73 | 0.73 | 0.69 | 0.48 |

[a] Median of 100 simulations

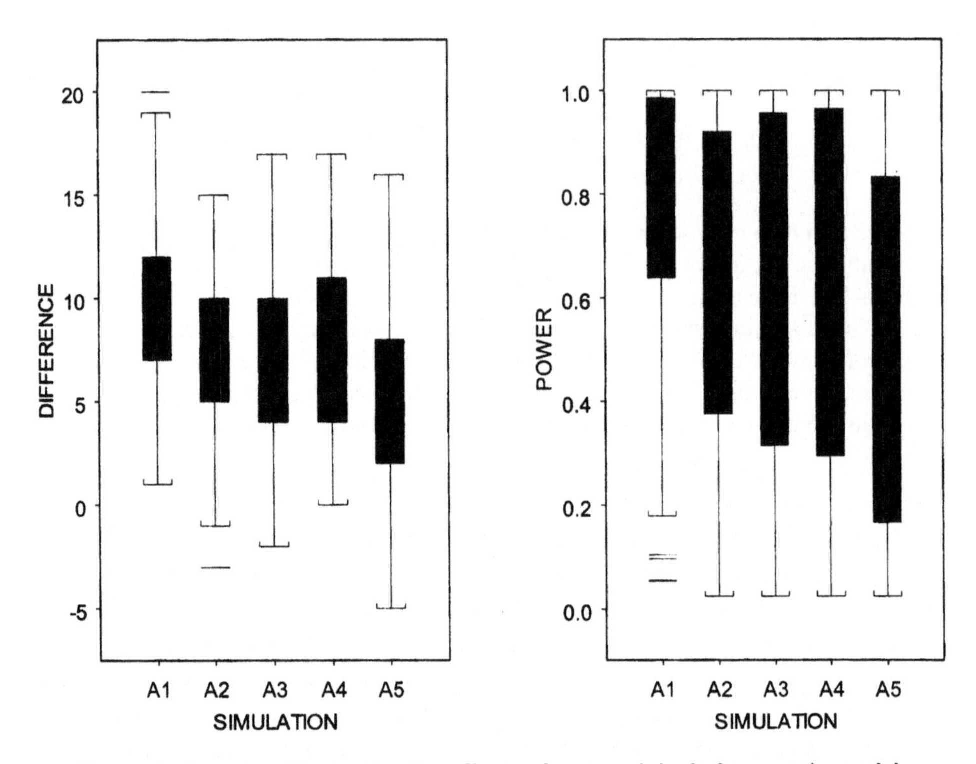


**FIGURE 4** Boxplots illustrating the effects of protocol deviations on the anticipated treatment effect (left panel) and on study power (right panel). Simulation conditions are detailed in the text.

tion models may be implemented in the nonlinear mixed effects modeling program, NONMEM (UCSF Regents, San Francisco, CA) which is commonly used for analysis of PK/PD data. For example, random assignment of a binary event *R* can be achieved using the following code:

```
$PROBLEM simulation example
$INPUT ID R DV            ;eg missing value for R
$DATA data


$PRED
"FIRST
"   REAL RSIM
"   INTEGER N
```

```
     IF(ICALL.EQ.4.AND.NEWIND.NE.2) THEN
"     CALL RANDOM(2,RSIM)   ;generates a random number ~ U[0,1]
"     N=1
"     IF (RSIM.LT.0.5) N=2 ;assigns subjects to event 1 or event 2
"     R=N
"     @R=N                 ;replaces missing value for R with N
     ENDIF


  Y=THETA(1)+ETA(1)        ;model definition


$THETA 10
$OMEGA 1
$SIMULATION(140978)(123456 UNIFORM)ONLYSIM
```

Dedicated CTS software has the advantage over other software packages in that integrated modules allow all components of the CTS project to be specified, including the protocol design, specification of the IO model, and covariate distribution models; the nature, extent, and combinations of protocol deviations to be simulated; and the execution models that will be used to describe each deviation; as well as modules for specification of how the results are to be evaluated and presented.

## 4.9 PUBLISHED CTS WITH PROTOCOL DEVIATIONS CONSIDERED

### 4.9.1 Quetiapine

Kimko and coworkers (17) simulated the anticipated results of a phase III clinical trial of the antischizophrenic drug, quetiapine, based on IO and covariate distribution models developed using data collected in earlier phase I and II trials. Model development was performed using the NONMEM program with first-order conditional estimation (18). The proposed trial design was a double-blind, placebo-controlled, randomized, parallel group study of fixed-dose quetiapine in hospitalized schizophrenic subjects. Subjects received 1 of 5 doses of quetiapine or placebo for a period of 4 weeks. Treatment was initiated after a placebo run-in period followed by a 2 week stepwise dose titration period.

The executed study design was replicated by excluding individuals wrongly included in the study who failed to meet the entry criteria. In addition, placebo responders identified during the placebo run-in period were replaced. A random dropout algorithm using a multiplicative congruential method (such that the random number generated is the remainder of a linear transformation of the previous number divided by an integer) was used to simulate the high dropout rate observed in the earlier phase II study (see Ref. 17). Based on the phase II study result, 70% of subjects assigned to the placebo group, 60% assigned to the lowest dose group, and 50% assigned to all other dose groups were withdrawn from the study.

Simulations were performed using ACSL Biomed (Pharsight Corp, Mountain View, CA) for 100 sets of 50 subjects per treatment group. Adequacy of the model to describe the original data was tested using sensitivity analysis and by comparing posterior parameter distributions and posterior predictions from the simulated trial design to parameters of the prior distribution and observed data. Dropout rates in the simulation and in the phase III trial were comparable. Comparison of the simulated results with actual results obtained in the phase III trial showed that the model adequately predicted responses to quetiapine; however, it was found to be inadequate in predicting the placebo response.

### 4.9.2 Docetaxel

Clinical trial simulation for docetaxel (19) was performed using PK/PD models previously developed from data obtained in earlier open-label, nonrandomized, phase II clinical trials of docetaxel in subjects with small cell lung cancer. The purpose of the simulation was to predict the influence of dose on survival time and time to disease progression in a high risk group in a planned phase III trial comparing doses of docetaxel of 100–125 mg/m$^2$ every 3 weeks.

IO and covariate distribution models were developed using the NONMEM program (18). Hazard models were used to simulate the primary and secondary clinical endpoints, death and disease progression, respectively. In addition, the execution model included a separate hazard model for patient dropout. Different models were tested and the Weibull distribution was selected based on goodness of fit assessed in the model-building phase of the analysis. A dose titration algorithm allowed for a 25% dosage reduction in the event of severe toxicity for each treatment cycle. To maintain consistency with study implementation, after two dosage reductions or if disease progression occurred, the patient was withdrawn from the study.

Simulations were performed using ACSL Biomed (Pharsight Corp, Mountain View, CA) for 100 sets of subjects and the results were analyzed using SAS (SAS Institute, Cary, NC). Adequacy of the model to describe the phase II data was tested using a posterior predictive check of the following test quantities: number of deaths and progressions, median survival time, 1 year survival, median time to progression, patient characteristics at baseline, number of side effects at the end of the first cycle, number of treatment cycles per patient, and total dose. Tabulated median and 95% confidence intervals of simulated test quantities agreed well with those obtained from the original data. In addition, 100 sets of

200 subjects per treatment group were simulated under the phase III trial design and test quantities were calculated. The results of the phase III trial simulation showed no clinical advantage of the higher docetaxel dose on survival or time to disease progression in high-risk subjects with small cell lung cancer. As a consequence of this analysis, it was determined that there would be no further clinical studies to evaluate the effect of dose intensification in subjects with small cell lung cancer.

## 4.10 SUMMARY

Execution models are used to examine the influences of protocol deviations on study outcomes. When implemented as a part of CTS, they allow "virtual" clinical trials to be run under varying conditions, from simple errors in data gathering to complex combinations of protocol deviations that emulate real-world situations. Thus, execution models are powerful tools for identifying weaknesses or limitations in a proposed study design, which may be anticipated, avoided, or resolved in order to increase robustness of the study design prior to implementation of the actual clinical study. As such, they are an integral component of CTS and an essential tool in clinical trial design.

## ACKNOWLEDGMENTS

This work was supported in part by the Hans Vahlteich Endowment Program for Faculty Research, College of Pharmacy, The University of Illinois at Chicago (HK).

## REFERENCES

1. J Urquhart. The electronic medication event monitor. Lessons for pharmacotherapy. Clin Pharmacokinet 32:345–356, 1998.
2. J Urquhart. Ascertaining how much compliance is enough with outpatient antibiotic regimens. Postgrad Med J 68:S49–S59, 1992.
3. J Urquhart, E de Klerk. Contending paradigms for the interpretation of data on patient compliance with therapeutic drug regimens. Statist Med 17:251–267, 1998.
4. H Kastrissios, T Blaschke. Medication compliance as a feature in drug development. Annu Rev Pharmacol Toxicol 37:451–475, 1997.
5. P Nony, M Cucherat, J-P Boissel. Revisiting the effect compartment through timing errors in drug administration. Trends in Pharmacological Sciences 19:49–54, 1998.
6. B Vrijens, E Goetghebeur. The impact of compliance in pharmacokinetic studies. Stat Methods Med Res 8:247–262, 1999.
7. W Wang, F Husan, S-C Chow. The impact of patient compliance on drug concentration profile in multiple doses. Statist Med 15:659–669, 1996.
8. W Wang, S Ouyang. The formulation of the principle of superposition in the pres-

ence of non-compliance and its applications in multiple dose pharmacokinetics. J Pharmacokinet Biopharm 26:457–469, 1998.

9. P Girard, LB Sheiner, H Kastrissios, TF Blaschke. Do we need full compliance data for population pharmacokinetic analysis? J Pharmacokinet Biopharm 24:265–282, 1996.
10. E Lesaffre, E de Klerk. Estimating the power of compliance—improving methods. Control Clin Trials 21:540–551, 2000.
11. P Girard, T Blaschke, H Kastrissios, L Sheiner. A Markov mixed effect regression model for drug compliance. Statist Med 17:2313–2333, 1998.
12. P Girard, F Varret. A Population Model for Compliance and Drop-out Model for Once a Day Regimen in Depressed Patients. In: *PAGE 1999*, Saintes, France.
13. M Karlsson, L Sheiner. The importance of modeling interoccasion variability in population pharmacokinetic analysis. J Pharmacokinet Biopharm 21:735–750, 1993.
14. S Laporte-Simitsidis, P Girard, P Mismetti, S Chabaud, H Decousus, J-P Boissel. Inter-study variability in population pharmacokinetic meta-analysis: when and how to estimate it? J Pharm Sci 89, 2000.
15. P Girard, S Laporte-Simitsidis, S Chabaud. Multilevel nested random effect model implementation. In: *PAGE 2000*, Salamanca, Spain.
16. J Lu, JM Gries, D Verotta, L Sheiner. Selecting reliable pharmacokinetic data for explanatory analyses of clinical trials in the presence of possible noncompliance. J Pharmacokinet Pharmacodyn 28(4):349–362, 2001.
17. H Kimko, S Reele, N Holford, C Peck. Prediction of the outcome of a phase 3 clinical trial of an antischizophrenic agent (quetiapine fumarate) by simulation with a population pharmacokinetic and pharmacodynamic model. Clin Pharmacol Ther 68:568–577, 2000.
18. S Beal, L Sheiner, NONMEM Users Guide. NONMEM Project Group, University of California at San Francisco, 1992.
19. C Veyrat-Follet, R Bruno, R Olivares, G Rhodes, P Chaikin. Clinical trial simulation of docetaxel in patients with cancer as a tool for dosage optimization. Clin Pharmacol Ther 68:677–687, 2000.

# 5

# Determination of Model Appropriateness

**Paul J. Williams**
Trials by Design, LLC, La Jolla and Stockton, and University of the Pacific, Stockton, California, U.S.A.

**Ene I. Ette**
Vertex Pharmaceutical Inc., Cambridge, Massachusetts, U.S.A.

## 5.1 INTRODUCTION

There are many reports published each year with results of the development of various pharmacometric (PM) models. These models are population (hierarchical) models that include but are not limited to pharmacokinetic, pharmacodynamic, disease progression, and outcomes link models. For population pharmacokinetic (PPK) and pharmacodynamic (PPD) models an FDA survey (1) reported that some PPK and PK/PD models developed as part of New Drug Applications (NDAs) did not add value or have any impact on the submission because the usefulness of such models was not addressed. Most PM models are developed but their applications are rarely addressed. When the purpose or intended use of the PM model is not stated, the applicability of the model is unclear. For PPK models all this continues in spite of the statement in the *Guidance for Industry: Population Pharmacokinetics* (The Guidance) (2), "A discussion of how the results of the analysis will be used (e.g., to support labeling, individualized dosage, safety, or to define additional studies) should be provided. In addition, the use of graphics, often based on simulations of potential responses under the final fitted model, to communicate the application of a population model (e.g., for dosage adjustment) is recommended." The nature of the application of a PM model must be taken into account during the entire modeling process. The intended use of a model should influence the attitude and modeling approaches used by the pharmacometrician at the various stages of the modeling process. This would determine what covariates are considered important, which parameters are of primary concern, and what are the extent and method of model evaluation and validation. Thus, it is the application or intended use of the PM model, which drives the modeling process from model development, through model evaluation, to model validation (as deemed necessary) that constitutes model appropriateness.

In the subsequent sections, we present the methodology for determining model appropriateness; approaches to model evaluation and validation; metrics used in model validation; and application examples to illustrate the principles involved in the determination of model appropriateness.

## 5.2 MODEL APPROPRIATENESS

### 5.2.1 Definition

A PM model is developed to help solve a problem. The identification and elucidation of this problem is necessary and preliminary to the modeling process. The establishment of model appropriateness requires that the intended use of the model be stated; model evaluation be done; and model validation be performed if the model is to be used for predictive purposes (evaluation and validation are addressed in detail later). For example, a descriptive model may be used to explain a higher incidence of adverse effects of a drug in one subgroup of subjects versus another. It may be that a drug is primarily eliminated by the liver and that the subgroup of subjects with impaired liver function has a lower clearance and therefore greater exposure to the drug than the subgroup without impaired liver function. It may also be that the group of subjects with impaired liver function had a greater incidence of adverse effects than the group with normal function. One possible explanation for the greater incidence of adverse effects in the impaired liver function group is that they had a greater exposure to the drug resulting in greater toxicity. This would be adequate as a key source of information that would help to explain subgroup differences in response and would also be a descriptive model because it is only applied to the subjects from which it was derived.

However, if one were interested in evaluating the effect of several dosing strategies on outcomes for a pivotal phase III study via simulation, then the distribution of the pharmacokinetic parameters becomes very important. In the former case evaluation of the estimated value for population clearance would be sufficient to establish appropriateness; in the latter both drug concentration predictability and the evaluation of the distribution of parameters would necessarily be done.

The purpose for which a model is developed has significant impact on the modeling process. Insufficient consideration is given to the purpose of the model in current practice. In most instances, a model is constructed using routine methods of identification and estimation and then used for any sort of application without regard to whether it was developed for description or prediction.

### 5.2.2 Descriptive Models

Sometimes a PM model is developed to explain variability in the PM of the drug. In this case, it is used as an empirical and numerical summary of information about PM variability in the population studied and would include all the features of the population that are found to be important covariates. When a model is used for descriptive purposes, it is important to assess its goodness of fit, reliability and stability. Model evaluation should be done in a manner consistent with the intended application of the PM model. The reliability of the analysis results can be checked by careful examination of diagnostic plots, key parameter estimates, standard errors, case deletion diagnostics (3, 4) and/or sensitivity analysis as may seem appropriate. Confidence intervals (standard errors) for parameters may be checked using nonparametric techniques, such as the jackknife and bootstrapping, or the profile likelihood method. Model stability to determine whether the covariates in the PM model are those that should be tested for inclusion in the model can be checked using the bootstrap (2, 3, 5).

When models are not checked for stability, it is possible to include spurious covariates in the model because of noise in the data that may have their source in only a few subjects or from multiple comparisons. Small changes in a data set may result in the selection of different covariates for a PM model when a model is not checked for stability.

### 5.2.3 Predictive Models

A second class of PM models are used for predictive purposes. Although these models contain descriptive components, they are used to answer "what if" questions about the effects of changes in the levels of covariates in the model. An example would be the effect of subpopulation differences on dosage regimen design. When PM models are developed for predictive purposes, much stronger assumptions are made about the relationship to the underlying population from which the data were collected. One is asking for correspondence of behavior outside the range over which one has actual empirical evidence. In the descriptive sense of a PM model one is concerned with what was observed; for prediction the behavior of the model is important.

For predictive purposes, a PM model should be validated in a way that is consistent with the intended use of the model. When doing this, it is always important to be reminded of the dictum from Box that "All models are wrong, some are useful" (6). Therefore, asking the question of whether one's model is true or false is not appropriate but rather one should ask whether the model's deficiencies have a noticeable effect on the substantive inferences. Model validation results in confidence that the model does not have deficiencies that will result in its not being applicable for its intended use.

In this chapter two examples of model appropriateness are presented, one a descriptive model and the other predictive. In the first example, a descriptive model is presented where there was a problem identifying which covariates ought to be included in a PPK model. In the second example we describe the modeling process for PM model development, model evaluation, validation (i.e., predictive performance), and application of the developed model to dosage regimen design to demonstrate the appropriateness of the model. Although there are several methods for determining the predictive performance of PM models (2), models developed for predictive purposes are seldom validated (1, 7) to establish their appropriateness.

When a model is to be employed in some type of simulation (e.g., dosage regimen design or clinical trial simulation) the model is by nature predictive, because simulations imply some correspondence of behavior to an external population. It has been pointed out that "A simulation can be no better than the quality of the model it uses" (8). Therefore, models that are developed with the intent of use in a clinical trial simulation must be validated.

## 5.3 APPROACHES TO MODEL EVALUATION AND VALIDATION

### 5.3.1 Model Evaluation

Model evaluation attempts to answer the question, "Are the structure and form of the model without significant error?" Model structure is a term that can be applied to both a deterministic model or a population model. In the case of a standard pharmacokinetic model (i.e., a deterministic model), structure most often deals with the compartmental character of the model. Thus, for a deterministic pharmacokinetic model, the structure may be defined as the number of compartments used to characterize the model. Population model structure is concerned with what covariates and random effects are included in the model and with which parameters have covariates and random effects been associated. Model structure can also be extended to pharmacodynamic and disease progression models. Model form, on the other hand, deals with the shape of the relationship between covariates and parameters. For example, if body weight were related to clearance, is the relationship linear or nonlinear? Model evaluation, therefore, encompasses goodness-of-fit assessments, checking the reliability and stability of the model.

#### 5.3.1.1 Goodness of Fit

Goodness-of-fit assessments require diagnostic plots of such things as the predicted (PDV) versus the observed (ODV) dependent variable, residuals versus PDV, weighted residuals versus PDV, weighted residuals versus time, and residuals versus covariates to examine for any type of systematic error. Pharmacometric models should be without systematic error.

#### 5.3.1.2 Model Reliability

Model reliability addresses the question, "Is the model worthy of confidence?" and thus deals with the issue of the degree of uncertainty associated with the estimated parameters. When characterizing reliability, the structure and form of the model remain unchanged and two elements must be addressed to assess reliability. First, the model is estimated with some degree of uncertainty and one would like that uncertainty to be "small." Percent relative standard error of less than 25% for point estimates and less than 35% for variability estimates would be considered small uncertainty (9). A model yielding estimated parameters with associated small uncertainties is deemed reliable. Second, if there are alternative approaches to computing parameter estimates, do these other approaches yield similar parameter estimates and conclusions? An example of addressing this second point would be estimating the parameters of a PPK model by the standard maximum likelihood approach and then confirming the estimates by either constructing the profile likelihood plot (i.e., mapping the objective function), using the bootstrap (5) or the jackknife (3). When the reliability of the estimates is established, confidence in the model is established.

#### 5.3.1.3 Model Stability

Model stability addresses the question, "How resistant is the model to change?" It is a partial confirmation of the structure of the population model and it also serves as a confirmation of the form of the population model as well. When stability is assessed one must ask if other plausible or probable data change the model structure or form. If the model form or structure does not change when the data are changed to other plausible data, then the model is stable. Bootstrapping can be employed to check for model stability by constructing other plausible data and determining if the model structure is unchanged for the majority of these data (4).

### 5.3.2 Model Validation

It has been previously pointed out that valid is defined as "having such force as to compel acceptance" and implies "being supported as objective truth" (7).

Regarding validation, it is most often defined as the evaluation of the predictability of the model developed (i.e., the model structure and form, together with the model parameter estimates) using a learning or index data set when applied to a validation (test) data set not used for model building and parameter estimation. Thus, we are concerned with the predictive performance of a PM model. This addresses the issue of generalizability of the PM model, that is, ascertaining whether predicted values from a developed PM model are likely to accurately predict responses in future subjects not used to develop the model. In a review of population PK, PK/PD, or PD models published from 1977 to 1996, 136 publications were identified. In this review only 19% were noted to have been validated as part of PM modeling (7). There are two broad categories of model validation, external and internal, which are discussed below.

#### 5.3.2.1 External Validation

External validation is the most stringent type of validation. With it the PM model from the index population is used to predict the dependent variable in an appropriate test (external) population. Thus, external validation addresses the wider issue of the generalizability or transportability of the model.

#### 5.3.2.2 Internal Validation

The methods of internal PM model validation include data splitting, resampling techniques (cross-validation and bootstrapping), and the posterior predictive check (PPC). The internal validation approaches are often useful when it is not practical to collect a new data set to be used as a test data set. Examples can be found in studies performed in pediatric populations and populations of patients with rare diseases (10).

*a. Data Splitting.* With this approach, a random subset of the data are chosen (e.g., two-thirds of the entire data set) for the index data and the model is developed, then the remaining data are chosen (e.g., one-third of the entire data set) for the test population. A model is initially fitted to the first portion of the data (the training data set) and its predictive performance evaluated in the second (test) data set. If the index model is validated, then the data may be recombined and the final model fitted to the combined data set. The disadvantage of data splitting is that the predictive accuracy of the model is a function of the sample size resulting from the splitting. An issue with data splitting is how to split the data into training and test data sets. Authors rarely consider what proportions of patients should be in the training and test sets, or fail to justify any recommendation made for splitting the data. Random splitting must lead to data sets that are the same other than for chance variation and is therefore a weak procedure (11). To maximize the predictive accuracy of data splitting, it has been recommended that the entire sample be used for model development and assessment (12, 13).

*b. Cross-Validation.* The resampling approaches of cross-validation (C-V) and bootstrapping (see the following subsection) do not have the drawback of data splitting in that all available data is used for model development so that the model provides an adequate description of the information contained in the gathered data. Cross-validation can be thought of as repeated data splitting and has the advantage over data splitting of preserving a much larger test data set and not relying on a single sample split. Therefore, all of the data is used in both the model development process and in the validation of the model. However, there is a high variation of estimates of accuracy and cross-validation is inefficient when the entire validation process is repeated (14).

C-V may be used to estimate how well a prediction rule predicts the response value of a future observation. It is often used for model validation or model selection since it is sensible to choose a model that has the lowest prediction error among a set of candidates. With C-V a part of the available data is used to fit the model, and the part of the data not used to fit the model is used to test the model. This process is done repeatedly; therefore C-V can be thought of as repeated data splitting. However, C-V is a generalization of data splitting that solves some of the problems of data splitting. When one employs the C-V process, the parent data set should be well mixed.

Figure 1 shows the overall flow of C-V. There are two types of C-V: *leave-*

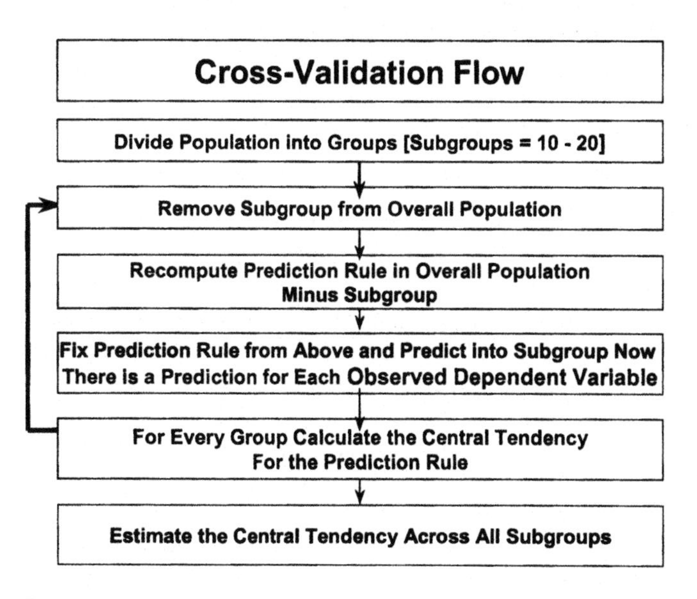


**Figure 1** Diagrammatic representation of the flow of the cross-validation process.

*out-one C-V* and *grouped C-V*. The leave-out-one C-V is the limit of C-V. With this approach one subject is omitted from the modeling process, and response from that observation is predicted using a model derived from the remaining $N - 1$ subjects. The modeling process is repeated $N$ times to obtain an average accuracy.

The grouped C-V *method* consists of:

1. Divide the data into $K$ approximately equal-sized groups. This would usually be 10–20 groups. Suppose that there is a group of 90 subjects who have been used to build a PPK model. Divide this population of subjects into 10 groups of 9 subjects each. Obtain a well-mixed population by choosing every tenth subject. Take all of the data for each subject on a subject by subject basis.
2. Remove the $K$th part of the data (9 subjects who belong to the $K$th group) and refit the model to the $-K$ parts of the data (the 81 remaining subjects) and recalculate the prediction rule. Here the $-K$ data serves as the training set.
3. Next, fix the prediction rule from step 2 then apply it to the $K$th group (9 subjects) to generate a set of predicted dependent variables, one for each observed dependent variable, in the $K$th part of the data. Thus, there is now a predicted dependent variable for every observed dependent variable in the Kth data based on the rule from $-K$ data.
4. For each $K$th replication estimate the central tendency of the prediction error parameter.
5. Do the above for the $k = 1, 2, \ldots, K$ and combine the $K$ estimates of prediction error by estimating the central tendency for the prediction error.

This process provides one with a revised estimate of the prediction error that one would expect to see if the rule was applied to a naïve test population.

*c. The Bootstrap.* Bootstrapping (15, 16) is the resampling with replacement method that has the advantage of using the entire data set. It is a resampling technique that was first presented by Efron in 1979 as a tool for constructing inferential procedures in modern statistical data analysis (16). It has been demonstrated to be useful in PM model validation (5), and has the same advantages as other internal validation methods in that it obviates the need for collecting data from a test population. Collecting test population data can be time consuming and difficult in some populations such as pediatric patients or rare diseases. In addition to being useful for model validation, it can also be applied to checking model reliability and stability (5). Bootstrapping has been applied to PPK model development, stability check or evaluation (5, 17, 18).

In using the bootstrap to check model reliability and stability, it is often informative to reestimate the distribution of parameters via the bootstrap and compare them to the originally estimated distribution of parameters. This is a valuable step in confirming that the developed model is an appropriate one. Of importance is the fact that the 95% confidence intervals (CI) for parameters can be estimated by the bootstrap in a nonsynthetic way without assuming asymptotic distributional properties of the parameters. When employing the bootstrap to estimate confidence intervals (CIs), no assumptions are made about the distribution of the estimation error of the parameters. This approach therefore provides improved estimates of the 95% CIs for these parameters. Depending on the intended use of the PM model, bootstrap estimates may be used to replace the parametric estimates of the CIs or one may use the bootstrap estimates to confirm the reliability of the underlying parametric estimates of the CIs and the stability of the PM model (2, 3, 5). The various bootstrap methods are described below.

The bootstrap is a method of repeatedly generating pseudo-samples distributed according to the same distribution as the original sample and is represented in Figure 2. The generation of the pseudo-sample is carried out by randomly selecting a subject's data. In the most general sense the bootstrap involves four steps.

1. First, a bootstrap pseudo-observation data set is generated from the same general region that contains all of the individual experimental data. This step generates equally plausible sets of data as that which has been observed. Each bootstrap sample is of the same sample size

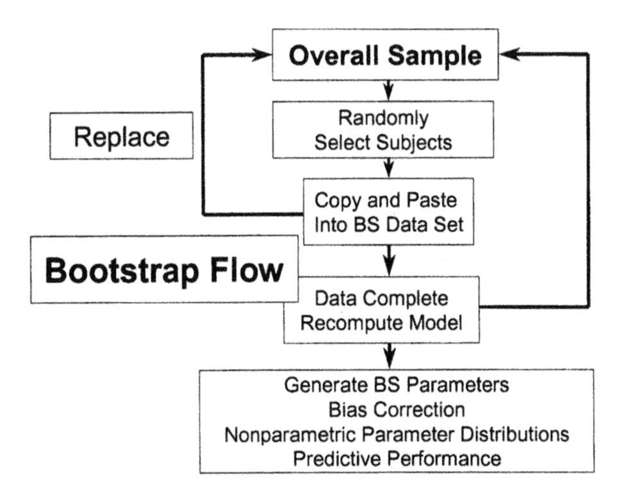


**FIGURE 2** Flow of the bootstrap process.

$N$ as the original sample. The subjects are resampled, and this constitutes the nonparametric bootstrap.

2. Second, each set of pseudo-data is used and the precedure of interest is applied to each pseudo-data set. This step maps a plausible region of the results of the precedure.
3. The results of the application of the procedure to each of the pseudo-samples is done.
4. The standard error is computed as the sample standard deviation of the $B$ replications.

There is another method of bootstrapping called *bootstrapping of the residuals*. When one develops a PM model, a residual can be estimated for each ODV:

$$\varepsilon_{i,j} = \mathrm{PDV}_{i,j} - \mathrm{ODV}_{i,j} \tag{1}$$

where $\varepsilon_{i,j}$ is the $j$th residual in the $i$th subject, $\mathrm{ODV}_{i,j}$ is the $j$th observed dependent variable in the $i$th subject, and $\mathrm{PDV}_{i,j}$ is a predicted dependent variable for $\mathrm{ODV}_{i,j}$. In this case one can estimate the approximate residuals $\hat{\varepsilon}_{i,j}$ which are also called errors. These residuals are assumed to be a random sample from an unknown distribution with an expectation 0.

When bootstrapping the residuals, one first selects a random sample from the residuals by resampling the residuals. At this point $\varepsilon^*,^*$ is any one of the $n$ values of $\varepsilon_{i,j}$ with a probability of $1/n$ of being chosen. Then a set of bootstrap concentrations are generated by $\mathrm{PDV}_{i,j} + \varepsilon^*.^*$. The bootstrap data set $x^*$ now equals $(x_1^*, x_2^*, \ldots, x_n^*)$, where $x_i^* = (\mathrm{ODV}_{i,j}, \mathrm{PDV}_{i,j} + \varepsilon^*.^*)$. Following the creation of $B$ data sets, steps 2–4 above are repeated.

Bootstrapping residuals assumes that the error between $\mathrm{PDV}_{i,j}$ and its mean doesn't depend on $\mathrm{ODV}_{i,j}$; that it has the same distribution no matter what the $\mathrm{ODV}_{i,j}$ would be; and that the model is known with certainty. These assumptions may not be met in PM modeling. Bootstrapping subjects is less dependent on assumptions than bootstrapping residuals. The parametric bootstrap involves resampling the parameter vector from its posterior distribution, and simulating the response variable from the distribution of each of the bootstrap samples.

The most commonly used parameter to assess predictability for the bootstrap has been the squared prediction error (SQPE). The SPE refers to the square of the difference between a future response and its prediction from the model:

$$\mathrm{SQPE} = E(\mathrm{ODV} - \mathrm{PDV})^2 \tag{2}$$

where ODV and PDV are the observed and predicted dependent variables and $E$ (the expectation) refers to repeated sampling from the true population.

Bootstrapping provides estimates of predictive precision that are nearly unbiased and that are of relatively low variance. When each of the models is applied to its own bootstrap data set, there is for each ODV a PDV that is generated.

The most common prediction error operator is the mean squared prediction error (MSPE). Thus, MSPE($M_1$, $D_1$) is an estimate of the MSPE when model 1 is applied to bootstrap data set 1. This is done until MSPE($M_1$, $D_1$) to MSPE($M_{200}$, $D_{200}$) are obtained using 200 bootstrap data sets are generated, as an example (5).

The next step in assessing predictive accuracy is to apply the frozen models ($M_1$ to $M_{200}$) to the data set $D_0$, which is the original data set (not from bootstrap resampling) with all individuals occurring once. In this process the coefficients and random effects are fixed for $M_1$ to $M_{200}$, so that at this step one has MSPE($M_1$:$D_0$) to MSPE($M_{200}$:$D_0$). The next step is to estimate the optimism (OPT) (or bias due to overfitting in the final model fit) for the prediction operator. This is executed for each model so that

$$\text{OPT}_i = \text{MSPE}(M_i{:}\text{D}_0) - \text{MSPE}(M_i, D_i) \tag{3}$$

where $\text{OPT}_i$ is the optimism for the $i$th model. It is expected that MSPE($M_i$, $D_i$) will be smaller than MSPE($M_i$:$D_0$) because MSPE($M_i$, $D_i$) is making predictions into the data set from which it was estimated and MSPE($M_i$:$D_0$) is making predictions into the original data set which serves as an "external" data set because $M_i$ is a PM from a bootstrap sample.

Next, the mean OPT across all the models is estimated and in this case is

$$\overline{\text{OPT}} = \frac{1}{n} \times \sum_{i=1}^{n} \text{OPT}_i \tag{4}$$

Once the mean optimism is known, it is added to the results of the prediction operator that was estimated from the original data set. This results in an improved estimate of the prediction operator as the prediction operator estimates generated when the process is applied to its own data will be overly optimistic compared to applying the prediction operator to the universe of possible external data sets.

$$\text{MSPE}_{\text{imp}} = \text{MSPE}(\text{M}_0{:}\text{D}_0) + \text{OPT} \tag{5}$$

where $\text{MSPE}_{\text{imp}}$ is the improved estimate of the MSPE provided by the bootstrap. The $\text{MSPE}_{\text{imp}}$ is estimated to provide an estimate of the MSPE that would be calculated if the model were applied to an external data set. The lower the $\text{MSPE}_{\text{imp}}$, the better the model.

*d. Posterior Predictive Check.* A new method, the posterior predictive check (PPC) may prove useful in determining whether important clinical features of present and future data sets are faithfully reproduced (19, 20, 21). The PPC can take either the form of internal or external validation depending on how the process is implemented. The underlying premise of the PPC is that simulated data based on the model should be similar to some set of observed data. The process of PPC can involve comparing the posterior distribution of (1) parameters to substantive knowledge or other data, (2) future observation to substantive knowledge or other data (external prediction), or (3) future observation to the data that have actually occurred (internal prediction). When this is done, then model replicates of data $y^{rep}$ that are comparable to the observed data $y$ are simulated by applying the same study design, and fixed and random effects values. The characteristics of $y^{rep}$ sets are compared with $y$. A test statistic is chosen and estimated for $y$ and each $y^{rep}$. Finally, the test statistic for $y$ is compared with the distribution of values for the $y^{rep}$ sets. This may be quantified as a posterior predictive $p$ value:

$$\text{Bayes } p \text{ value} = \text{probability } [T(y^{rep}, \theta) \geq T(y, \theta | y)] \tag{6}$$

where $T$ denotes a scalar-valued function (a statistic) of the observations or estimated parameters ($\theta$).

Two clear examples of the PPC have been presented where the model was used to simulate a set of dependent variables, drug concentrations, and the pharmacodynamic marker of heart rate (20, 21). For these simulations, the population parameters were fixed and concentration-time profiles with the heart rate–time profiles were simulated for 200 subjects. The expected 10th and 90th percentile concentrations and heart rate from the simulations were plotted and overlayed on the observed concentrations, overlayed on an external set of observed concentrations, and observed heart rate for each time point. In a further and very informative analysis, the authors compared the cumulative density functions for the ob-

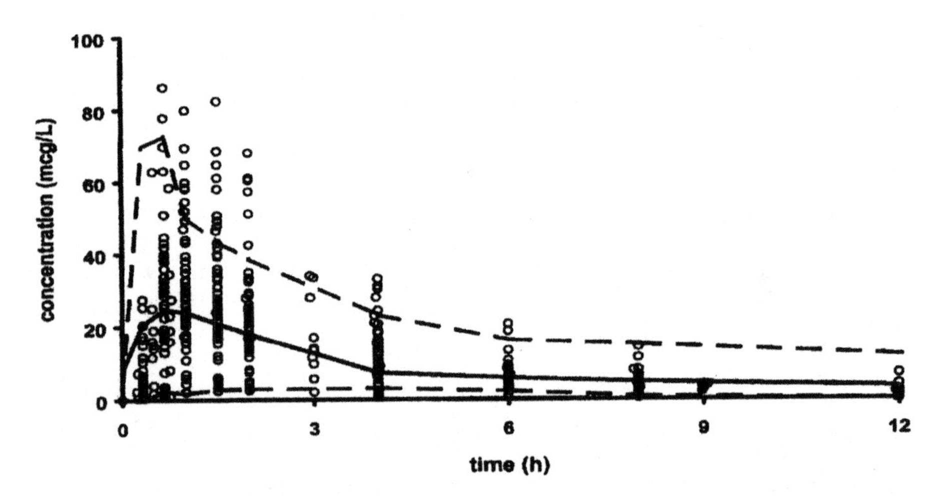


**FIGURE 3** Plot of concentrations from internal data overlay on 10th and 90th percentile lines (dashed lines). [From Duffull et al. [20].]

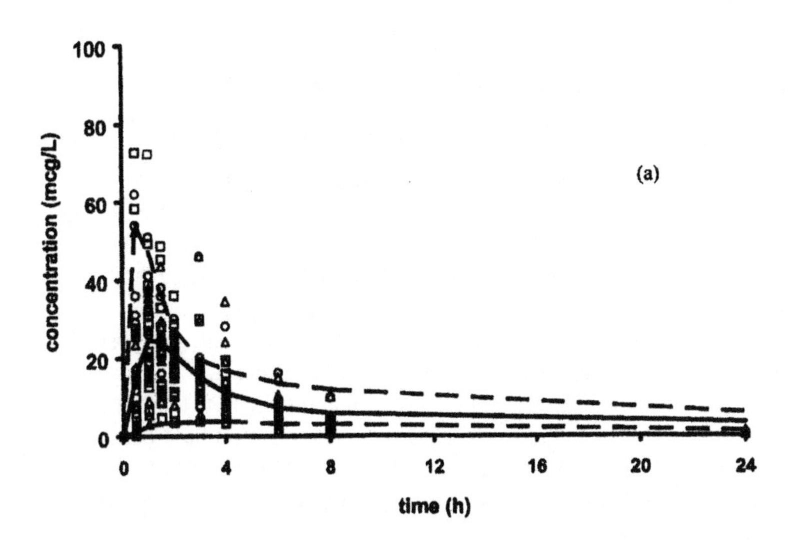


FIGURE 4 Plot of concentrations from external data overlay on 10th and 90th percentile lines (dashed lines). [From Duffull et al. [20].]

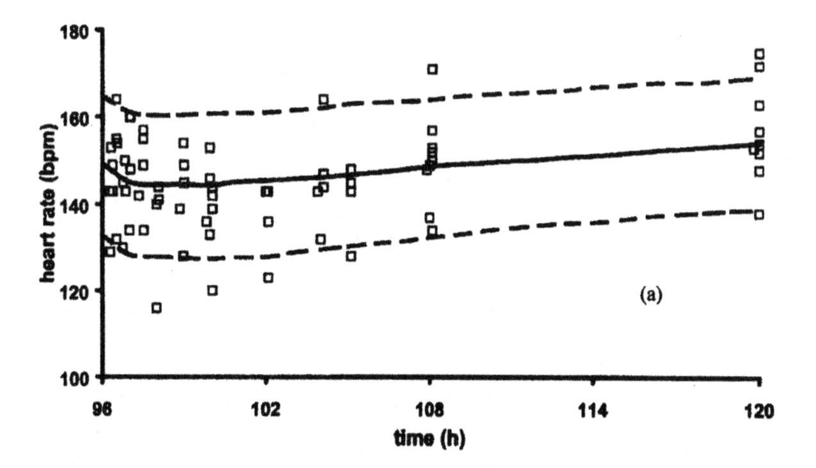


FIGURE 5 Plot of heart rate from internal data, a pharmacodynamic endpoint, overlay on expected 10th and 90th percentiles generated from model simulation. [From Duffull et al. [21].]

served versus the simulated dependent variables using a Kolmogorov-Smirnov (KS) test. The plots of percentiles and dependent variables versus time are presented in Figures 3 to 5. These plots indicated that the model was without significant error because about 10% of the observations fell above the 90th percentile line and 10% of the observations fell below the 10th percentile line. The simulated density functions of the predicted and observed density functions were not significantly different per the KS test. The values of the KS statistic for the drug concentrations from index and test data sets were 0.27 ($p = 0.83$) and 0.18 ($p > 0.9$), respectively, and 0.15 ($p > 0.90$) for heart rate at a dose of 10 mg. These final results indicated that the simulated dependent variable and the observed dependent variable distributions were not significantly different.

However, few papers have been published that apply the PPC to PM modeling (22–24). In a limited evaluation of PPC in a study the authors concluded that the failure of the PPC to invalidate a model offers little assurance that the model is correct (25). Further evaluation of the utility of the PPC especially in hierarchical models is needed.

## 5.4 METRICS APPLIED TO MODEL VALIDATION

Several approaches to quantitating dependent variable predictability have been proposed. The metric should be chosen by its usefulness, and there are instances when one of several metrics could be applied.

### 5.4.1 Graphical Displays

Graphical displays can also be very useful for model evaluation and validation. Often used plots include $ODV_{ij}$ versus $PDV_{ij}$, $PE_{ij}$ versus $PDV_{ij}$, $PE_{ij}$ versus time, $PE_{ij}$ versus covariates, and weighted residuals (WRES) versus covariates, where $ODV_{ij}$ is the $i$th observed dependent variable in the $j$th subject and $PDV_{ij}$ is the corresponding $i$th population predicted dependent variable in the $j$th subject based on the dosing history, sampling history, and covariates, and PE stands for prediction error. The individual PEs, which are residual differences, are calculated as

$$PE_{ij} = ODV_{ij} - PDV_{ij} \tag{7}$$

When the $ODV_{ij}$ is plotted versus the $PDV_{ij}$, the line of identity (unity) should be included and the scatter about the line observed.

In Figure 6 the plot of WRES versus subject identification is presented (26). The WRES are the residuals ($ODV_{ij} - PDV_{ij}$) that have been normalized by their standard deviations. The WRES are nearly independent even within the same individual. Thus, when one views the data from many individuals, the correlation that one would expect to see from several measurements within a single individual should not be seen when observing the WRES (26) (S.L. Beal, "Vali-

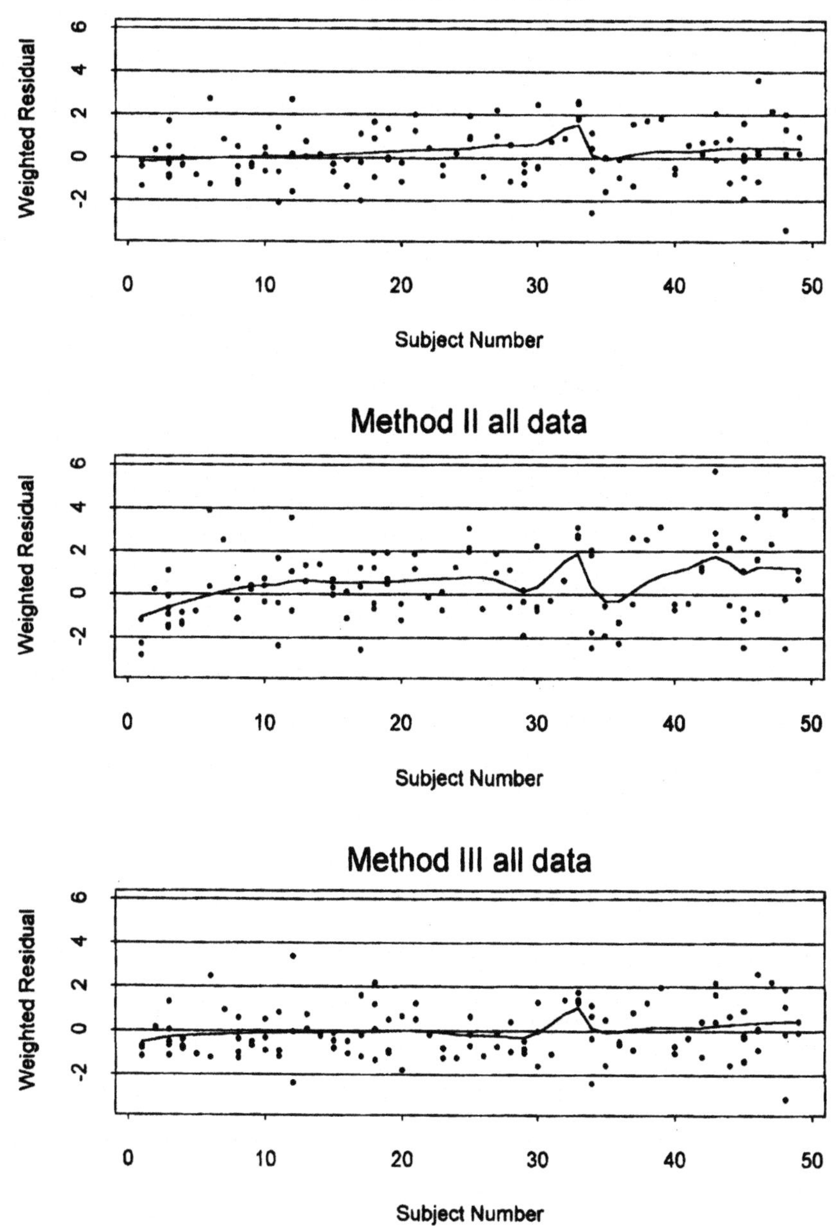


**FIGURE 6** Plot of weighted residuals versus subject identification number. [From Williams et al. [26].]

dation of a Population Model," e-mail to NONMEM Usernet Participants, Feb. 2, 1992). For an adequate model (i.e., a stable model yielding reliable parameter estimates with no severe deficiencies) the mean of the WRES should be scattered evenly about zero on the WRES axis and most observations should be within ±4. Figure 6 was taken from a comparison of several methods for generating predictions of digoxin serum concentrations. Method III appears to have the least bias (26).

### 5.4.2 Prediction Error

Prediction errors (PE) on dependent variables are a commonly calculated metric for validating PM models (26, 27). The mean PE (MPE) and the mean absolute error (MAE = mean |PE|) are often estimated. Under the assumption of normality, confidence intervals (CI) are constructed around the MPE.

$$CI = MPE \pm 2 * SE_{MPE} \tag{8}$$

where $SE_{MPE}$ is the standard error of the MPE. If the CI contains 0, then the model is said to have adequate predictability and the model is assumed to be without significant error (27).

One problem with the PE metric is that when there is more than one observation per subject the method is inadequate because the PEs within an individual are not independent (11). However, recently an approach has been suggested to overcome this problem where the CI is constructed under the statistical model

$$PE_{ij} = MPE + \eta_j + \varepsilon_{ij} \tag{9}$$

where $\eta_j$ is a random effect representing between subject variability and a persistent shift from the MPE for subject $j$ and $\varepsilon_{ij}$ is a random effect representing the residual variability (26). This technique accounts for the lack of independence of several measures made within the same subject as the model contains two sources of random effects, between and within subject random effects. This approach is easily implemented in several software programs (SAS, NONMEM, Splus, etc.) using mixed effects methods.

### 5.4.3 Standardized Prediction Error

The standardized prediction error (SPE) takes into account the variability and correlation of observations within an individual. The $SPE_{ij}$ is the $i$th standardized prediction error in the $j$th subject and is defined as the $PE_{ij}/SD_{ij}$, where the $SD_{ij}$ is the standard deviation associated with the $PDV_{ij}$. The mean SPE (MSPE) and the standard deviation of the SPE are estimated with their respective 95% confidence interval (CI). The CI for the MSPE should include 0 and the CI for the standard deviation of the SPE should include 1 for a validated PM model (28). However, when applied to a validation data set, this method and the method from

Section 5.4.2 may be overly conservative as uncertainty in parameter estimation is not taken into account (7, 13).

### 5.4.4 Prediction Through Parameters

Another method involves validation through parameters. In this case the PM model from the index data set is used to generate predicted parameters of the model in the test population. The parameters must be estimated in the test data, which is a straightforward process when the data in the test data set are rich. When sampling is sparse in the test data set then one could obtain Bayesian maximum a posteriori individual parameter estimates from the population. As for the prediction of the dependent variables, the bias and precision are estimated for the parameters usually under the assumption of normality. Bruno et al. (29) have provided an example of predictions through parameters. The predictive performance of the model for drug clearance was evaluated in a validation data set. For this study

$$PE_j\% = (Cl_{indv} - Cl_{pop})/Cl_{pop} \times 100 \tag{10}$$

The mean $PE_j\%$ was 8.2% for the overall population clearance and the mean absolute error was 21%. An interesting approach that these authors pursued was the estimation of the percent of subjects whose model prediction of clearance was better than the naïve prediction (prediction error at least 10% less) or worse (prediction error at least 10% greater). When this was done it, was noted that the model prediction error was smaller in 53% of subjects but was larger in 36% of subjects indicating that although the model did a bit better overall than the naïve predictor, the gain was modest.

Several approaches to model validation have been discussed. It would be unreasonable to expect that there would be a "one size fits all" method for model validation and it may be that a given model could be appropriately validated in one of several ways. The approach to a validation may be dictated by the proposed use of the model. Failure to do this can lead to a misleading validation. For predictive purposes, the focus of the validation has to be on the predictive ability of the model. In this case the success of the model depends solely on the closeness of the forecasts to the observed values. In this section, we will present examples of the appropriateness of descriptive and predictive model examples.

## 5.5 APPLICATION EXAMPLES

### 5.5.1 A Descriptive Model Example

In one example (5) of an anti-infective agent administered to 88 full-term infants, a problem confronted those evaluating a PPK model. The problem: A covariate, HIV status, was included in a PPK model at the $\alpha = 0.01$ level when evaluated by univariate approach. However, when a full model was constructed, backward elimination indicated that HIV status should be deleted as a predictor of clearance. There was a concern that HIV status may be a predictor of clearance but that the adopted PM approach indicated that it should be deleted from the model. Therefore, a bootstrap approach was constructed to rigorously evaluate the importance of HIV status as a predictor of population clearance for the drug.

One hundred nonparametric bootstrap replicate data sets were created and the deterministic structural model for each bootstrap data set was ascertained. This was to ensure that the structural model that best described the bootstrap data was not different from the basic structural model used for developing the PPK model. Once a structural model was confirmed, a selection method was used to identify the significant covariates. The bootstrap replicate data sets were used here because important covariates should be included in the model from most bootstrap data sets. Here it is assumed that each replication data, being a random sample from the patients under study, should reflect the underlying structure of the data. Therefore, the frequency of inclusion of a covariate in the model is a criterion for the predictive importance of a covariate, and the authors were looking for a frequency of inclusion of a covariate in the model of 50%. If a covariate was not included in at least 50% of the models from the bootstrap replicate data sets, it was excluded from further investigation.

The process was:

1. Generalized additive modeling (GAM) was applied to the results of each of 100 bootstrap replications with a selection level of $\alpha = 0.05$ and a frequency cutoff value of 50%. That is, each covariate found to be a significant predictor at $p \leq 0.05$ in at least 50 bootstrap replicates is selected.
2. Those covariates that do not attain the cutoff value are eliminated from further consideration for inclusion in the model.
3. With the appropriate pharmacostatistical models, NONMEM PPK model building was performed using the covariates retained in step 2 with the covariates selection level of $p < 0.005$. Essentially backward elimination for covariate selection was applied to each of the 100 bootstrap samples. The covariates found to be important in explaining the variability in the parameter of interest were used to build the final population pharmacokinetic model.
4. The PPK model was applied to the original full (not bootstrapped) data to obtain parameter estimates.

When the above process was applied to all the covariates under consideration, reciprocal serum creatinine (RSC) and HIV status were identified as the predictors of clearance. However, for the final PPK model only RSC was retained in the model as a linear predictor of clearance.

In this example, GAM and bootstrapping combined to serve as a powerful tool to create a PPK model with stability. The result of a stepwise selection procedure is usually a single model without any information about its stability. Stepwise selection depends very much on covariates selected in the early steps and on influential data points. With the bootstrap approach using a fixed selection level, an attempt was made to estimate the whole distribution of importance for covariates under consideration.

For this agent the problem was identified, and it was concluded to eliminate HIV status as a predictor of clearance. After model evaluation was performed, the author was confident that HIV status was not spuriously deleted from the model. It was important to know if HIV status should have been included in the PPK model because if HIV status had been included, this may have demonstrated the need of a dosing strategy specifically for HIV patients. When HIV status was eliminated from the model, it was evident that no modification in dosing strategy would be needed for this patient subpopulation on a pharmacokinetic basis.

### 5.5.2 A Predictive Model

5-Fluorocytosine (5FC) is an antifungal agent that is useful in the treatment of infections due to *Cryptococcus* neoformans and *Candida* species (30–32). It has a low therapeutic index with resistance developing at low concentrations (<25 mg/l) and bone marrow toxicity occurring at higher concentrations (>75–100 mg/l) (33). There are two common competing dosing strategies for 5FC which involve 2 degrees of difficulty of administration and it would be desirable to employ the simpler strategy (UCSD) in clinical practice, if both methods result in similar concentrations. The Sanford dosing method (35) recommends a 16 h dosing interval for patients with creatinine clearances of 15–49.9 ml/min, whereas the UCSD method (35, 36) recommends a 12 h dosing interval for this same patient group. At the creatinine clearance of 15–49.9 ml/min the 16 h Sanford dosing interval is inconvenient and problematic because dosing intervals that are not evenly divided into a 24 h schedule often result in skipped or missed doses and nurses or pharmacists are inconvenienced by the necessity of calculating and scheduling odd dosing times. The objective was to develop a PPK model for 5FC that would be useful in comparing these two competing dosing strategies to determine if the simpler (UCSD) dosing strategy was as good as or better than the more complex (Sanford) approach to dosing 5FC. If the expected concentrations from the UCSD dosing strategy were in the target concentration range as often or more often than the Sanford approach then it would be the preferred approach based on scheduling convenience.

#### 5.5.2.1 Pharmacokinetic Model Development

PPK modeling began with data structure analysis so that the data was examined for any hidden structure, outliers, or leverage observations (37). Hidden structure was investigated by generating a locally fit smooth scatter plot matrix of parameters and covariates and outliers were sought by histogram plots with density lines (SPlus 2000, Insite Corporation, Seattle, Washington). Once the outliers were identified, a decision could be made as to whether they constituted leverage observations which were overly influential in estimation of model parameters. No leverage observations were identified. Model development proceeded in a stepwise manner employing NONMEM software version IV (38). Regarding the development of the structural population model, model comparisons were conducted by observing the objective function, residual variability, and between-subject variability.

A PPK model was developed using the approach outlined in the previous paragraph. The base model was estimated and the NONMEM post hoc process executed to estimate individual deviations (etas) from the typical PPK parameters (clearance and apparent volumes). Variables tested for inclusion in the model were weight, age, ideal body weight, serum creatinine, creatinine clearance, gender, and height. Ideal body weight (ibw) was calculated from the formulas of Devine (39), and CLCR was calculated from the method of Cockroft and Gault (40). Ka was set at 1.5 $h^{-1}$ and was determined from the data presented by Schonebeck et al. (41) by applying the method of Franke and Ritschel (42). Between and within individual variability terms were modeled as proportional error terms. This way of expressing variance results in a lack of scale which facilitates between parameter comparisons of variance.

The final irreducible PPK model developed for CL and Vd was:

$$CL = \theta_1 * CLCR + \theta_2 * Wt \quad (11)$$

$$Vd = \theta_3 * Wt \quad (12)$$

where CL was the 5FC clearance, Vd was the apparent volume of distribution, CLCR was the creatinine clearance, Wt was the patient weight in kg, and $\theta_1$ to $\theta_3$ were the coefficients that related the parameter of interest to a covariate.

When the population model was developed, weight was the only covariate that could be used to explain variability in clearance and volume of distribution. The parameter estimates of the above model are presented in Table 1.

#### 5.5.2.2 Model Evaluation Using the Bootstrap

From the original data set, 1000 bootstrap data sets were constructed by resampling with replacement (43, 44). The sampling was repeated until the bootstrap sample consisted of *N* subjects where in this case *N* was 78, the same size as the original data sample. The structure of the model from the initial model develop-

**Table 1** A Summary of Population Pharmacokinetic Model Parameters of 5FC from Initial NONMEM Analysis

| Parameter | Estimate | CV%[a] | SEE[b] | 95% CI |
|---|---|---|---|---|
| $\theta_1$ | 0.041 | | 0.008 | |
| $\theta_2$ | 0.028 | | 0.009 | |
| $\theta_3$ | 1.46 | | 0.204 | |
| $Cl_{pop}$ | | 33 | | 19,42 |
| Vd | | 46 | | / |
| Cp | | 27 | | 19,33 |

[a] Coefficient of variation. / = solution intractable (solution is intractable because the lower CI $\eta$ for the CV is a negative and we would have to be able to take the square root of a negative number)
[b] Standard error of the estimate.

ment process was retained. That is, the coefficients that related the pharmacokinetic parameters (i.e., clearance and apparent volume) to covariates (i.e., weight, height, creatinine clearance) were allowed to be estimated from each of the 1000 bootstrap data sets. To estimate the distribution of population pharmacokinetic parameters, a total of 1000 bootstrap pseudosamples were generated and the structural pharmacokinetic model was fit to each of these 1000 bootstrap data sets. For one method, a normal distribution was assumed and standard methods, which conform to normal theory, were employed to construct the 95% confidence intervals. This is the standard bootstrap confidence interval (standard bootstrap 95% CI normal). Then using the results from all 1000 fitted models the percentile bootstrap 95% CI method was also used (15, 18, 44).

The summary and comparison of the estimates from the 1000 bootstrap pseudosamples (data sets) are presented in Tables 2 to 4. The PM model was to be used in evaluating two competing dosing strategies by simulating the expected range of serum concentrations; it was important to evaluate not only the typical values for PM parameters but also the variability associated with those typical values. When employing the bootstrap to assess reliability of a PM model, the model parameter estimates from the bootstrap should be within ±15% of the irreducible PM model (5). If the bootstrap parameter estimates meet these criteria, then one may employ the originally estimated model because it can be considered reliable. Tables 2 to 4 confirm that the PM model met these criteria and that it was therefore reliable and could be used for simulation purposes with confidence. Of notable importance for the current model was that the fixed effects were within ±5% and the random effects were within ±10% of the final irreducible model (see Table 1), which resulted in a high degree of confidence in the final model.

**TABLE 2** Estimates of Central Tendency and Distribution of the Pharmacokinetic Parameters for 5FC from Several Approaches

| | $\theta_1$ | $\theta_2$ | $\theta_3$ |
|---|---|---|---|
| Mean | 0.041 | 0.029 | 1.49 |
| Median | 0.041 | 0.029 | 1.45 |
| Standard bootstrap 95% CI[a] | 0.021, 0.060 | 0.006, 0.052 | 1.02, 1.96 |
| Percentile bootstrap 95% CI[a] | 0.023, 0.063 | 0.003, 0.050 | 1.12, 2.06 |
| NONMEM | 0.041 | 0.028 | 1.46 |
| NONMEM 95% CI[a] | 0.025, 0.057 | 0.010, 0.046 | 1.05, 1.87 |

[a] CI is the confidence interval.

#### 5.5.2.3 Predictive Performance: The Bootstrap Approach

Predictive performance, as an essential element in the determination of model appropriateness, was assessed by estimating the mean absolute error (MAE). Estimation of the MAE can be done using the bootstrap.

The bootstrap method has been proposed and used to assess the predictive accuracy or performance of a PPK model (5). In the case of the current analysis, 200 of the 1000 total bootstrap replicate data sets were used. The "optimism" (OPT) of the MAE was calculated and was then added to the prediction error metric that was estimated when the final pharmacokinetic model was applied to

**TABLE 3** Estimates of Random Effects and 95% CI for the Random Effects

| | $\eta_1$[b] | $\eta_2$[c] | $\varepsilon$[d] |
|---|---|---|---|
| Mean | 0.111 | 0.198 | 0.076 |
| Median | 0.109 | 0.177 | 0.075 |
| Standard bootstrap 95% CI[a] | 0.033, 0.189 | −0.059, 0.455 | 0.041, 0.111 |
| Percentile bootstrap 95% CI[a] | 0.046, 0.180 | 0.000, 0.458 | 0.042, 0.109 |
| NONMEM | 0.109 | 0.211 | 0.072 |
| NONMEM 95% CI[a] | 0.038, 0.180 | −0.009, 0.431 | 0.035, 0.109 |

[a] CI is the 95% confidence interval.
[b] $\eta_1$ is the intersubject random effect associated with the Cl.
[c] $\eta_2$ is the intersubject random effect associated with the Vd.
[d] $\varepsilon$ is the residual random effect.

**TABLE 4** Estimates of Random Effects and 95% CI for the Random Effects

| | 95% CI CV Clearance | 95% CI CV Volume |
|---|---|---|
| Mean | 33.3% | 44.5% |
| Median | 33.0% | 42.0% |
| Standard bootstrap 95% CI[a] | 18.0%, 43.5% | *i*, 67.5% |
| Percentile bootstrap 95% CI[a] | 21.0%, 42.4% | 0.000, 67.6% |
| NONMEM | 33.0% | 45.9% |
| NONMEM 95% CI[a] | 19.4%, 42.4% | *i*, 65.7%[b] |

[a] CI is the 95% confidence interval.
*i* = solution intractable.

the original data set. These 200 bootstrap data sets were denoted as $D_1$ to $D_{200}$. See Table 5 for a detailed explanation of the notation that follows. The structure $S_0$ of the model $M_0$ [$M_0 = F(S_0, D_0)$] was retained; i.e., the coefficients that related the pharmacokinetic parameters (i.e., clearance and apparent volume) to covariates (i.e., weight, height, creatinine clearance) were estimated from each of the 200 bootstrap data sets as

$$\mathrm{CI} = \theta_1 * \mathrm{CLCR} + \theta_2 * \mathrm{Wt} \tag{13}$$

$$\mathrm{Vd} = \theta_3 * \mathrm{Wt} \tag{14}$$

When $S_0$ was fit ($F$) to $D_i$, the model that resulted would be noted as $M_i$ so that when $S_0$ was applied to $D_{11}$ the model that resulted was noted as $M_{11}$ [$M_i = F(S_0, D_{11})$]. Thus, there were 200 bootstrap models ($M_1$ to $M_{200}$) that were fit, one for each bootstrap data set.

The next step in assessing predictive accuracy was to apply the models ($M_1$ to $M_{200}$) to the data set $D_0$, which was the original data set (not from bootstrap resampling) with all individuals occurring once. In this process the coefficients and random effects are fixed for $M_1$ to $M_{200}$ so that at this step there are 200 sets of predicted outcomes $P(M_1 : D_0)$ to $P(M_{200} : D_0)$ providing $\mathrm{PE}_1$ to $\mathrm{PE}_{200}$.

The next step was to estimate the optimism (OPT) for the prediction metric. This was executed for each model so that

$$\mathrm{OPT}_j = \mathrm{PE}_j - \mathrm{AE}_j \tag{15}$$

where $\mathrm{AE}_i$ was the apparent error that was estimated when model $M_i$ was fit to data $D_i$, $\mathrm{PE}_i$ was the prediction error when model $M_i$ was fixed and fit to data

**TABLE 5** Mathematical Notation Utilized in Bootstrapping for Computing the Prediction Error

| Notation | Representation of notation |
|---|---|
| $D_0$ | Study data or original data |
| $S_0$ | The structure of the developed population pharmacokinetic model with the $D_0$, which can be expressed as follows: $CL = \theta_1 * CLCR + \theta_2 * Wt$ $Vd = \theta_3 * Wt$ |
| $F$ | The fit metric: $M = F(S, D)$ means $M$ is the model produced when structure $S$ is fit to data set $D$. |
| $P$ | The prediction error metric: $PE = P(M, D)$ means the prediction error when model $M$ is applied to data set $D$. |
| $D_i$ ($i = 1, 2, 3, \ldots, 200$) | Bootstrapped data or samples; each has the same size (78 subject's data) as the data $D_0$, which the developed population pharmacokinetic model was based on. They were drawn with replacement from the observed data $D_0$; Observed data could either appear in the bootstrap samples ($D_i$) once, more than one time, or not at all. For each bootstrap data set the structure was retained but the coefficient and the intercept were reestimated. |
| $M_i = F(S_0, D_i)$ ($i = 1, 2, 3, \ldots, 200$) | Models generated from fitting the structural model $S_0$ to the bootstrapped samples; bootstrapped models. |
| $AE_i = P(M_i, D_i)$ ($i = 1, 2, 3, \ldots, 200$) | Bootstrap apparent error; prediction errors obtained when bootstrapped models $M_i$ were applied to bootstrapped samples $D_i$. |
| $PE_i = P(M_i, D_0)$ ($i = 1, 2, 3, \ldots, 200$) | Bootstrap prediction error; prediction errors obtained when bootstrapped models $M_i$ were applied to the original data $D_0$. |
| $OPt_i = PE_i - AE_i$ ($i = 1, 2, 3, \ldots, 200$) | Bootstrap optimism; the difference between the prediction error and the apparent error. |

$D_0$. and $\mathrm{OPT}_j$ is the optimism for model $M_i$. Next, the mean OPT for the prediction metric was estimated as follows:

$$\overline{\mathrm{OPT}} = 1/200 \times \sum_{j=1}^{200} \mathrm{OPT}_j \tag{16}$$

Once the mean optimism was known, it was added to the results of the MAE estimated from the original data set. This resulted in an improved estimate of the prediction metric, as the prediction metric estimates generated when a model is applied to its own data would be smaller compared to applying the prediction metric to an external data set.

$$P_{\mathrm{imp}} = P(M_0 : D_0) + \overline{\mathrm{OPT}} \tag{17}$$

where in the current investigation $P_{\mathrm{imp}}$ was the improved estimate of the MAE ($\mathrm{MAE}_{\mathrm{imp}}$) provided by the bootstrap.

The OPT of the MAE was 0.4 mg/l, while the improved MAE was 16.2 mg/l. It should be noted that when the optimism of the MAE is small, say, less than 15% of the improved MAE; this is an indication that the model is without substantial deficiencies (5).

In Table 6 the MAE was that obtained when the final model of the development stage was applied to its own data. The mean OPT was added to the mean absolute error to inflate the values so that they would approach the MAE of applying the PPK model to an external data set. It was observed that the optimism for the MAE increased it by 2.6%, implying confidence in the model. Other results for predictive performance are presented in Table 6.

**Table 6** Summary of Results for Bootstrap of Concentrations

| Mean Absolute Error[a] $P_2(M_0, D_0)$ (mg/l) | $\overline{\mathrm{OPT}}$ | $\mathrm{MAE}_{\mathrm{imp}}$[b] (mg/l) |
|---|---|---|
| 15.8 | 0.4 | 16.2 |

[a] Mean absolute error is the mean absolute error when the original model was applied to the original complete data set with all of the 78 subjects appearing once in the data.

[b] $\mathrm{MAE}_{\mathrm{imp}}$ is the improved estimated of the mean absolute error obtained by adding the $\mathrm{OPT}_2$ to the mean absolute error.

#### 5.5.2.4 Model Application: Simulation

Simulation methods can be applied to PM modeling to aid in designing studies that are well powered, efficient, informative, and robust (44). Simulations can also aid in determining the influence of the PPK on other study design features (2) and in interpreting the impact of the PPK model on dosing strategies.

Once the model was evaluated with the estimates deemed reliable and the predictive performance determined (i.e., it could predict concentrations), it was considered appropriate and could be applied toward its intended use.

Pharsight Trial Designer simulation software (Pharsight Corp.) was used to produce sets of concentrations that would reflect the expected range of concentrations given that the underlying PPK model was determined to be appropriate. The parameters and variables from the developed PM model were used for the simulation. The simulations were undertaken to evaluate two commonly used dosing methods for 5FC. This simulation was conducted to assess the expected range of serum 5FC concentrations that would result from two currently used dosing methods (the Sanford method and the UCSD method) (34–36). Both of these methods recommend a constant dose, either 37.5 mg/kg (Sanford) (34) or 25 mg/kg (UCSD) (35, 36) with increasing dosing interval as renal function declines. The Sanford method has an inconvenient dosing interval of 16 h for patients with creatinine clearance of 15–49.9 ml/min. Therefore, the UCSD method was recommended as the dosing strategy of choice. A statistical description for some of the ranges of expected concentrations are presented in Table 7. Some of these concentrations were outside of the targeted range for the Sanford dosing strategy.

#### 5.5.2.5 Implications

In determining the predictive performance of the PM model for 5FC, it was found that the MAE was less than 0.4 of the average measured concentration and the optimism for the MAE was very small. These findings resulted in confidence that the final PPK model was adequate and appropriate for its intended use.

For a predictive model to be deemed appropriate, it must have practical worth. If a model meets the statistical criteria of reliability and the predictions are unbiased, its generalizability will be limited if it does not contain adequate covariate information. Thus, two related questions must be answered in order for one to arrive at a generalizable or transportable PM model:

1. With the available covariates and data, is the model the best that can be developed? A small sample size can exacerbate difficulties in data-dependent covariate selection since small perturbations in the data can affect the apparent statistical significance of covariate relationships. When the sample size is small, there is a low signal-to-noise ratio, which may result in an increased risk of selecting unimportant covari-

**Table 7** Sanford Regimen 37.5 mg/kg at Steady State

| | CICR = 15–49.99 ml/min Dosing interval = 16 h | | CICR = 50–79.99 ml/min Dosing interval = 12 h | | CICR ≥ 80 ml/min Dosing interval = 6 h | |
|---|---|---|---|---|---|---|
| Percentile | Peak | Trough | Peak | Trough | Peak | Trough |
| 95th | 82.7 | 56.7 | 83.1 | 54.5 | 115.2 | 89.9 |
| 75th | 66.2 | 40.9 | 63.8 | 40.2 | 89.4 | 67.6 |
| 50th | 55.7 | 31.7 | 52.7 | 32.7 | 72.9 | 54.6 |
| 25th | 44.4 | 25.9 | 43.1 | 25.4 | 59.4 | 43.1 |
| 5th | 28.9 | 16.6 | 29.7 | 15.1 | 41.8 | 28.4 |

ates while failing to include important ones. Even if the study design is impeccable, is the patient base broad enough to permit the transportability of the model? Thus, for a predictive PM model to be adopted by others, confidence in its reliability and predictive performance is required.

2. For its intended purpose, does the model predict accurately; and if it does, is it transportable? With regression modeling, there can be a tendency for the production of overly optimistic models. The bootstrap approach for the determination of predictive performance tends to overcome this problem because the optimism is accounted for in predicting model accuracy (5, 15, 16). Therefore, the bootstrap approach was used in this study in the absence of a test data set that would have made external validation of the model possible. Thus, for a predictive PM model to be adopted by others, confidence in its reliability and predictive performance is required. Empirical demonstration of transportability or generalizability is a way to gain such credibility.

In the 5FC data set used for the second example, the ranges of the covariates were wide. This wide distribution of the covariates with a reasonable sample size provided a broad patient base, muting the problem of low signal-to-noise ratio and lack of transportability of the PM model. The PM model was developed after data examination for hidden structure, thus enabling the development of a parsimonious model that contained descriptive and predictive information. That is, the PM model developed for the drug was shown to be reliable and to yield accurate predictions with minimal bias (i.e., with reduced overoptimism). Therefore, the demonstration of predictive performance of the model using the bootstrap provided the empirical evidence of the generalizability of the model and the confidence needed for the model to be used for evaluating two competing 5FC dosing strategies. The results of the simulations enabled the selection of one dosing strategy when compared with the other.

## 5.6 SUMMARY

Model appropriateness has been defined and explained using a descriptive and predictive model. It involves clearly identifying and stating the problem, stating the application or the intended use of the model, evaluating the model, and determining the predictive performance of the model (via the bootstrap, for instance) if the model is to be used for a predictive purpose.

PM models are either descriptive or predictive. The first example presented a descriptive model, where the purpose was to determine if HIV status was a covariate associated with drug clearance. In this case HIV status was eliminated from the model which from the pharmacokinetic perspective would mean that dosing strategies specifically constructed for studies of HIV infected patients would not be necessary and it may be possible to have large data sets for future PPK model development by combining data sets from HIV patients and non-HIV patients.

The second example presented a predictive model. The range of expected concentrations along with other factors determined the dosing method of choice. Thus, the question of how 5FC should be dosed was addressed. It must be remembered that in answering this question, the approach to model development, evaluation, and predictive performance is important. When a PM model is developed for predictive purposes, it is assumed that the population from which the model was derived will be similar to (or representative of) the population to which the model is applied. Because of the presumed generalizability or transportability of the PM model developed, the steps delineated above should be given careful attention in the development of predictive models. Implied in the generalizability or transportability of the PM model is the fact that it can be applied to a similar set of patients in another location. This concept of model generalizability can also be referred to as model validity. The inclusion of PM models for dosage adjustment in drug labels is an implicit statement about their generalizability.

Models whose intended use is simulation are by nature predictive, therefore model evaluation and the determination of predictive performance are the necessary components for the determination of model appropriateness for this type of intended use. Implied in models used for the simulation of clinical trials are their generalizability, therefore characterization of the appropriateness of such models is a sine qua non to good clinical trial simulation.

## REFERENCES

1. EI Ette, R Miller, WR Gillespie, et al. The population approach: FDA experience. In: L Aarons, LP Balant, G Danhof, et al., eds. The Population Approach: Measuring and Managing Variability in Response, Concentration and Dose. Brussels: Commision of the European Communities, European Cooperation in the Field of Scientific and Technical Research; 1996, pp 272–275.
2. Guidance for Industry: Population Pharmacokinetics. U.S. Department of Health and Human Services, Food and Drug Administration. February 1999.
3. EI Ette, T Ludden. Population pharmacokinetic modeling: the importance of informative graphics. Pharm Res 12(12):1845–1855, 1995.
4. J Mandema, D Verotta, LB Sheiner. Building population pharmacokinetic-pharmacodynamic models. In: DZ D'Argenio, ed. Advanced Pharmacokinetic and Pharmacodynamic Systems Analysis. New York: Plenum Press, 1995, pp 69–86.
5. EI Ette. Stability and performance of a population pharmacokinetic model. J Clin Pharmacol 37:486–495, 1997.

6. GEP Box. Robustness in the strategy of scientific model building. In: RL Laurer, GN Wilkinson, eds. Robustness in Statistics. New York: Academic Press, 1979, pp 201–236.
7. F Mentre, ME Ebelen. Validation of population pharmacokinetic-pharmacocynamic analyses: review of proposed approaches. In: LP Balant and L Aarons, eds. The Population Approach: Measuring and Managing Variability in Response Concentration and Dose (COST B1). Brussels: Office for Official Publications of the European Communities, 1997, pp 146–160.
8. NHG Holford, M Hale, HC Ko, JL Steimer, LB Sheiner, CC Peck. Simulation in drug development: good practices. Washington, DC: Center for Drug Development Science, Georgetown University, 1999.
9. EI Ette, H Sun, TM Ludden. Balanced designs in longitudinal population pharmacokinetic studies. J Clin Pharmacol 38:417–423, 1998.
10. MF Delgado Iribarnegaray, B Santo, MJ Garcia Sanchez, MJ Otero, AC Falcao, A Dominguez-Gil. Carbamazepine population pharmacokinetics in children: mixed-effect models. Ther Drug Monit 19:132–139, 1997.
11. RP Hirsch. Validation samples. Biometrics 47:1193–1194, 1991.
12. EB Roecker. Prediction error and its estimation for subset-selected models. Technometrics 33:459–468, 1991.
13. H Sun, EO Fadiran, CD Jones, L Lesko, SM Huang, K Higgins, C Hu, S Machado, S Maldonado, R Williams, M Hossain, EI Ette. Population pharmacokinetics: a regulatory perspective. Clin Pharmacokinet 37:41–58, 1999.
14. B Efron. Estimating the error rate of a prediction rule: improvement on cross-validation. J Am Stat Assoc 78:461–470, 1983.
15. B Efron. Bootstrap methods: another look at the jackknife. Ann Stat 7:1–26, 1979.
16. B Efron, G Gong. A leisurely look at the bootstrap, the jackknife and cross-validation. American Statistician 37:36–48, 1983.
17. CA Hunt, G Givens, S Guzy. Bootstrapping for pharmacokinetic models: Visualization of predictive and parameter uncertainty. Pharmaceutical Research 15:690–697, 1998.
18. EI Ette. Comparing non-hierarchical models: Application to non-linear mixed effects modeling. Comput Biol Med 26:505–512, 1996.
19. A Gelman, XL Meng, H Stern. Posterior predictive assessment of model fitness via realized discrepancies. Statistica Sinica 6:733–807, 1996.
20. SB Duffull, S Chabaud, P Nony, L Chritian, P Girard, L Aarons. A pharmacokinetic simulation model for ivabradine in healthy volunteers. Eur J Phamaceut Sci 10:285–294, 2000.
21. SB Duffull, L Aarons. Development of a sequential linked pharmacokinetic and pharmacodynamic simulation model for ivabradine in healthy volunteers. Eur J Pharmaceut Sci 10:275–284, 2000.
22. JF Bennet, JC Wakefield, LF Lacey. Modeling of trough plasma bismuth concentrations. J Pharmacokinet Biopharm 25:79–106, 1997.
23. EH Cox, C Veyrat-Follet, S Beal, S Fuseau, S Kenkare, LB Sheiner. A population pharmacokinetic-pharmacodynamic analysis of repeated measures time-to-event pharmacodynamic responses: the anti-emetic effect of ondansetron. J Pharmacokinet Biopharm 27:625–644, 1999.

24. LE Friberg, A Freijs, M Sandstrom, M Karlsson. Semiphysiological model for the time course of leukocytes after varying schedules of 5-fluorouracil in rats. J Pharmacol Exp Ther 295:734–740, 2000.
25. H Yano, SL Beal, LB Sheiner. Evaluating pharmacokinetic/pharmacodynamic models using the posterior predictive check. J Pharmacokinetic Pharmacodynam 28:171–191, 2001.
26. PJ Williams, JR Lane, EC Capparelli, YH Kim, RB Coleman. Direct comparison of three methods for predicting digoxin concentrations. Pharmacotherapy 16:1085–1092, 1996.
27. LB Sheiner, SL Beal. Some suggestions for measuring predictive performance. J Pharmacokinetic Biopharm 9:503–511, 1991.
28. L Aarons, S Vozeh, M Wenk, M Weiss, PH Weiss, F Follath. Population pharmacokinetics of tobramycin. Br J Clin Pharmacol 28:305–314, 1989.
29. R Bruno, N Vivier, JC Vergniol, SL De Phillips, G Montay, LB Sheiner. A population pharmacokinetic model for docetaxel (Taxotere): model building and validation. J Pharmacokinet Biopharm 24:153–172, 1996.
30. JP Utz, BS Tynes, HJ Shadom, RJ Duma, MM Kannan, KN Mason. 5-Flourocytosine in human cryptococcosis. Antimicrob Agents Chemother 344–346, 1968.
31. JP Utz: Flucytosine. N Eng J Med 286:777–778, 1972.
32. CO Record, JM Skinner, P Sleight, DCE Speller. Candida endocarditis treated with 5-fluorocytosine. Brit Med J 1(5743):262–264, 1971.
33. R Schlegel, GM Bernier, JA Bellanti, DA Maybee, GB Osborne, JL Stewart, et. al. Severe candidiasis associated with thymic dysplasia, IgA deficiency and plasma antilymphocyte effects. Pediatrics 45:926–936, 1970.
34. JP Sanford, DN Gilbert, MA Sande. The Sanford Guide to Antimicrobial Therapy, 1995, p 112.
35. DN Wade, G. Sudlow. The kinetics of 5 Fluorocytosine elimination in man. Aust NZ J Med 2:153–158, 1972.
36. TH Ittel, UF Legler, A Polak, WM Glockner, HG Sierberth. 5 Fluorocytosine kinetics in patients with acute renal failure undergoing continuous hemofiltration. Chemotherapy 33:77–84, 1987.
37. EI Ette, P Williams, H Sun, E Fardiran, F Ajayi, LC Onyiah. The process of knowledge discovery from large pharmacokinetic data sets. J Clin Pharmacol 41:25–34, 2001.
38. YH Kim, PJ Williams, JR Lane, EV Capparelli, MJ Liu. 5 Flourocytosine: population pharmacokinetic model development, nonparametric bootstrap validation and dosing strategy evaluation [abstract 2008]. Quest for the magic bullet: an astounding century, Nov 14–18, 1999, New Orleans, LA. AAPS Pharma Sci Supplement 1:S4, 1999.
39. BJ Devine. Gentamicin therapy. Drug Intell Clin Pharm 8:650–655, 1974.
40. D Cockcroft, M Gault. Creatinine clearance from serum creatinine. Nephron 16:41, 1976.
41. J Schonebeck, A Polak, M Fernex, HJ Scholer. Pharmacokinetic studies on the oral antimycotic agent 5 fluorocytosine individuals with normal and impaired renal function. Chemotherapy 18:321–336, 1973.

42. EK Franke, WA Ritschel. Quick estimation of the absorption rate constant for clinical purposes. Drug Intell Clin Pharm 10:77–82, 1976.
43. Bryan FJ Manly. Randomization, Bootstrap and Monte Carlo Methods in Biology. 2nd ed. London: Chapman and Hall, 1998, pp 34–68.
44. Efron B, Tibshirani R. An Introduction to the Bootstrap. New York: Chapman and Hall, 1993.

# 6

# Computational Considerations in Clinical Trial Simulations

**John C. Pezzullo and Hui C. Kimko***
Georgetown University, Washington, D.C., U.S.A.

## 6.1 INTRODUCTION

Simulation is a standard component of the analytical arsenal of many disciplines, including engineering, physics, astronomy, and statistics. Its theoretical foundations and techniques are well developed, but the application of this technology to the field of drug development is relatively new (1). This chapter aims to explain the process underpinning simulation as it applies to clinical trial design.

Simulation involves creating a mathematical representation of some part of the real world, and observing how this "virtual world" behaves under a set of specified conditions. Real-world entities (subjects, dosage regimens, drug concentrations, clinical outcomes, adverse events, etc.) are represented as numbers stored in a computer. These numbers are manipulated according to various rules—formulas that are assumed to describe real-world activities. Then, the entities in this simulated world are observed to play out a more-or-less realistic scenario. Virtual subjects get enrolled and randomized into treatment arms according to the rules of the protocol. Drugs are administered, distributed among various body compartments, and ultimately eliminated. Subjects react to the drug and/or its metabolites; they may experience sporadic adverse events, drop out,

* *Current affiliation*: Johnson & Johnson Pharmaceutical Research, Raritan, New Jersey, U.S.A.

and cross over to different dosage regimens. The simulation program records and saves all the outcomes experienced by all the virtual subjects in its virtual world. This simulated activity takes place in the computer much faster than it would in the real world.

Any realistic simulation will incorporate the effects of random fluctuations and variability due to the myriad of uncontrollable factors that influence everything that happens in real life. Some processes may be deliberately randomized, such as the assignment to treatment groups; other events are intrinsically sporadic, such as the occurrences of adverse events; pharmacokinetic (PK) and pharmacodynamic (PD) parameters vary from one subject to the next; and all measurements (body temperature, blood pressure, drug concentration, etc.) are subject to random measurement errors. Computer simulations incorporate this unpredictable variability by means of random number generators, so successive simulations of the same model will produce somewhat different outcomes.

Consideration of variability is, in fact, the most useful aspect of computer simulation, revealing the extent to which unavoidable variability and fluctuations will affect the results. Clinical trials are generally carried out to estimate parameters (PK, safety, efficacy) or to draw conclusions (show that different treatment groups do or do not have different outcomes), and their results are always subjected to statistical analyses to account for the effects of random variability. The primary goals of a clinical trial simulation are to evaluate whether, and to what extent, a proposed trial design can achieve suitable power and/or precision, and to optimize the trial properties (sample size, dose, PK sampling times, etc.) to achieve the greatest power and precision possible within the financial and operational constraints of the study.

## 6.2 CHOICE OF SOFTWARE PLATFORM

Choices have to be made as to the software environment in which the simulation will be specified and executed. The basic computational steps involved in a simulation are generally straightforward, so simulations can be (and have been) carried out in almost every imaginable computing environment.

Any *general-purpose programming language* that supports basic arithmetic operations and simple random number generation (either built-in or user-added) will do, e.g., Fortran, BASIC, Java, JavaScript, C, Pascal, and ALGOL, among a host of others. A general-purpose programming language would not be expected to provide the amenities found in more specialized simulation languages, such as the generation of random numbers from specific types of distributions, the management of discrete entities, the solution of differential equations, the calculation of statistical analyses, or the automatic tabulation of the results of many simulation runs; these would have to be programmed explicitly.

*Mathematical programming systems*, such as MATLAB, Mathematica,

Maple, Omatrix, and MuPad, provide well-validated, high-quality random number generators, automatic equation solvers, and statistical analysis as built-in commands, so less work would be required to program a simulation.

*Spreadsheets* such as Excel have all the computational capability required, even without using their embedded macro language capabilities such as Visual Basic for Applications (VBA). These also have the advantage of providing a visual layout of the simulated entities (i.e., subjects), with all of their characteristics, behaviors, and outcomes; this approach might therefore be appropriate for those just learning the concepts of simulation. The universal availability of spreadsheet software is another factor: If someone only rarely has occasion to perform a simulation, it might not be feasible to invest in more expensive specialized software. However, since stability and quality of the random number generator in Excel are in question, cautious use should be warranted.

*Statistical languages*, such as R, S-plus, and the programming facility within SAS would make it much easier to simulate data and perform advanced statistical analysis. They also offer a larger selection, and a higher quality, of random number generators.

*Simulation languages* would, of course, provide a much higher level of built-in support for the many special computational tasks involved in a real-world simulation: random number generation, management of discrete events, statistical testing, repetitive running of simulated trials, and tabulation of the outcomes of the multiple simulations. They may also provide a simple way to define the process being simulated, often by a graphical interface in which components can be selected from a toolbar and dragged into a diagram that describes the components, parameters, processes and relationships that make up the system.

At the top of the specialization scale are the *programs specifically designed to simulate clinical trials* (e.g., Pharsight Trial Simulator). These incorporate much of the relevant knowledge [e.g., pharmacokinetics (PK), pharmacodynamics (PD)], specialized techniques (e.g., AUC estimation, nonlinear regression, the solution of stiff systems of nonlinear differential equations), and trial execution irregularities (e.g., missed doses, dropouts), so the investigator can concentrate entirely on the problem to be solved, without the burden associated with the computational details.

## 6.3 BASIC CONCEPTS OF SIMULATION

Simulation involves generating a "model" of a virtual world, or at least a small part of a world, that can be expressed in terms of a set of rules (usually expressed by formulas and tables) that describe how that world works. If the simulation of a clinical trial is to be in any way realistic, it should incorporate:

1. *What is known about the subjects* that might affect how they react to the drug (pharmacokinetics, disease status, demography, etc.)

2. *What is known about the drug* under study that might affect the subjects in the trial (effectiveness, adverse effect, etc.)
3. *Unpredictable random elements* that characterize the real world in which the actual study will take place (between-subject, within-subject, between-occasion variability, etc.)
4. *Uncontrollable factors in patient or clinician behavior*, such as protocol adherence (dropouts, dose administration adherence, missing records, etc.)

Simulating a planned clinical trial involves generating a cohort of virtual subjects to participate in the trial. A virtual subject exists in the computer as a set of parameters—numbers that characterize the relevant aspects of that person; e.g., characterization may include genetic, demographic, physiologic and pathophysiologic information. This could be a set of cells in a spreadsheet, a set of records in a database, or a collection of data structures in the computer's memory.

These entities embody all that is known or can reasonably be guessed about the subjects and their characteristics, about their behavior in the trial, and about the way they react to the study drug. More readings can be found in Chapters 2 to 4. Sections 6.3.1 to 6.3.9 list a logical sequence of steps for computer-based simulation of a clinical trial.

### 6.3.1 Give the Virtual Subjects Genetic, Demographic, Physical, and Pathophysiological Characteristics

A virtual subject should be assigned individual characteristics according to covariate distribution models. If subjects between 18 and 50 years old are enrolled in the study, then each virtual subject must be given a randomly generated age somewhere within that range. This requires certain assumptions about how the ages of the subjects will be distributed. Will they be spread uniformly across that range, or might a normal distribution with a mean of 34 and a standard deviation of 8 be more realistic? Alternatively, the ages could be selected randomly from the distribution actually observed in a previous trial. Other patient characteristics might have to be taken into account when generating random ages for the simulated subjects—the age distribution of healthy volunteers might be quite different from that of patients with a specific illness.

Similarly, each virtual subject will have some particular height, weight, gender, and a whole collection of other relevant characteristics—degree of liver function, genomic makeup (presence of various mutations that may affect the way this subject responds to the treatment)—all of which will need to be represented as numeric or categorical parameters. This can introduce complexities; height and weight, for example, are likely to be correlated, and the randomly generated values will have to reflect this correlation. All available resources (the medical literature, similar prior studies, etc.) should be brought to bear to get a sense of how the subjects' characteristics are likely to be distributed.

The characterization of covariates must also include those physiological attributes that affect the important outcomes of the trial. A pharmacokinetic simulation must specify how parameters such as volume of distribution and clearance can vary from one individual to another because of the covariates. More details about covariate distribution models can be found in Chapter 3.

Once the covariate distributions have been specified, a set of random values consistent with these distributions must be generated for each subject. This utilizes a critically important capability of a computer—its ability to generate random (or at least pseudo-random) numbers. Section 6.4 describes methods for generating random numbers that conform to most of the common distribution functions needed when generating virtual subjects for a virtual clinical trial.

An alternative to generating virtual people by randomly generating all their physical characteristics would be to select them randomly from pools of real people (2). There might exist a large database of individuals containing most, if not all, of the variables needed for the simulation. A random subset of records could be selected from this database that meets the study's inclusion criteria. A database of this type is very valuable for simulation work, since it eliminates the need to establish the distributions of the variables of interest and to generate random values for these variables. In addition, all intercorrelations between variables are automatically taken care of. Such a database is likely to be missing some variables that might be needed for a realistic simulation; in this case a "hybrid" approach could be used, in which the missing variables could be imputed (3).

Estimates of the population distribution of the parameters will have to be obtained, perhaps from prior studies, or if no prior data is available, simply the best intuitive guesses that can be made. So from the difference in covariates and unexplained between-subject variability, one set of parameters will be derived for each individual.

### 6.3.2 Determine How the Virtual Subjects Will Behave During the Trial

Once a cohort of virtual subjects have been simulated, complete with realistic demographic and pathophysiological characteristics, additional attributes are assigned that govern how the subjects will behave during the simulated clinical trial. For example, from earlier comparable trials a 15% dropout rate might be expected. To simulate this in the virtual subjects, another variable could be created and set to 0 or 1 in such a way that a randomly selected 15% of the subjects have a 1, and the other 85% have 0. Then, later in the simulation, these subjects would be eliminated from the analysis. Of course, these randomizing decisions could be made *during* the simulation run instead of at the outset. The choice depends on which is more convenient: Static predestined attributes might be appropriate if using a spreadsheet, but dynamic decision making might be more convenient in a procedural programming environment. Nonadherence, crossovers, and the occurrence of adverse events would be simulated in the same way. The incidences may be correlated with some existing covariates. More details about protocol deviation models can be found in Chapter 4.

### 6.3.3 Simulate the Observed Outcomes of the Virtual Trial

Once the collection of virtual subjects have been created and endowed with random, but realistic, characteristics, predestined behavior and outcome parameters, then the appropriate calculations are executed to evaluate the outcome measures that form the basis of the analyses. For example, in a pharmacokinetic study the plasma concentration of the drug at specified times after dosing would be calculated, using each subject's randomly simulated kinetic parameters; each subject would therefore have different calculated concentrations. For a realistic simulation, additional random "noise" would be added to these values to simulate the combined effects of uncontrolled factors.

### 6.3.4 Analyze the Data Generated by the Simulated Trial

After generating the simulated outcome variables, the same analysis would be performed on this simulated data that would be used during the "real" study. For example, if the statistical analysis plan for the study calls for analyzing area under the concentration-time curve (AUC), then the AUC would be evaluated for the virtual subjects. If the plan calls for a between-group comparison of AUCs by a Student t-test, then this test would be performed on the simulated AUCs.

### 6.3.5 Repeat the Simulation Many Times

A single simulated trial will provide only a single set of outcomes, and therefore only a single estimate for each population parameter, and only a single $p$ value for each planned statistical significance test. Sometimes that one trial will provide enough useful information to make an intelligent decision about the adequacy of the proposed study design. For example, the outcome of the primary hypothesis test may be so far from significant (such as $p = 0.8$) that there could be no doubt that the proposed study is seriously underpowered. Also, many statistical analyses provide asymptotic uncertainty estimates for the parameters (standard errors), and the standard errors of the parameters of interest may be so large that the proposed study design could not possibly provide the needed precision.

But it is risky to draw too many conclusions from a single trial—this just might be the one trial in a hundred where all the random fluctuations combine in the worst possible way to produce results that are quite unrepresentative of the true situation. More likely, the simulation will produce a result that seems more or less reasonable—a $p$ value near 0.05 or a precision estimate of approximately the hoped-for magnitude. Important decisions should not be made on the basis of just one simulation that produced a result that indicated neither a severely underpowered nor a flagrantly overpowered study. Instead, the same model should be rerun with a different set of virtual subjects, that is, a different set of random fluctuations. This would give another independent set of outcomes, which could be combined with the outcomes from the first run to get a more realistic overall picture of what is likely to happen when the real trial is run. With very little human effort—but perhaps enormous computer effort—the trial simulation can be replicated dozens, hundreds, or even thousands of times.

### 6.3.6 Calculate Precision, Accuracy, and Power

One of the main reasons for simulating a clinical trial is to determine whether the sample size, experimental conditions, planned observations, and proposed analyses provide a reasonable likelihood of detecting some effect. The *power* of a study is the likelihood that the study will produce a significant result (at some arbitrary significance level, such as $p < 0.05$) if the actual size of the effect is at least some postulated value. If the study is well designed, there should be a good chance of getting a significant $p$ value. If not, the design and parameters of the proposed trial will have to be adjusted until the simulations indicate that it can provide the required statistical performance (power and precision)

The power of a proposed trial design is obtained by counting the fraction of simulated studies that turned out significant for each statistical test performed. Power depends on the magnitude of the true (population) effect: Large effects can be detected with small studies, but small effects require large studies. If, for example, everyone who receives a drug lives and everyone who does not receive that drug dies, it will not take a very large (or very well-designed) study to get a significant result on a Chi Square test. If, on the other hand, a drug lowers the true mortality rate of a disease from 50 to 49% (and assuming that such a miniscule effect is really worth detecting at all), it is going to be *very* difficult to obtain statistical significance unless the study enrolls an enormous number of people and is very well designed. The beauty of the simulation method is that for any particular study design (number and size of each study group, dosing strategy, number and timing of PK and PD observations, etc.), and for any particular set of postulated parameters, and for any particular statistical analysis strategy (types of statistical tests, critical significance levels, etc.), a realistic simulation will show the percent of runs that produce significant results. Extensive power tables and graphs can be prepared showing the interrelationships among sample size, alpha level, dosing strategies, and PK/PD sampling strategies, and the corresponding statistical power, sensitivity, specificity, and all the other characteristics of the trial's outcome.

Compared to an actual clinical trial, a simulation offers the tremendous advantage of knowing what the virtual subject's "true" PK parameters are, making it possible to assess how precisely and how accurately those parameters can be estimated from the time course of concentration data. It will show, for example, whether or not the proposed experimental design will produce *biased* estimates (consistently too large or consistently too small). It can be determined, before a single human is enrolled into an actual trial, whether the choice of sampling intervals and the proposed sample size can produce sufficiently precise parameter estimates, or whether it is likely to produce a significant result.

### 6.3.7 Evaluate the Proposed Trial and Revise Its Design

After multiple simulations of the proposed trial have been performed, the effectiveness of the design can be evaluated. Did it provide the desired level of precision for the parameters of interest? Were the results sufficiently free of bias? Did the design have sufficient statistical power to detect effects of a meaningful size? If the answers are "no," then the study is poorly designed and/or underpowered. It may need to be redesigned, perhaps with more measurements, or with differently spaced measurements, or with more precise analytical methods, or with more subjects, or with more closely matched subjects. If the answers are "yes," the study would be considered well designed and sufficiently powered. In fact, it may be *overpowered*—it may have more subjects and/or more measurements than needed for the desired precision and power. Overpowered studies waste resources (time, money, subjects) that could be used to support other studies. Smaller and/or simpler designs should be evaluated to see if they could provide adequate power and precision. Subsequent application chapters will address in greater depth the matter of interpreting the results of simulation runs.

## 6.4 COMPUTER-GENERATED RANDOM NUMBERS

The remainder of this chapter will address the numerical techniques underlying the computer simulation of clinical trials. If the simulation will be carried out using a program that has been designed specifically to perform clinical trial simulations, all of these techniques will have been programmed into the system, and all that will be needed is to specify the particulars of the study design and provide postulated population models and parameters that characterize the subjects. But if the simulation is to be undertaken using general-purpose software, then the random number generation and differential equation solving, and pharmacodynamic relationships will have to be programmed explicitly.

Random numbers used in computer simulations are almost always generated by means of a computation that produces a *pseudo-random* sequence of numbers that have most of the statistical characteristics of truly random numbers. That is, individually they do not appear to follow any pattern that could be used to guess the next value in the sequence, but collectively they tend to be distributed according to some mathematically defined distribution function. That is, a histogram prepared from thousands of generated numbers will have some particular shape such as a horizontal line (for uniformly distributed numbers) or a bell curve (for normally distributed numbers).

Different parts of the simulation may require different kinds of random numbers:

- Ages might be approximately *uniformly* distributed within a narrow interval if, for example, the study called for healthy adults between 25 and 35 years of age.
- Clearance values might be *log-normally* distributed.
- The instrument errors in a lab assay might be *normally* distributed.
- The time to occurrence of a specific adverse event might be *exponentially* distributed.

The simulation software will have to generate random numbers from these probability distribution functions. In this chapter, we will cover basic probability distribution functions; information on other distributions can be found in literature (4). Chapter 3 illustrates several discrete probability functions. The following sections will use an informal pseudo-code that is similar to the syntax of typical modern programming languages.

### 6.4.1 Adequacy of Random Number Generators

*Is the computer's random function good enough*? This topic has been intensively studied over the years. For extremely critical work, it is important not only that the random numbers conform to the distribution they claim to represent (uniform, normal, etc.), but that for all intents and purposes they appear to be uncorrelated. That is, there should be no discernable relationship between two consecutively generated random numbers. Experts in computational mathematics have been able to detect subtle failures in most of the computerized random number generators in common use. This slight correlation is not likely to cause trouble in most simulations, and the issue is not discussed here further. But it should be noted that for rigorous work, a specialized random number generator might be required (5–7).

Almost all computer–random number generators will eventually, after producing some number of random values, start generating the same sequence over again. This can be troublesome in simulation work because even a simple trial simulation may require thousands of random numbers, and the trial may have to be simulated many thousands of times. The more primitive generators (often found in the common implementations of many programming languages) will repeat after generating fewer than 100,000 distinct numbers, and would be unsuitable for all but the simplest and most informal simulations. Better methods will generate billions or trillions of numbers before repeating, and some recent ones will, for all practical purposes, simply *never* repeat. Methods for generating random numbers are briefly introduced in Chapter 3.

### 6.4.2 Generating Uniform Random Numbers

Uniform random numbers follow one of the simplest of all mathematical distributions. They are fractional numbers that can have any value between 0 and 1. They are usually generated by the computer with many digits after the decimal point. So a uniform random number generator might produce the sequence 0.391175, 0.926644, 0.079791, etc. The term *uniform* means that the numbers tend not to bunch up near the middle, or near either end of the 0-to-1 interval, but rather have an equal chance of being found anywhere within that range. In a set of 10,000 uniform random numbers, about 1000 of them would fall between 0.0 and 0.1, and about 1000 of them would fall between 0.62 and 0.72, or within any 0.1-wide interval.

Almost every numerical computing environment has a built-in function that returns a different uniformly distributed random number between 0.0 and 1.0 every time it is invoked. This function will be represented in this chapter as $U_{0\text{-}1}$, and depending on the particular software environment, it may be called RANDOM, RAND, RND, UNIFORM, RUNIF, URAND, etc., and it may or may not take arguments. If it does take arguments, they will usually specify the range as something other than 0 to 1. Sometimes the random number generator function is implemented as returning an integer from 1 to some specified integer, such as Random(100) returning an integer between 1 and 100. Such functions can be converted to the more common 0-to-1 generators by calling them with a large argument, such as 10,000, and then dividing the resulting random integer by 10,000. When doing this, it is best to subtract 0.5 from the integer before dividing, so that the resulting fractional number will never be exactly 1.0 (a $U_{0\text{-}1}$ random number exactly equal to 0 or 1 might cause computational problems in subsequent calculations). Sometimes the function takes an optional argument to reset the "seed" value used by the random number algorithm. And some random number functions can generate an entire collection of random numbers with a single call.

It is important to understand that most random number–generating functions return a different value every time they are invoked. So if the random function reference appears in two places in a formula, two different random numbers will be generated. So, the expression

$$U_{0\text{-}1} - U_{0\text{-}1}$$

generally will *not* produce zero, even though one would naturally think that any number subtracted from itself is always zero. But in the above expression, the two references to $U_{0\text{-}1}$ do not produce the same number; each invocation of $U_{0\text{-}1}$ will produce an independent random number between 0 and 1, so the result of the subtraction could be any number between $-1$ and $+1$.

The following formula would generate a random number that is uniformly distributed between $a$ and $b$:

$$U_{a\text{-}b} = a + (b - a) \cdot U_{0\text{-}1}$$

### 6.4.3 Generating Categorical Random Numbers

When a categorical attribute must be assigned to a subject or a process, such as "lived or died," "treatment or placebo," or "poor or extensive metabolizer," a random integer can be generated to represent the different possible categories. Depending on the programming environment, it may be possible to store the category as a text value ("Lived" or "Died") or a logical (true/false) value instead of as a number. Several "discrete" random distributions are commonly used in simulation.

#### 6.4.3.1 Simulating a Coin Flip (Or Any Situation with Two Outcomes)

The outcome of flipping a coin can be represented by the word *Head* or the word *Tail*. Because $U_{0\text{-}1}$ will be less than 0.5 about half the time, and greater than 0.5 about half the time (Figure 1), a random categorical value can be generated in the following way:

If $U_{0\text{-}1} < 0.5$
  Then Outcome = "Head"
  Else Outcome = "Tail"

Similarly, the following procedure can be used to randomize virtual subjects into two treatment groups, say Drug and Placebo, with three times as many Drug subjects as Placebo subjects (Figure 1):

If $U_{0\text{-}1} < 0.75$
  Then Group = "Drug"
  Else Group = "Placebo"

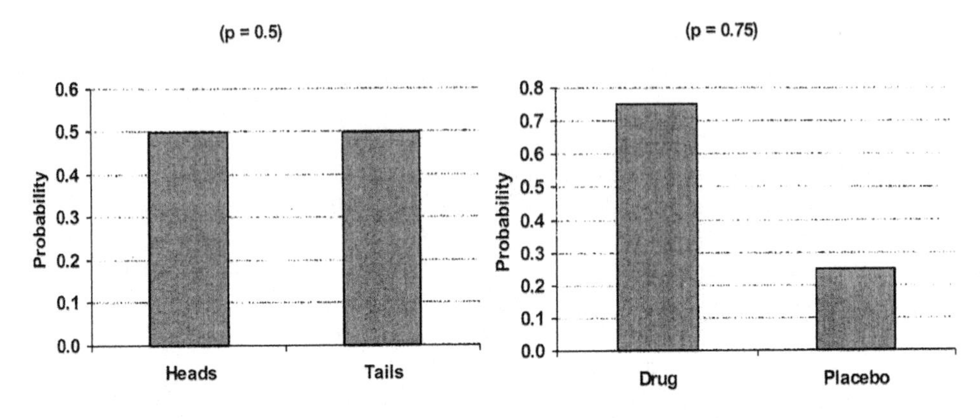


**FIGURE 1** Binomial distribution. (a) p = 0.5, (b) p = 0.75.

The same technique can be generalized to simulate events that have a specified probability of occurring. For example, if prior experience suggests that 10% of subjects are likely to drop out of a study, this can be simulated by the following expression:

If $U_{0\text{-}1} < 0.1$
  Then Completion_Status = "Dropout"
  Else Completion_Status = "Complete"

### 6.4.3.2 Simulating *n*-way Branches

A further generalization of the above branching situation occurs when there are more than two events. If tests can be nested to create a sequence of decisions. To randomize subjects into three groups, Controls (20%), Treatment 1 (35%), and Treatment 2 (45%) (Figure 2A), two "cut points" must be determined that divide the 0-to-1 interval into three segments with relative lengths 0.2, 0.35, and 0.45. A simple way to do this is to compute cumulative cut points:

$$0 + 20 = 20$$
$$20 + 35 = 55$$
$$55 + 45 = 100$$

The first two of these running-total numbers (corresponding to the fractions 0.20 and 0.55) are the two cut points for transforming a random uniform 0-to-1 number into one of the three categories. If $U_{0\text{-}1}$ is between 0 and 0.2 (which should occur 20% of the time), "Control" would be selected; if it is between 0.2 and 0.55

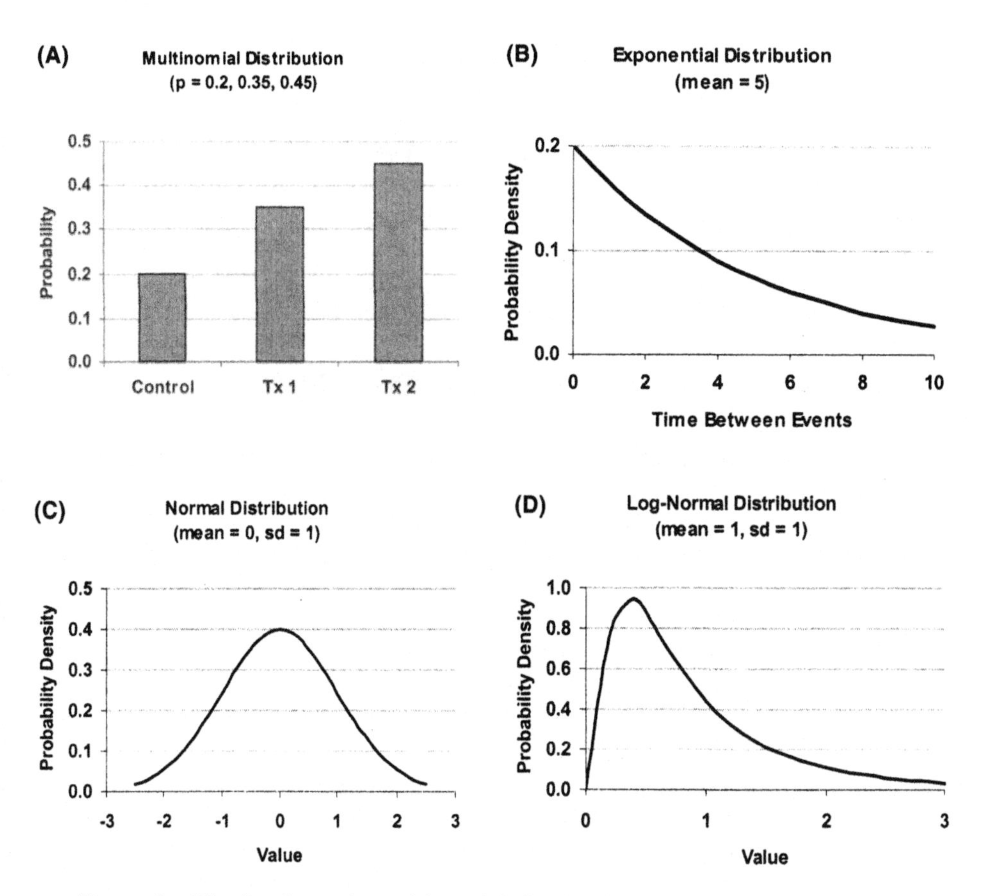


**FIGURE 2** Distributions: A, multinomial distribution; B, exponential distribution; C, normal distribution; D, lognormal distribution.

(which should occur about 35% of the time), "Treatment 1" would be selected; otherwise (the other 45% of the time, when it is between 0.65 and 1), it would be "Treatment 2." The following statements do this:

```
X = U0-1
If X < 0.2
    Then Group = "Control"
    Else If X < 0.55
        Then Group = "Treatment 1"
        ELSE Group = "Treatment 2"
```

Here, the Else part of the If construct contains another complete If construct. Note that it is not necessary to do the third test; if the random number fails the first two tests, it must be in the third region. This process can be extended to handle multiway branches of considerable complexity.

It is crucially important to invoke the random number generator only once and save its value in a variable (X in the example) for use in the two If tests. If the random number generator appeared in each If test, then each invocation would return a different random number, and the comparison with the running-total cut points would produce incorrect proportions of treatments 1 and 2.

### 6.4.4 Generating Exponentially Distributed Random Numbers

The exponential distribution is useful in simulating the time intervals between which sporadic events occur. An exponentially distributed random variable can have any value between zero and infinity, but small values will occur more frequently than large values. The exponential distribution is characterized by a single adjustable parameter—the mean event rate, which is the average number of events expected per unit of time (Figure 2B).

Exponentially distributed random numbers can be generated directly from uniformly distributed random numbers using the formula

$$\text{Event Interval} = -\text{Mean} \cdot \text{Ln}(U_{0\text{-}1})$$

where Ln is the natural logarithm function. The occurrence times for a set of random events are generated by a two-step process: First, random numbers are generated from an exponential distribution to simulate the intervals between successive events; then a cumulative summation (running total) of these times is calculated, giving the actual times of the successive events.

### 6.4.5 Generating Normally Distributed Random Numbers

There are several ways to generate normally distributed random numbers from uniformly-distributed numbers. Some are approximate; others are exact. Some methods are more efficient than others, and some are easier to program. Perhaps the simple and most convenient of the exact methods involves the inverse of the distribution function. In Excel this function is called NORMSINV (Excel also supports NORMINV which allows the user to specify the mean and SD of the distribution). This function may not be present in all programming environments, but if it is available then normally distributed random numbers from a population having a mean of zero and a standard deviation of one can be generated from the following formula:

$$N_{0,1} = \text{Inverse\_Normal}(U_{0\text{-}1})$$

where Inverse_Normal returns the value for which the integral, from minus infinity to that value, is equal to the argument passed to the subroutine (Figure 2C). One efficient and commonly used method for generating normally distributed variables, the Box Muller method, is described in Chapter 3.

#### 6.4.5.1 Generating Normally Distributed Random Numbers with a Specified Mean and Standard Deviation

The "standard" normally distributed random numbers generated above can easily be scaled to have any mean (m) and standard deviation (SD):

$$N_{m,SD} = m + SD * N_{0,1}$$

The following example simulates the body weights of a set of virtual subjects, using prior data indicating that subjects tend to have normally distributed weights that average 170 with a SD of 30.

$$\text{Body\_Weight} = 170 + 30 * \text{Inverse\_Normal}(U_{0\text{-}1})$$

#### 6.4.5.2 Generating a Pair of Correlated Normally Distributed Random Variables

When creating virtual subjects for use in a simulation, it may be necessary to simulate characteristics that are known to be correlated. For example, height and weight tend to be strongly correlated. In an X-Y scattergraph of the two variables, the correlation (or lack of it) is indicated by the degree to which the numbers tend to cluster along a line, rather than in a round "shotgun" pattern. The strength of the correlation is measured by the correlation coefficient—a number whose value can range from $-1$ when the two variables are perfectly inversely correlated (lie exactly on a line that slopes downward) through 0 when the variables are totally uncorrelated (form a round shotgun pattern) to $+1$ when the variables are perfectly positively correlated (fall exactly on an upward-sloping line).

A pair of normally distributed random variables (c1 and c2) having a correlation coefficient of $\rho$ can be generated as follows:

$$c1 = N_{0,1}$$
$$c2 = c1 * \rho + N_{0,1} * \text{Sqrt}(1 - \hat{\rho}\ 2)$$

The two appearances of $N_{0,1}$ above refer to two *independent* normally distributed random numbers, not the same number (Figure 3).

To produce two variables having specific means and SDs, the Cs can be scaled in the usual way. The following procedure simulates subjects having normally distributed heights with a mean of 150 cm and SD of 20 cm, and normally distributed weights with mean of 80 kg and SD of 25 kg, where the correlation coefficient between height and weight is 0.65:

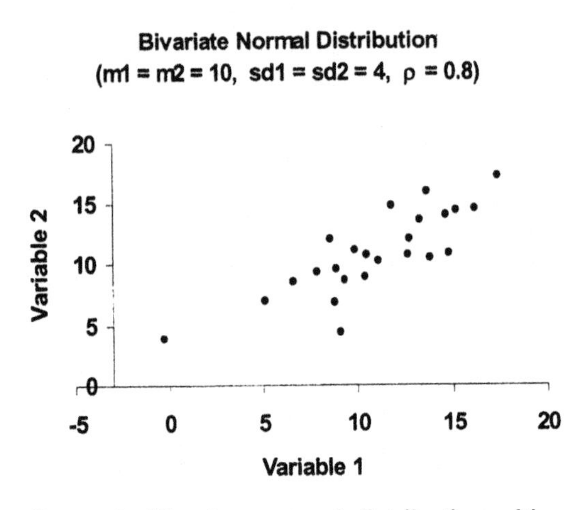


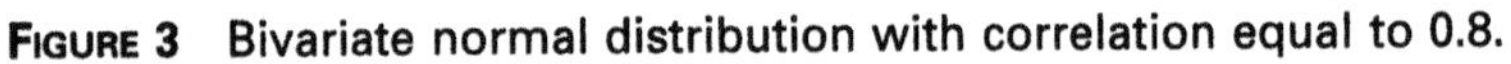
**FIGURE 3** Bivariate normal distribution with correlation equal to 0.8.

$$N1 = N_{0,1}$$

$$N2 = N_{0,1}$$ (remember, N1 and N2 are two independent random numbers)

$$\text{Height} = 150 + 20 * N1$$

$$\text{Weight} = 80 + 25 * (N1 * 0.65 + N2 * 0.76)$$

where 0.76 is Sqrt $(1 - 0.65\hat{}\ 2)$.

It is possible to generate a set of three or more random numbers that are mutually intercorrelated, but the method is more complicated. See Chapter 3.

### 6.4.6 Generating Log Normally Distributed Random Numbers

The log normal distribution is almost as commonplace as the normal distribution. Many physiological properties (e.g., drug concentrations) and operational quantities (e.g., lengths of hospitalization) tend to be log normally distributed. Quantities that by their very nature can never be negative numbers (concentrations, durations, etc.) often tend to be log normally distributed.

This distribution is characterized by many occurrences of relatively small values, a few occurrences of large values, and an occasional occurrence of a very large value. The frequency distribution tends to be skewed—it has a steep left side with a very short left tail, but a much more gradual trailing off to the right (Figure 2D). The SD is often as large as the mean itself. Taking the logarithms of these individual numbers produces numbers with a normal distribution that has its own mean and SD.

Because the logarithms of log normally distributed numbers are normally distributed; it follows that the exponentials of normally distributed numbers will be log normally distributed. So, the following expression will generate a log normally distributed random number.

$$\text{LogNormal} = \text{Exp}(N_{0,1})$$

where Exp is the function for raising *e* to a power.

But this expression generates random numbers from only one very specific log normal distribution: the one that happens to have a mean of 1.65 and a SD of 2.16, which will almost certainly not be the mean and SD needed for a particular simulation. Generating random numbers from a log normal distribution with a specific mean and SD is a not an intuitively obvious process; it is not as simple as multiplying by the desired SD and adding the desired mean, and it is not a simple matter of taking the logarithms of the means and SD's either, as shown below.

### 6.4.6.1 Generating Log Normally Distributed Random Numbers Having a Specific Mean and Standard Deviation

The procedure to generate log normal random numbers with a specified mean *m* and standard deviation SD is:

$$Q = \text{Sqrt}(\text{Ln}(1 + (SD/m)^2))$$

$$\text{LogNormal}_{m,SD} = m \cdot \text{Exp}(Q \cdot (N_{0,1} - \hat{Q}\ 2/2))$$

So, to generate random numbers from a log-normal distribution with a mean of 10 and SD of 8, first compute Q:

$$Q = \text{Sqrt}(\text{Ln}(1 + (8/10)^2)) = \text{Sqrt}(\text{Ln}(1.64)) = 0.7033$$

$$Q^2/2 = 0.7033^2/2 = 0.24735$$

Then, log normal random numbers can be generated from the expression

$$\text{LogNormal}_{10,8} = 10 \cdot \text{Exp}(0.7033 \cdot N_{0,1} - 0.24735)$$

### 6.4.6.2 Generating Log Normally Distributed Random Numbers Whose Logarithms Have a Specific Mean and Standard Deviation

The situation is much simpler if the mean m and standard deviation SD of the (normally distributed) logarithms of the log normally distributed numbers are available. The procedure in that case is simply to generate the appropriate normal variable and then exponentiate it.

$$X = \text{Exp}(m + SD \cdot N_{0,1})$$

So, if it is known from prior studies that the natural logarithm of the clearance of a drug is normally distributed with a mean of 2.5 and standard deviation of 0.9, a set of simulated clearance values can be generated by the expression.

$$X = \mathrm{Exp}(N_{2.5,0.9})$$

which is equivalent to

$$X = \mathrm{Exp}(2.5 + 0.9 \cdot N_{0,1})$$

These numbers will be log normally distributed with a mean of 18.2 and SD of 20.4, and the logarithms of these numbers will be normally distributed with a mean of 2.5 and SD of 0.9.

When utilizing prior data that is known to be log normally distributed, always take the logarithms of each number, calculate the mean and SD of these logarithms, then use the simple formula above to generate the random numbers. The more complicated formulas in the previous section should be used only if individual data points are not available, but only the summarized mean and SD of the log normally distributed numbers.

## 6.5 SIMULATING THE PHARMACOKINETICS OF A DRUG TRIAL

After the virtual patients, along with their demographic and physiological characteristics and their predestined behaviors have been generated, the results of the study can be simulated. The first part of this process is calculating the behavior of the drug in the subject's body. For example, a pharmacokinetic study plan may call for sampling drug concentrations every 2 h over a 24 h period after dosing. The simulation must calculate, for each patient, what the drug concentration would be at the end of each 2 h interval, based on that subject's dose level, covariates, and postulated values of the PK parameters.

### 6.5.1 Solving Pharmacokinetic Models Analytically

Pharmacokinetic models can be fairly simple, or incredibly complex, or anything in between. It is fairly straightforward to solve simple models analytically (that is, by means of algebraic solution of the kinetic equations). In the case of a single bolus infusion with first-order elimination from a single compartment, the plasma level rises (almost) immediately to a maximum value determined by dose and volume of distribution, and then declines exponentially, with a rate constant that depends on the clearance (CL) and volume of distribution (Vd):

$$\mathrm{Concentration} = (\mathrm{Dose/Vd}) \cdot e^{-\mathrm{Time \cdot CL/Vd}}$$

Because the differential equation that gives rise to the above formula happens to be linear, the *superposition principle* applies. This means that if a complex situation can be broken down into a combination of first-order models, the individual systems can be solved separately and then the resulting concentrations at corresponding times can be added. So, if a study calls for a bolus injection, followed by an 8 h infusion, followed by an oral dose every 4 h for 5 doses, all of these individual administrations have simple solutions in terms of exponential functions. The seven systems can therefore be solved individually, to get what the concentration-versus-time curve would have been if each of the seven administrations had been the only one, and then the seven sets of concentrations can be added together. In doing this, it is necessary to align the corresponding time points of the individual solutions so that the concentrations being added are those that would apply at each actual point in time.

### 6.5.2 Solving Pharmacokinetic Models Numerically

More complicated processes, such as those involving multiple compartments, may have analytical solutions, although they may be very complicated. It is possible to use mathematical software such as Mathematica or MATLAB to assist in finding analytical solutions to complex systems of differential equations, if such solutions do in fact exist. But for more complicated models it is usually necessary to resort to numerical integration of the ordinary differential equations.

Here the advantages of a mathematical programming environment or simulation language become evident—they usually have built-in capabilities for numerically integrating systems of simultaneous linear or nonlinear differential equations. But if the chosen implementation environment does not provide this capability, and if the simple analytic methods described above cannot realistically describe the system being simulated, the numerical solution will have to be programmed explicitly. This is almost impossible in a spreadsheet environment without resorting to a script or macro language, such as VBA in Excel. But it can be done in any procedural programming environment, such as FORTRAN or BASIC.

Many methods exist for integrating systems of differential equations of the type that occur in pharmacokinetics. They all proceed stepwise through time, taking a set of known concentrations at some particular time point, t, estimating those concentrations at some slightly later time point, $t + dt$, and then repeating the process for subsequent times. The methods differ in how they come up with these incremental estimates. The following sections present some of the more commonly used numerical integration methods here; much more elaborate discussions can be found in any good reference on numerical analysis (8).

#### 6.5.2.1 Euler's Method

The simplest of the numerical methods for integrating one or more linear or nonlinear differential equations is based on the fact that the differential equations directly provide the slope of a straight line tangent to the Concentration-vs-Time curve for each species in each compartment at any point in time. Euler's method is based on the assumption that, for sufficiently small time increments (very small dt), the true Conc-vs-Time curve can be approximated by its tangent line:

$$\text{Conc}(t + dt) = \text{Conc}(t) + d\text{Conc}/dt \cdot dt$$

The derivatives of each concentration are provided directly by the differential rate equations themselves. Euler's method is based on the expansion of the true Conc-vs-Time function in a Taylor series and retaining only the first (linear) term (See Section 6.5.2.2).

Although Euler's method is easy to implement (because the required first derivatives are so readily available), it is almost never used in clinical trial simulation. This is because it can require an extremely small step size to achieve reasonable accuracy, and therefore is susceptible to very long running times and an unacceptable accumulation of round-off errors.

### 6.5.2.2 Higher-Derivative Methods

An obvious extension of Euler's method would involve the use of higher derivatives, that is, higher-order terms in the Taylor expansion:

$$\text{Conc}(t + dt) = \text{Conc}(t) + d\text{Conc}/dt \cdot dt + d^2\text{Conc}/dt^2 \cdot (dt)^2/2! + d^3\text{Conc}/dt^3 \cdot (dt)^3/3! + \ldots$$

This method retains much of the stability of Euler's method, and it can achieve much higher accuracy (approaching machine accuracy) while retaining reasonably large step sizes and therefore a reasonable running time.

This method does require the direct evaluation of higher derivatives, but these are (in theory) fairly easy to obtain by recursively differentiating the original rate expressions. All of the commonly used methods for solving systems of differential equations attempt to approximate these higher-order derivatives without the need to evaluate them algebraically.

### 6.5.2.3 Runge-Kutta Methods

The Runge-Kutta (R-K) family of methods achieves higher-order accuracy, comparable to the Taylor series approach, without the need to explicitly evaluate the higher derivatives. It does this by repeatedly evaluating the differential equations at various increments, and combining them to obtain estimates of the higher derivatives. There are a number of specific R-K formulas, corresponding to different orders of approximation; the most commonly used is the fourth-order R-K method, which is generally considered to provide a good trade-off between precision and programming simplicity. It uses the first four terms of the Taylor expansion of $dy/dx = f(t, y)$, without actually requiring higher derivatives.

Consider the case of a single differential equation of the form

$$d(Conc(t))/dt = F(t, C)$$

that is, the rate of change of concentration with respect to time can be written as some function of the concentration and the time. The fourth-order Runge-Kutta formula is

$$Conc(t + dt) = Conc(t) + (k1 + 2 \cdot k2 + 2 \cdot k3 + k4)/6$$

where

$$k1 = dt \cdot F(t, Conc)$$

$$k2 = dt \cdot F(t + dt/2, Conc + k1/2)$$

$$k3 = dt \cdot F(t + dt/2, Conc + k2/2)$$

$$k4 = dt \cdot (t + dt, Conc + k3)$$

Each step requires that the function defining the differential equation be evaluated four times, at slightly different combinations of t and Conc values.

The Runge-Kutta algorithm is fairly easy to implement, and often provides a good combination of adequate accuracy and reasonably short run times. It is often considered to be the "workhorse" of ordinary differential equation integrators. But other methods have been developed that offer improved performance in terms of faster execution, higher precision, and better stability.

#### 6.5.2.4 Other Methods

*a. Extrapolation.* In these methods the computed result is extrapolated to estimate what would have been obtained had we used a very much smaller step size than we actually used. The general idea of Richardson extrapolation has been applied to differential equation solving to produce the Bulirsh-Stoer methods.

*b. Predictor-Corrector.* In these methods the time-advancement proceeds in two passes—a prediction pass, often using Runge-Kutta or other methods to obtain a first estimate of the function at the later time points, followed by correction pass in which derivatives are reevaluated at the new time point and used to obtain better estimates of the increments to be applied to the function. The Adams-Bashforth-Moulton family of algorithms are probably the best-known of the predictor-corrector methods.

The choice of method is no trivial matter: It requires a good knowledge of numerical analysis and a good understanding of the idiosyncrasies of pharmacokinetic rate equations. No one should attempt to implement these advanced methods without consulting a good reference work on the topic, such as Press et al. (8).

### 6.5.2.5 Ill-Conditioned Systems of Equations

As if the above complexities were not enough to deter one from grappling with pharmacokinetic equations, some sets of differential equations can contain another pitfall that may render them difficult or impossible to integrate numerically: When the values of the various rate constants in a system vary over many orders of magnitude, different components of the system can undergo changes according to vastly different time scales: One component may, within seconds, attain steady state within a system that, overall, requires hours (or days or weeks) to settle down. This special type of ill-conditioning has been given the colorful term "stiffness," and there are well-established criteria for diagnosing this condition—the eigenvalues of the characteristic matrix of coefficients display a wide (orders of magnitude) range of values.

When integrating stiff differential equations numerically, the problem is in the choice of time interval for each integration step. If the interval is small enough to capture the rapidly changing parts of the system, it will require an astronomically large number of steps to span the slowly changing behavior, with inordinate runtimes and unacceptably large accumulations of round-off errors. But if the time scale is based on the slowly varying processes, then the numeric integration will be completely erroneous for the rapidly varying parts, and is therefore likely to be wrong for the entire system.

Algorithms are available for dealing with stiff differential equations. Most of the general routines are based on the idea of backward differencing, in which the derivatives are evaluated at the new time-points. If the differential equations are linear (representing first-order kinetics), the implementation of this approach is fairly easy, but for nonlinear systems it can involve having to solve a complex set of nonlinear simultaneous equations at every step of the iteration. For the special kinds of differential equations that arise in pharmacokinetic modeling, it may be possible to develop an ad hoc procedure by considering the faster process as not "kinetic" at all, but rather an equilibrium process, reducing the number of simultaneous differential equations that need to be solved. In any case, it remains important to be aware that this problem may be lurking in the simulation, to be able to recognize its presence, and to realize that it is something that needs to be dealt with by special programming. An excellent description of modern methods for dealing with stiff systems of differential equations can be found in Press et al. (8).

## 6.6 SIMULATING PHARMACODYNAMIC EFFECTS

Once the pharmacokinetic equations have been solved to obtain drug concentrations over time for the simulated patients, the next step is to simulate the effects these drug levels will produce on the body, if drug concentrations are the driving force for the PD effect.

### 6.6.1 Simulating Continuous Pharmacodynamic Effects

A continuous PD effect would be a measurable change in some physiologic property that varies as drug level changes. We may be interested in how this quantity changes during the time when the drug concentration rises and falls, or may only be interested in the maximum size of the effect.

Some formula will be required that relates the observed effect to the drug concentration. A simple example is: A drug might prolong the QT interval in an ECG by an amount that is directly proportional to the drug concentration within a certain range:

$$\text{QT} = \text{k} \cdot \text{Drug Concentration}$$

The value of k would have to be estimated, perhaps from earlier pilot studies, or from the behavior of similar compounds. Of course, a realistic simulation will have to take between-subject and within-subject variability into account as considered in Chapter 18, and the expression for QT will therefore need a random component. If the drug is thought to prolong QT by 3 ms for every 1.0 ng/ml of concentration, and the patient-to-patient variability (the SD of the prolongations among a large number of patients with the same drug concentration) is 5 ms, the following formula would be used to simulate a particular patient's QT prolongation:

$$\text{QT}_i(\text{ms}) = 3 \cdot \text{Drug Concentration (ng/ml)} + \text{N}_{0,5}$$

where $\text{N}_{0,5}$ is a normally distributed random variable with a mean of 0 and SD of 5.

### 6.6.2 Simulating Discrete Outcomes

A discrete pharmacodynamic outcome would be an event like fainting, vomiting, cardiac arrest, etc., which may or may not occur, and whose probability of occurrence is assumed to vary with changing concentration. To simulate the occurrence of discrete events, we need to do two things in succession:

1. Calculate the probability of a subject experiencing this event, based on drug concentration.
2. Simulate the occurrence or nonoccurrence of the event for that subject.

The first step requires a function capable of converting the subject's responses (e.g., drug concentrations or some biomarker of response) to a probability of having the event, i.e., can convert a real number into a probability bounded by a 0, 1 interval. One commonly used formula for predicting the probability of an event's occurring is the logistic formula:

$$\text{Probability} = 1/(1 + e^{-\text{Drug effect}})$$

where

$$\text{Drug effect} = c0 + c1 \cdot x1 + c2 \cdot x2 + \text{etc.}$$

where x1, x2, etc. are the predictors such as drug concentration and disease status, and c0, c1, c2, etc. are parameters obtained by means of logistic regression from prior studies.

Once a logistic formula has been found, it can be used to calculate, for each subject, the probability of that subject experiencing the event. But in the real world, an individual subject either *will* or *will not* experience the event, so to simulate the occurrence or nonoccurrence of a headache, the subject's calculated probability (a number that can have any fractional value between 0 and 1) must be transformed into a dichotomous "yes" or "no." This can be done easily as described in Section 6.4.3. So if the fitted logistic model is:

$$\text{Probability of Headache} = 1/(1 + e^{(-(-1.23+0.456 \cdot \text{Conc}))})$$

then this formula could be used to compute each subject's probability of headache, which would then be converted to the simulated dichotomous event by the formula:

If $U_{0\text{-}1}$ < Prob_of_Headache
Then Headche = "Yes"
Else Headache = "No"

### 6.6.3 Simulating Survival Outcomes

The term "survival analysis" refers to a collection of statistical techniques for analyzing the rate at which a group of individuals experience some "terminal" event such as death, disease recurrence, or an adverse event. The results are often presented graphically as a survival curve, showing the estimated fraction of the population that remains "event-free" at the end of any particular amount of time. These curves are generally calculated by the "life table" method (for grouped data) or the Kaplan-Meier method (for ungrouped data). Survival data is often summarized by quoting the mean survival time (the average of the survival times of all the subjects) or the median survival time (the time at which 50% of the population has experienced the terminal event), or the fraction of the population remaining event-free at some specified time point (such as 5 years for cancer studies). To determine if survival is influenced by certain factors (treatment group, age, gender, etc.), the most commonly used method is Cox proportional hazards (CPH) regression (9). The CPH model is:

$$\text{Surv}(t) = \text{Base}(t)^{e^{c1 \cdot x1 + c2 \cdot x2, \text{etc.}}}$$

where Surv(t) is the survival curve, showing the probability of a subject's surviving for time t; Base(t) is the "base" survival function, which can have any shape; x1, x2, etc. are covariates; and c1, c2, etc. are coefficients, typically obtained from CPH analysis of data from earlier studies.

Simulating survival outcomes requires a postulated survival function from which random "survival times" can be generated for the virtual subjects. In the absence of any precise knowledge of what the survival curve looks like, a reasonable assumption might be an exponential function. For this model, there are simple relationships between mean survival time, median survival time, *n* month survival rate, and the exponential constant appearing in the model. Having this constant, exponentially distributed random survival times can be generated using the formula in Section 6.4.4.

## 6.7 CONCLUSIONS

A complete discussion of computer simulation would fill many volumes; this introduction could provide no more than a cursory presentation of the basic principles and techniques. But even a very simple simulation, if done properly, can be of enormous help in the design of an experiment.

It is often not necessary to simulate an entire trial in order to obtain information that is useful in optimizing some aspect of that trial. Suppose that one objective of a trial is to determine the clearance of a drug. It is well known that the precision of an estimated parameter is strongly dependent on sampling times. One might wonder whether a proposed set of PK sampling times will provide the best possible precision in the estimated clearance. To help answer this, a very simple simulation could be run with a PK model to calculate the expected Concentration-vs-Time curve, adding some random fluctuations to generate data points with a reasonable amount of scatter (based on prior experience), and fitting a model to the simulated data. Several hundred repetitions of this simulation will produce a reasonable estimate of the uncertainty (and bias) of the estimated clearance. The simulation could then be rerun using a different set of sampling times, and perhaps additional (or fewer) PK samples, to see how these design alterations affect the precision of the estimated parameter. Even this very simple simulation could lead to an optimized study design (at least an optimized set of sampling times) that might not have been envisioned originally.

Simulations can range from the very simple to the very complex. If even the simplest simulation of only one component of a planned trial, as described above, can lead to meaningful improvements in its design, then it is evident that a more thorough and realistic simulation, while requiring considerably more time and effort, would be an indispensable tool in the design of clinical trials.

## REFERENCES


1. NHG Holford, HC Kimko, JPR Monteleone, CC Peck. Simulation of clinical trials. Ann Rev Pharmacol Thera 40:209–234, 2000.
2. The Third National Health and Nutrition Examination Survey 1988–1994 (NHANES III). US Centers for Disease Control and Prevention Division of Health Examination Statistics. National Center for Health Statistics (NCHS). http://www.cehn.org/cehn/resourceguide/nhanes.html
3. JL Schafer. Analysis of Incomplete Multivariate Data, Monographs on Statistics and Applied Probability 72. London: Chapman and Hall/CRC, 1997.
4. AM Law, WD Kelton. Simulation Modeling and Analysis, 2nd ed. New York: McGraw-Hill, 1991.
5. WJ Kennedy, JE Gentle. Statistical Computing. New York: Marcel Dekker, 1980.
6. SK Park, KW Miller. Random number generators: good ones are hard to find. Comm. ACM 31:1192–1201, 1988.
7. EJ Dudewicz, TG Ralley. The Handbook of Random Number Generation and Testing with TESTRAND Computer Code, Vol. 4 of American Series in Mathematical and Management Sciences. Columbus, OH: American Sciences Press, 1981.
8. WH Press, SA Teukolsky, WT Vetterling, BP Flannery. Numerical Recipes in Fortran, and Numerical Recipes in C Cambridge University Press, 1999.
9. JF Lawless. Statistical Models and Methods for Lifetime Data New York: John Wiley & Sons, 1982.

# 7

# Analysis of Simulated Clinical Trials

**Ene I. Ette, Christopher J. Godfrey, and Stephan Ogenstad**
Vertex Pharmaceuticals, Inc. Cambridge, Massachusetts, U.S.A.

**Paul J. Williams**
Trials by Design, LLC, La Jolla and Stockton, and University of the Pacific, Stockton, California, U.S.A.

## 7.1 INTRODUCTION

Maximization of knowledge gained during the drug development process with the intent of increasing the probability of "success" in a clinical trial has received considerable attention recently (1). Clinical trial simulation has been recognized as being potentially useful for streamlining clinical drug development and evaluation with a view to maximizing information content of clinical trials (2–5).

Simulation of a clinical trial based on pharmacokinetic, pharmacodynamic (PK/PD), and clinical outcome link models is a technique that is being newly developed. When there is a considerable amount of information about the drug, simulation provides the means of synthesizing the information into a coherent package that indicates the drug developer (sponsor) has good control over the pharmacology and, eventually, the therapeutics of the drug.

The objective of the clinical trial simulation paradigm in drug development is to increase the efficiency of drug development, i.e., minimizing the cost and time of the development of a drug while maximizing the informativeness of data generated from a trial or trials. Simulation of clinical trials offers the means of generating complex data sets, which may include the influence of prognostic factors, sample size, and dropouts, for testing new analysis methods (6, 7). It offers the possibility of improving the chances of an efficacy trial succeeding by allowing the user to ask questions which would involve perturbing different aspects of the input model (5). For instance, what would be the impact of a 20% noncompliance rate on the study outcome? What would be the minimal size of a study that would be commensurate with a reliable detection of the treatment effect? Thus, clinical trial simulation as an abstraction of the clinical trial process is used to investigate assumptions built into a proposed study, study designs, perform sensitivity, and power analysis in order to maximize the information content that can be gained from a study.

Hale et al. (8) used simulation to choose experimental design, sample size, and study power for a pivotal clinical trial of mycophenolate mofetil, a prodrug of mycophenolic acid—an immunosuppressant, in combination with cyclosporin and corticosteroids in transplantation. This was a randomized concentration-controlled trial (RCCT) in which subjects were randomized to one of three target drug exposure [area under the plasma concentration-time curve (AUC)] levels instead of three dose levels. An RCCT was chosen because modeling and simulations based on phase II data had indicated a study using AUC rather than dose would have much greater power to detect treatment effect.

Bias and precision of the estimates of quantitative descriptors that reflect treatment effect size, time to peak effect, drug disposition, and effect with associated variability can be evaluated using simulation. In addition, simulation has been used as a tool for investigating the performance of various sampling designs and design factors employed in population pharmacokinetics (PPK) and population PK/PD studies (9–16). It has also been used for dose selection for clinical trial structure and design (17, 18).

Clinical trial simulation can be likened to meta-analyses of clinical trials. Meta-analysis has been defined as statistical analysis of a large collection of analysis results from individual studies for the purpose of integrating the findings (19). Without meta-analysis useful data can be left fallow or are at least not utilized to their maximum extent (20). The results of meta-analysis of clinical trials can point to specific areas that need to be addressed, either because there are few available data or because the available data suggest a particular hypothesis that requires additional attention (e.g., an age group of patients that may be at an increased risk, or an unexpected exposure from a low dose may appear harmful). These issues can be addressed by simulating the clinical trials, but unlike meta-analysis that is retrospective, clinical trial simulation is prospective. However, a simulated trial is similar to meta-analysis in that it consists of many replications of the trial. It differs, however, in that the replications of the trial by simulation are usually of greater homogeneity than the trials included in meta-analyses. Hence, statistical analysis methods used in analyses of simulated clinical trials usually incorporate a random effects component which is generally unnecessary for the meta-analysis of actual clinical trials. Nevertheless, they are analogous systems.

One of the challenging tasks for a pharmacometrician/statistician is to convey findings from the analyses of a simulated trial to clinicians and other members of the drug development team. Communication of the results can be effected through words, tables, or graphics. The use of high-quality graphics can effectively enhance the communication of the outcome of a simulation project to a drug development team. Graphics, in particular, are essential for conveying relations and trends in an informal and simplified visual form (21).

The remainder of the chapter deals with analysis of simulated data from single and replicated trials with some examples from PPK and efficacy trials, multiplicity in simulated efficacy trials, and the communication of the results of a clinical trial simulation (CTS) project.

## 7.2 ANALYSIS OF SIMULATED DATA

The procedure for analyzing a simulated trial should be specified in a simulation plan. A simulated trial should be analyzed using the same data analysis method as the actual trial. Insights into a single replication of the trial or integrative and comparative insights can be gained by analyzing a single-trial replication or the group of trial replications performed. Thus, we consider these two levels of analysis next.

### 7.2.1 Single-Trial Replication Analysis

The experimental unit of a replication of a single clinical trial is each individual in the trial. Thus, analyzing one replication of a simulated trial allows the responses across subjects within that replication to be summarized. In this case, the analyst would be interested in individual subject outcome measures, for example, the duration of drug effect, before proceeding to perform the complete simulation experiment.

### 7.2.2 Simulated Experiment Analysis

The replicated in silico clinical trial is akin to the meta-analysis of clinical trials as discussed previously. In this case, a response across trials is defined. A summary of the analysis of the outcome measures obtained from each clinical trial replication yields the outcome of each simulated experiment. For a specific clinical trial design the power of the study is determined by the number of trials that reject the null hypothesis. This is often the case in a confirmatory trial. It is good practice in performing CTS to assess study power that the type I error rate ($\alpha$) be estimated to verify the size of the test (e.g., $\alpha = 0.05$). An error rate is a probabilistic measure of erroneous inference in a given set of hypotheses from which some common conclusions are drawn. Bias and precision of parameter estimates such as drug clearance in a population pharmacokinetic study or maximum drug effect in a PK/PD study are typically computed.

#### 7.2.2.1 A Population Pharmacokinetic Study Example: Determination of Power

Kowalski and Hutmacher (22, chapter by Hutmacher and Kowalski) performed clinical trial simulations to assess the power to detect subpopulation differences in apparent drug clearance (CL/F) and examine sample size requirements for a population pharmacokinetic substudy (3) of a phase III clinical trial. The simulations were based on a population PK model developed from a phase I healthy volunteer study. Taking into account the practical constraints of the proposed study, a sparse sampling design was developed for the study. Due to the narrow range of the sampling times, it was expected that a sparse sampling design would not support a two-compartment model that was used to describe the phase I data. Thus, a minimal model, a one-compartment model, was used to fit the data from the simulated experiment. The key parameter of interest in the simulated study was CL/F, and a 40% reduction was considered to be of clinical significance. That is, this degree of reduction in CL/F would result in a need for dosage adjustment.

Three hundred hypothetical clinical trials were simulated to determine the sample size necessary to detect 40% reduction in CL/F in a subpopulation of proportion $p = 0.05$ or $p = 0.10$ with at least 90% power. Sample sizes of 150 and 225 were investigated. The power of the study was estimated as the percentage of trials out of 300 in which statistically significant ($\alpha = 0.05$) difference in CL/F was observed using the likelihood ratio test.

To obtain the empirical estimates of $\alpha$, Kowalski and Hutmacher (22) simulated 300 clinical trials for each combination of sample size and $p$, where the proportional reduction in CL/F ($\phi$) was constrained to zero. Covariate and base models were fitted to each of the trials and the likelihood $\chi^2$ ratio tests were performed at the 5% level of significance. The percentage of trials where a statistically significant difference in CL/F was observed provided an empirical estimate of $\alpha$ (i.e., $H_0$: $= 0$ is rejected when $H_0$ is true). The data were analyzed with the NONMEM (23) software. The results suggested that approximately a 9 point change in the objective function should be used to assess statistical significance at the 5% level rather than the commonly used $\chi^2$ critical value of 3.84 for 1 degree of freedom. Also, the proposed sampling design with the minimal model yielded accurate and precise estimates of apparent drug clearance (CL/F) and apparent volume of distribution at steady state. This study was executed in accordance with the spirit and methods of the consensus document on clinical trial simulation (24).

*a. Use of the Likelihood Ratio Test.* The work of Kowalski and Hutmacher (22) also provides a poignant example of the risk inherent in the use of the likelihood ratio test (LRT) in the analysis of simulated trials, particularly in the context of mixed effects modeling.

If minus twice the log likelihood associated with the fit of a saturated model $A$ with $p + q$ parameters is designated $\ell_A$, and a reduced version of this model (model $B$) with $p$ parameters has minus twice the log likelihood $\ell_B$, the difference in minus twice the log likelihoods $(\ell_A - \ell_B)$ is asymptotically $\chi^2$ distributed with $q$ degrees of freedom. This formulation is widely used to assess the statistical significance level of the parameters associated with the $q$ degrees of freedom.

For the determination of the significance level of fixed effects, the LRT is known to be anticonservative, i.e., the empirical $p$ value will be greater than the nominal $p$ value (25). Generally, as the number of parameters (degrees of freedom) being tested increases, the more liberal the test.

Conversely, Stram and Lee (26) noted that the LRT tends to be asymptotically conservative for the assessment of random effects significance level. In this context, the conservative nature is attributable to the null hypothesis consisting of setting the variance term at a boundary condition, i.e., zero. While the inaccuracy in $p$ value is modest when the number of random effects being tested is small, the conservativeness increases with an increase in the number of random effects being tested.

Wählby et al. (27) explored via simulation a number of factors influencing the disparity between nominal and actual significance level of tests for covariate (fixed) effects in nonlinear mixed effects models using the software, NONMEM. Approximation method [first order (FO) versus first-order conditional estimation (FOCE) with interaction between interindividual variability $\eta$ and residual variability $\varepsilon$], sampling frequency, and magnitude and nature of residual error were determined to be influential on the bias associated with the $p$ value. An important finding was that the use of the FOCE method with $\eta$-$\varepsilon$ interaction resulted in reasonably close agreement of actual and nominal significance levels, whereas the application of the LRT after estimation using the FO approximation generally resulted in marked bias in $p$ values.

The implications of the disparity between actual and nominal significance levels of the likelihood ratio test in mixed effects modeling and simulation are clear; however, simple algorithms for $p$ value correction are not readily available. The bias in likelihood ratio test–determined $p$ value for fixed effects could be very influential on trial simulation findings. Ultimately, simulation exercises should provide for determination of empirical $p$ values to avoid faulty conclusions about power and sample size, for example. This is what Kowalski and Hutmacher did in the example discussed in the previous section.

*b. Reliability and Robustness of Parameter Estimates.* Determining the reliability of parameter estimates from the pharmacokinetic/pharmacodynamic (input-output) models is necessary as this may affect study outcome. Not only should bias and precision associated with parameter estimation be determined, but also the confidence with which these parameters are estimated should be examined. Confidence interval estimates are a function of bias, standard error of parameter estimates, and the distribution of parameter estimates. Paying attention to these measures of parameter estimation efficiency is critical to a simulation study outcome (11, 12).

Simulation is useful for evaluating the merits of competing study designs (3, 4). Competing study design should be evaluated for power, efficiency, robustness, and informativeness. In evaluating the power of a study with a particular design, the ability to reject a null hypothesis or to estimate a parameter such as drug clearance with the design is examined. Efficiency examines the ratio of effort/cost to expected result. Efficiency is sacrificed but power is improved when more subjects are enrolled, sampled more often (when an efficacy variable is involved) and for longer periods of time.

It is also important to evaluate the quality of the results of a simulated population pharmacokinetic, pharmacokinetic/pharmacodynamic, or confirmatory efficacy study for robustness. Robustness addresses the question, "If my assumptions underlying the study design are wrong, am I still able to meet the objectives of the research project?" Evaluation for robustness may be approached by sensitivity analysis. Evidence of robustness renders the results acceptable and independent of the analyst. Informativeness addresses the question of how much can be learned from the study. Very often informativeness can both be increased in a study simply by collecting extra data on each subject without enrolling additional subjects (10).

The intended objective of a study will dictate which of the criteria are most important for any given design. For studies where it is important to continue to learn about the effects of factors such as renal function, size, gender, or race on drug pharmacokinetics, dynamics, or other effects, one would be most interested in an informative study design. When one is interested in showing a drug effect is superior to a placebo, power becomes a major issue and informativeness may be sacrificed. A study design should be selected that balances the four criteria for evaluation.

### 7.2.2.2 An Efficacy Study Example: Dosage Optimization

Another well-executed example of CTS is a report on dosage optimization for docetaxel in a proposed oncology trial (17). This application is also consistent with the spirit and methods of the Simulation in Drug Development: Good Practices (24) consensus document. The investigation was performed to determine whether non–small-cell cancer patients might benefit from dose intensification. Prior PK/PD analysis performed during the development of the drug showed that those non–small-cell cancer patients who had high baseline $\alpha$1-acid glycoprotein (AAG) levels had shorter time to disease progression and death. Thus, PK/PD models were developed for time to disease progression, death, and dropout and validated with data from 151 non-small-cell lung cancer patients from phase II studies. Separate hazards were estimated for each event type. Different types of hazard models (exponential, Gompertz, and Weibull) were tested, and the Weibull model was found to provide the best explanation of the data. Prognostic factors (dose at first cycle, cumulative dose, and cumulative AUC) were also included in the models for death and disease progression. Posterior predictive check (3, 28) was used to evaluate whether the models provided adequate description of the data.

The simulation process was evaluated by simulating 100 complete trials of previous phase II trials from which the PK/PD models were derived. Kaplan-Meier analyses were performed on each simulated trial. Several statistics, such as median time to progression, 1 year survival, number of deaths, and disease progression, etc. were computed and compared with those obtained from the real trials. The statistics from real and simulated trials were in good agreement. For instance, from the Kaplan-Meier analysis a similar time course of death was observed in the simulations and phase II data. Thus, the simulation process was evaluated as the basis for simulating a phase III trial of docetaxel in non-small-cell lung cancer.

The primary objective of the simulated phase III trial was the comparison of overall survival between two doses (100 and 125 mg/m$^2$) of docetaxel. A secondary objective was the comparison of the time to progression and the safety of the two dosage regimens. This was a proposed randomized phase III trial in non-small-cell lung cancer patients with high AAG levels (i.e., AAG levels higher than 1.92 g/L). It was thought that dose intensification would result in a clinical benefit to this group of patients. Patients were randomized to one of the two doses, and the drug was administered as a 1 h intravenous infusion every 3 weeks. Two hundred patients were randomized to each treatment arm, and 100 trials were simulated. This sample size was assumed to be adequate and would permit the detection of a survival advantage of 8 weeks in the 125 mg/m$^2$ treatment arm with an $\alpha$ of 5% and 80% power. A comparison of the overall survival and time to progression for the two treatments was done using the log-rank test at the 5% level of significance.

The difference in the median time to progression between the two treatments (i.e., 9.18 and 9.48 weeks for the 100 and 125 mg/m$^2$, respectively) was only significant in 11 of 100 trials. A slightly longer median survival time of

5.49 months was observed for patients on the 125 mg/m$^2$ regimen as opposed to a 5.31 month median survival time for patients on the 100 mg/m$^2$ regimen. However, the difference in the median survival times was only significant in 6 of 100 trials. The 1 year survival rate (14% for patients on the 100 mg/m$^2$ regimen and 15% for those on the 125 mg/m$^2$ regimen) was similar in both treatment arms. From this finding it was concluded that dose intensification would not result in any significant clinical benefit to patients with high AAG. Thus, a real trial was not performed because of the outcome of the simulated study.

These examples have been chosen to highlight the importance of appropriate analysis of simulated trials. It is worth noting that the number of replications in a simulated trial should be justified by the objectives of the study and the precision required (3). The presence of more than one endpoint raises the issue of multiplicity.

### 7.2.3 Multiplicity

Response variables in clinical trials are usually classified as primary or secondary endpoints. The multiplicity problem arises when performing many hypothesis tests on the same data set or having more than one endpoint. The presence of two or more primary endpoints in a clinical trial usually means that some adjustments of the observed $p$ values for the multiplicity of tests may be required for the control of the type I error rate. To promote good clinical trial simulation practice, multiplicity should be addressed in simulated trials that lend themselves to it.

Consistency in the design and analysis of simulated and actual clinical trials is essential for CTS to be useful. The conclusions drawn from the interpretation of the results of a clinical trial are dependent on factors such as the disease under study, patient population, endpoints, study design, study conduct, appropriateness of statistical analysis for the given design, and the sensitivity of the chosen statistical test(s) to the scientific question(s) the study is designed to answer. The presence of multiplicity in a clinical trial that is unaccounted for in the design and ensuing analysis is one of the factors that may make interpretation of results difficult if not impossible.

#### 7.2.3.1 Sources of Multiplicity

Multiplicity can arise from the presence of multiple active treatment arms in a study as in dose ranging studies, multiple endpoints because of the nature of the disease, multiple analyses and/or tests. Multiple analyses of clinical trials are usually the rule and not the exception and are executed in an effort to understand the trial outcome. There are usually per protocol analysis, evaluable subset analysis, protocol-defined subset analysis, and the all randomized (intent-to-treat) or all patients treated analysis in the case of confirmatory trials. Also, the need to minimize the cost of obtaining data (interim analyses), the exploration of alternative statistical methods, and the desire to discover new aspects to the data (subset analyses) are necessary ingredients in the conduct of a clinical trial that will also introduce multiplicity. Thus, there is an inherent multiplicity component in most clinical trials, and it is important to consider and address multiplicity in the analysis of simulated clinical trials.

#### 7.2.3.2 Error Rate and Control

Performing multiple tests is often reasonable in a clinical trial because of the cost of obtaining data. However, a negative feature of multiple testing is the greatly increased probability of declaring false significance (type I error). To prevent this, some adjustments for the observed $p$ values for the multiple tests is required for the control of the type I error rate. By an error rate, one means the probability of false rejection of either the null hypothesis (referred to as the type I error rate), or the alternative hypothesis (referred to as the type II error rate). A type I error is committed when an ineffective drug gains entrance into the market, and a type II error is committed when an effective drug fails to reach the market. Both of these errors are equally serious.

The probability of false rejection of at least one of $K$ null hypotheses in a multiple hypothesis testing problem can be controlled via one of two approaches: comparisonwise error rate or experimentwise error rate. If one opts to control the individual type I error rates at a given nominal $\alpha$ level for each of $K$ hypotheses, then one is controlling what is called the comparisonwise error rate. On the other hand, controlling overall type I error rate at a given nominal $\alpha$ level for all possible $K$ hypotheses is controlling the experimentwise error rate. If, for instance, $K = 4$ with a prespecified nominal $\alpha$ level of 0.05, a comparisonwise error rate method of control would imply testing each of the four corresponding null hypothesis at the 0.05 nominal level to ensure a 0.05 significance level for each test. On the other hand, an experimentwise type I error rate method of control would imply testing each of the four null hypotheses at some $\alpha$ level so as to ensure an overall significance level of 0.05 or less for all four tests (i.e., testing each of the null hypothesis at $\alpha = 0.013$ using the Bonferroni procedure, for instance).

Increased chance of a higher type I error rate is a well-known disadvantage associated with the comparisonwise error rate control approach. The probability of committing at least one type I error increases as the number of comparisons increases. If $n$ independent tests are separately performed at $\alpha = 0.05$ level of significance, the probability that at least one will be significant is $1 - (1 - 0.05)^n$. As the type I error rate increases, the type II error rate decreases giving rise to more powerful tests (an experimenter's delight?). On the contrary, a type I error rate less than or equal to the designated $\alpha$ level would result from controlling the experimentwise error rate. From an experimenter's perspective, the downside of this approach is that the resulting type I error rate is usually smaller than the designated $\alpha$ level which may lead to less sensitive tests (an experimenter's dismay?). Controlling experimentwise error rate instead of comparisonwise error rate does not depend on which null hypotheses are true. Controlling the experimentwise error rate seems to be a more appropriate option since it is not practical to require that an improvement be realized for every component of a global test in a given clinical trial. In addition, a multiple test procedure that controls the experimentwise error rate also controls the global error rate. The reverse is not true (29). A global test (which is outside the scope of this chapter) focuses on providing an overall probability statement about the efficacy of treatments by considering simultaneously all endpoints (30–40). It does not focus on any individual endpoint effect, which may sometimes be an important question of research.

#### 7.2.3.3 Multiplicity Adjustment Approaches

The Bonferroni procedure is the most commonly used approach for multiplicity adjustment. There are improvements to this approach and they are referred to as stepwise procedures (41–44).

*a. The Bonferroni Procedure.* The simplest multiplicity adjustment procedure that is applicable to most multiplicity adjustment problems is the Bonferroni test. No assumptions are made about the correlation structure among the endpoints and/or data distributions. Thus, it is too conservative when there are many endpoints and/or the endpoints are highly correlated. If there are $K$ endpoints, one accepts as statistically significant all those $p$ values $\leq \alpha/K$, where $\alpha$ is the overall error type I error rate. $K \times p_k$ are the adjusted $p$ values, where $p_k$ are the observed $p$ values and $k = 1, 2, \ldots, K$. This methodology would be applied to each replicate of the simulated trial.

There are other "less conservative" and more powerful improvements to the Bonferroni procedure that are also used to control experimentwise error rate. Among these are the Holm's procedure (39) and modifications and improvements on the Holm's procedure [i.e., the Hochberg (41) and Hommel (42) procedures] that are less conservative than the Holm's procedure. However, Holm's procedure requires fewer assumptions. The basis of these approaches is the realization that of the $K$ null hypotheses to be tested, the only ones to protect against rejection at any given step are those not yet tested.

*b. Bonferroni Modified (Stepwise) Procedures*

HOLM'S PROCEDURE. A step-down adjustment procedure introduced by Holm (41) was an improvement on the Bonferroni method by providing additional power while maintaining the experimentwise error rate. The testing is done in a decreasing order of significance of (ordered) hypotheses. Testing for significance is continued until a null hypothesis is accepted. Thereafter all remaining (untested) null hypotheses are accepted without further testing.

The following is the algorithm for Holm's procedure:

1. Let $p_1 \leq p_2 \leq \ldots \leq p_K$ be the ordered $p$ values and $H_{01}, H_{02}, \ldots, H_{0K}$ be the corresponding ordered null hypothesis.
2. Reject $H_{01}$ if $p_1 < \alpha/K$ and go to the next step; otherwise stop and accept all null hypotheses.
3. Reject $H_{02}$ if $p_{(2)} < \alpha/(K - 1)$ and go to the next step; otherwise stop and accept all remaining $K - 1$ null hypotheses.
4. In general reject $H_{0k}$ if $p_k < \alpha/(K - k + 1)$, k = 1, 2, . . . , $K$, and go to the next step; otherwise stop and accept the remaining null hypotheses. The adjusted $p$ values are $p_{\text{adj},k} = \max \{(K - j + 1) \times p_k\}$, $j = 1, 2, \ldots, k$ and $k = 1, 2, \ldots, K$.

Consider, for example, $K = 4$ endpoints and the following (ordered) $p$ values were observed: $p_1 = 0.005 < p_2 = 0.020 < p_3 = 0.024 < p_4 = 0.081$. With the Holm procedure:

1. Reject $H_{01}$ since $p_1 = 0.005 < 0.0125 = 0.05/4$ and go to the next step.
2. Given that $p_2 = 0.020 > 0.017 = 0.05/3$, stop and accept $H_{02}$ and all remaining untested null hypotheses, $H_{0k}$ $(k = 2, 3, 4)$ since $p_2 > \alpha/3$. The resulting adjusted $p$ values are $p_{\text{adj},1} = \max \{4 \times p_1\} = 0.020$, $p_{\text{adj},2} = \max \{4 \times p_1, 3 \times p_2\} = 0.060$, $p_{\text{adj},3} = \max \{4 \times p_1, 3 \times p_2, 2 \times p_3\} = 0.060$ and $p_{\text{adj},4} = \max\{4 \times p_1, 3 \times p_2, 2 \times p_3, 1 \times p_4\} = 0.081$.

HOCHBERG'S AND HOMMEL'S PROCEDURES. With the Hochberg procedure (41) testing is done in an increasing order of significance of the (ordered) hypotheses. Once a null hypothesis is rejected, significance testing is discontinued. Thereafter all remaining untested null hypothesis are rejected.

The following is the algorithm for the Hochberg procedure:

Let $p_1 \geq p_2 \geq \ldots \geq p_K$ be the ordered $p$ values and $H_{01}, H_{02}, \ldots, H_{0K}$ be the corresponding null hypotheses. Reject $H_{0k}$ and $H_{0j}$ for $j > k$ and $k = 1, 2, \ldots, K$ if $p_k\ \alpha/k$. The adjusted $p$ values are $p_{\text{adj},k} = \min \{j \times p_j\}$ for $j = 1, 2, \ldots, k$ and $k = 1, 2, \ldots, K$.

Using the same example as with the Holm procedure, the Hochberg procedure would proceed as follows:

1. Accept $H_{01}$ since $p_1 = 0.081 > 0.05$ and go to the next step.
2. Given that $p_{02} = 0.024 < 0.025 = 0.05/2$, stop and reject $H_{02}$ and all remaining null hypotheses $H_{0k}(k = 3, 4)$ that are untested. The resulting adjusted $p$ values are $p_{\text{adj},1} = \min \{p_1\} = 0.081$, $p_{\text{adj},2} = \min \{p_1, 2 \times p_2\} = 0.048$, $p_{\text{adj},3} = \min\{p_1, 2 \times p_2, 3 \times p_3\} = 0.048$, and $p_{\text{adj},4} = \min\{p_1, 2 \times p_2, 3 \times p_3, 4 \times p_4\} = 0.020$.

The algorithm for the Hommel procedure (43) is as follows: Let $p_1 \leq p_2 \leq \ldots \leq p_K$ be the ordered $p$ values, and $H_{01}, H_{02}, \ldots, H_{0K}$ the corresponding ordered null hypothesis. Assume an a priori $p$ value of 0.05. The procedure is performed by starting in succession with $m = 1, 2, \ldots, K$ until the maximum m that has $p_{K-m+j} \geq j\ \alpha/m$ for $j = 1, 2, \ldots, m$ is found. Suppose the maximum m is equal to t. Then we reject all $H_{0k}$, $k = 1, \ldots, K$, for which $p_k \leq \alpha/t$.

Continuing with the example above, and starting with $m = 1$ and $j = 1$, $p_{K-m+j} = p_4 = 0.081$ which is greater than 0.05 ($j \times \alpha/m$). Proceed to $m = 2$, $j = 1$. $p_3 = 0.024 < 0.025$ ($j \times \alpha/m$). Therefore, the maximum $m$ is 1. Reject all null hypotheses for which $p_k < 0.05$; i.e., reject $H_{01}$, $H_{02}$, $H_{03}$ as with the Hochberg procedure. The resulting adjusted $p$ values are $p_{\text{adj},k} = m \times p_k = p_k$, $k = 1, 2, 3, 4$; i.e., $p_{\text{adj},1} = 0.081$, $p_{\text{adj},2} = 0.024$, $p_{\text{adj},3} = 0.020$, and $p_{\text{adj},4} = 0.005$.

It is worth noting from the example that the Hochberg and Hommel procedures lead to two more null hypothesis rejections ($H_{02}$ and $H_{03}$) than the Holm procedure. The Holm procedure is uniformly more powerful than the Bonferroni procedure, the Hochberg procedure is uniformly more powerful than the Holm, and the Hommel procedure has been shown to be only slightly more powerful than the Hochberg (42–43).

Some Ad Hoc Procedures. Some ad hoc procedures that make use of correlation information among endpoints without any distributional assumptions have been developed. Tukey et al. (45) have suggested that for strongly correlated endpoints and for a given nominal $\alpha$ level the adjustments $p_{\text{adj},k} = 1 - (1 - p_k)^{\sqrt{K}}$ and $\alpha_k = 1 - (1 - \alpha)^{1/\sqrt{K}}$ be used, where $p_k$ and $p_{\text{adj},k}$ are the observed and adjusted $p$ values and $\alpha_k$ is the adjusted critical level for the $k$th hypothesis for $k = 1, \ldots, K$. Dubey (46) and Armitage and Parmar (47) have suggested the use of $p_{\text{adj},k} = 1 - (1 - p_k)^{m_k}$ and $\alpha_k = 1 - (1 - \alpha)^{1/m_k}$, where

$$m_k = K^{1-r_k}$$

and

$$r_k = \frac{1}{K-1} \sum_{j \neq k}^{K} r_{jk}$$

replaces $\sqrt{K}$ in the Tukey et al. (45) formula for $p_{\text{adj},k}$; $r_{jk}$ is the correlation coefficient between the $j$th and the $k$th endpoints. A nice feature of the Dubey (46) and Armitage and Parmar (47) procedure is that when the average of the correlation coefficients is 1, the adjusted and the unadjusted $p$ values are the same; and when it is zero, the adjustment is according to the Bonferroni test. For equicorrelated endpoints with 0.5 correlation coefficient, the Dubey (46) and Armitage and Parmar (47) procedure is equivalent to the Tukey et al. (45) procedure.

Some of the approaches for addressing multiplicity in clinical trials have been discussed with their applicability. It is important to use the appropriate multiplicity adjustment procedure in the analysis of simulated clinical trials. This would lead to a meaningful interpretation of the outcome of the simulation experiment.

## 7.3 COMMUNICATION OF ANALYSIS RESULTS

The results of the analysis of a simulated clinical trial should be presented in a manner that can be readily understood by the intended audience. Communication of the results of a simulation project can be effected through words, tables, or graphics. The efficient conveyance of the message from a simulation project should involve the use of all three communication media, which has been described respectively as infantry, artillery, and cavalry of the pharmacokinetic/pharmacodynamic defense force (21). These methods should be used to supplement one another, although the effectiveness of each depends on the contents of the message. Tables should be used to communicate information that can best be conveyed using this means of presentation. Graphics, in particular, are essential for conveying relations and trends in an informal and simplified visual form.

Graphs are analogous to written language; they communicate quantitative and categorical information. Written language communicates thoughts, ideas, observations, emotions, theories, hypotheses, numbers, etc. Graphical language is used extensively to convey information because it does so, effectively. Quantitative patterns and relationships in data are readily revealed by graphs because of the enormous power of the eye-brain system to perceive geometrical patterns. The eye-brain system can quickly summarize vast amounts of quantitative information on a graph, perceiving salient features, or focusing on specific detail. The power of a graph is its ability to enable one to take in the quantitative information, organize it, and see patterns and structure not readily revealed by other means of studying and presenting data (21, 48).

Numerical data can be displayed in different formats, but only some are well suited to the information processing capacity of the human vision. The phrase *graphical perception* has been coined by Cleveland and McGill (49) to refer to the role of visual perception in analyzing graphs. These authors have studied several elementary visual tasks (such as discrimination of slopes or lengths of lines) relevant to graphical perception. They attribute the great advantage of graphical displays (e.g., scatterplots) over numerical tables to the capacity of the human vision to process pattern information globally at a glance. In comparing the effectiveness of data display using graphs and numerical tables, Legge et al. (50) found perceptual efficiencies to be very high for scatterplots, ≥60%. Efficiencies were much lower for numerical tables, 10%. Efficiency in the study referred to the performance of a real observer relative to the ideal observer. The ideal observer makes an optimal use of all available information (50). Performance with scatterplots was reported to have the hallmark of a parallel process:

weak dependence on viewing time. Processing of tables of numbers was found to be performed in a much more serial fashion. Efficiencies dropped roughly with increasing information content in the tables and increased in rough proportion to viewing time. They concluded that entries in tables are processed sequentially at a fixed rate. Given enough viewing time, efficiency of information processing from tables could approach that of graphics (50). Thus, information content of tables should be kept to a minimum to allow efficient extraction of such information by the reader, or the audience in the case of oral presentation of the outcome of a simulation project.

However, it is worth noting that the merits of a graphical display depend on the information chosen for display and the amount of effort that will be expended by the reader in deciphering what is encoded in the graph (51). In summarizing how sample size affects the power of a study, an integrated display such as a line graph would be a superior method of display. This is because in decoding the graph the reader would have to compare the power of the study for the different sample sizes and integrate that information to form his opinion. Judging change requires comparing quantities and integrating that information (51). A line graph is more effective in conveying change than other types of display (51). This is because the eye is focused on the physical slope of the line. Bar plots are also effective in conveying change (trends) in that the eye, in decoding change (or a trend) with bar plots, is tracing a perceived slope (51). The effectiveness of a graph, therefore, depends on the amount of work that is to be performed by the reader in decoding the information contained in the graphical display.

In summarizing the results of a population pharmacokinetic study in which the effect of sample size on the bias and precision with which population pharmacokinetic parameters were estimated, Ette et al. (52) used line plots. A similar line plot display (Figure 1) of the effect of sample size and intersubject variability on the estimation of population pharmacokinetic parameters was created from an aspect of data generated in a simulation study performed to determine the performance of mixed designs in population pharmacokinetic studies (53). The plot shows the influence of intersubject variability on parameter estimation as sample size was varied. In the study, the effect of three different levels of intersubject variability, ranging from 30 to 60% coefficient of variation, and different sampling designs on the sample size required for efficient population pharmacokinetic parameter estimation were investigated. However, in Figure 1 data for only one of the designs are plotted to illustrate the effectiveness of a line plot.

The *coplot* (21, 54) is a powerful tool for studying how a response depends on two or more factors. It presents conditional dependence in a visually powerful manner. Two variables are plotted against each other in a series of overlapping ranges. This enables one to see how a relationship between two variables ($y$ and $x$) changes as a third variable $z$ changes, i.e., $y \sim x \mid z$. Thus $y$ is plotted against

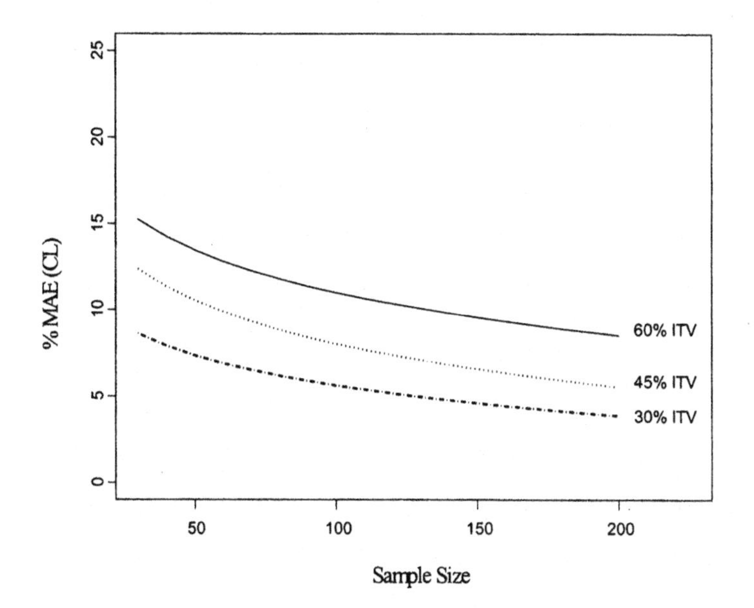


**FIGURE 1** A line plot of the effect of sample size and intersubject variability on the precision (mean absolute error: %MAE) with which clearance (CL) was estimated in a simulated population pharmacokinetic study. [Created from data reported in Fadiran et al. (53).]

$x$ for a series of conditioning intervals determined by $z$. Coplots may have two simultaneous conditioning variables, i.e., $y \sim x \mid z_1 * z_2$.

In studying the performance of mixed design in population pharmacokinetic studies Fadiran et al. (53) used conditioning plots (coplots) to summarize their findings (Figure 2). They investigated the influence of total cost of a population design (i.e., the combined cost of individual designs expressed as the total number of samples for a sample size given the sampling design considered), sample size, and intersubject variability on bias and precision associated with population pharmacokinetic parameter estimation. The presentation of the results from that study using coplots is a good example of using multipanel display to summarize the results of a simulated study; and this is recommended in the consensus document on clinical simulation (24). In this example, the use of the coplot allowed information from four variables to be presented using a two-dimensional graphical display.

Figure 2, extracted from the work of Fadiran et al. (53), is a coplot illustrating the effect of cost on the precision associated with the estimation of drug

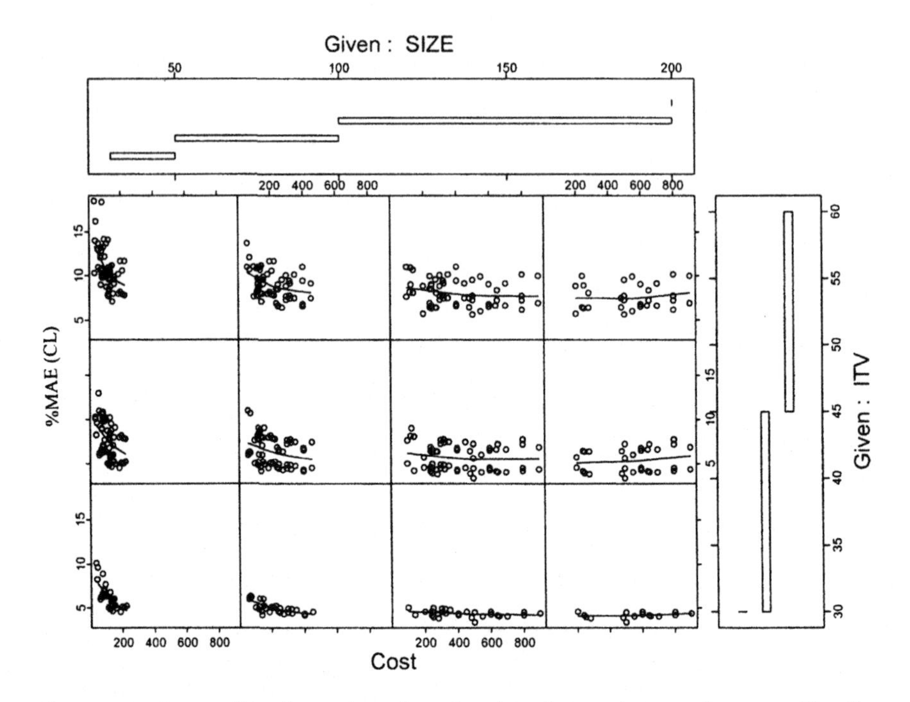


**FIGURE 2** A conditioning plot of a simulated population pharmacokinetic study showing the dependence of precision (expressed as percent mean absolute error (%MAE) in clearance (CL) on cost (i.e., total number of samples) conditioned on sample size (SIZE, top panel) and levels of intersubject variability (ITV, right panel). The study evaluated the performance of mixed designs in population pharmacokinetic study. [Excerpted from Fadiran et al. (53).]

clearance (CL) as sample size and intersubject variability are altered (53). The panel at the top is the *given panels* (*or conditioning intervals*), the panels below is the *dependence panel* (cost). The panel at the right is also the *given panels* (*or conditioning intervals*). Locally weighted regression (loess) lines are added to the plots to aid perception. The rectangles of the top conditioning panel specify the sample size, while those on the right specify the intersubject variability. On the corresponding dependence panel precision in clearance estimation (expressed as percent mean absolute error, %MAE) is graphed against the cost of a population design for those estimates whose value lie in the interval. The graph is read from left to right of the bottom row, then from left to right of the next row, and so forth. It can be seen from the figure that the CL was reasonably well estimated with mean absolute error (MAE) ranging from less than 10% at the 20% level of intersubject variability to less than 10% at the 60% level of intersubject variability, irrespective of cost, sample size, and sampling design that was investigated. The scatter of points around the loess regression fitted lines reflect the residual variation about the fits.

In presenting data in a graphical display that requires attention to be focused on one variable, performance is better served by the use of more separated displays. The histogram and the boxplot are examples of separated displays. The histogram (Figure 3) as used by Hale (55) in presenting the results of simulated randomized concentration controlled trial with mycophenolate mofetil is a good example of the use of separated displays to convey information on a simulated study outcome. This plot compares simulation predicted trial outcomes and the actual trial result. The bars represent complete simulated trials using a developed simulation model. Outcomes to the right of the cutoff line is statistically significant, and the actual study outcome is shown. The actual trial value fell between

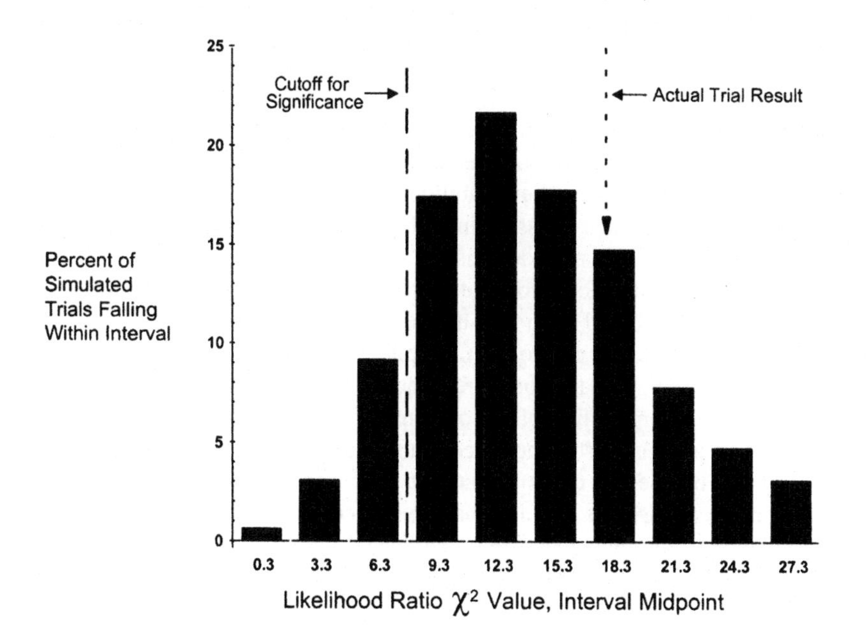


**FIGURE 3** Frequency distribution of the test statistic for the primary analysis resulting from 500 completed simulated RCCTs with mycophenolate mofetil under "worst case" trial conditions, completed before real study initiation. The actual study outcome is shown, falling in the central portion of the distribution in the interval centered at 18.3 [Excerpted and adapted from Hale [55].]

the 80th and 90th percentile of the simulated results, which means that the actual trial outcome is not unusual based on the simulation model, which reflects that the simulation model was a reasonable description of the trial process.

The boxplot has proven to be a popular graphical method for displaying and summarizing univariate data, to compare parallel batches of data and to supplement more complex displays with univariate information. Its appeal is due to the simplicity of the graphical construction (based on quartiles) and the many features that it displays (location, spread, skewness, and potential outliers). Boxplots are useful for summarizing distributions of treatment outcomes. A good example would be the comparison of the distribution of response to treatment at different dose levels.

## 7.5 SUMMARY

A simulated clinical trial has been likened to a meta-analysis of clinical trials. While meta-analysis is retrospective, a simulated trial is prospective. Like meta-analysis, careful attention should be paid to the analysis of a simulated trial as this may affect the interpretation of the outcome.

The procedure to be used in the analysis of a simulated trial should be specified in the simulation plan, and the method used in analyzing the simulated trial should be the same as that to be used for the analysis of the actual trial data. The importance of determining the reliability of parameter estimates in PPK and PPK/PD trials has also been stressed. In addition, assessment of the robustness of parameter estimates and study designs has been discussed. The importance of an empirical determination of error rate has been discussed and its application in the use of the likelihood ratio test has been presented. Attention has been drawn to error rate and its control in multiplicity testing where there are two or more primary endpoints, two or more treatment groups, etc. This is of great importance in the interpretation of the outcome of a simulated clinical trial.

Finally, it is of utmost importance that appropriate displays (numeric or graphic) should be used to communicate the outcome of a simulation project. Different graphical displays that can be used to communicate effectively and efficiently some of the outcomes from a simulated clinical trial have been presented. Failure to communicate the outcome of a simulated trial successfully puts at risk all the energy spent in analyzing a simulated trial, irrespective of its quality.

## REFERENCES

1. CC Peck. Drug development: improving the process. Food and Drug Law J 52(2): 163–167, 1997.

2. M Hale, WR Gillespie, SK Gupta, B Tuk, NH Holford. Clinical trial simulation: streamlining your drug development process. Appl Clinic Trials 5:35–40, 1996.
3. US Food and Drug Administration. Guidance for Industry: Population Pharmacokinetics. February 1999. http://www.fda.gov/cder/guidance/index.htm
4. H Sun, EO Fadiran, CD Jones, L Lesko, SM Huang, K Higgins, C Hu, S Machado, S Maldonado, R Williams, M Hossain, EI Ette. Population pharmacokinetics: a regulatory perspective. Clin Pharmacokinet 37:41–58, 1999.
5. NHG Holford, HC Kimko, JPR Monteleone, CC Peck. Simulation of clinical trials. Annu Rev Pharmacol Toxicol 40:209–234, 2000.
6. DR Jones. Computer simulation as a tool for clinical trial design. Int J Bio-Med Comp 10:145–150, 1979.
7. KL Lee, M Frederick, CF Starmer, PJ Harris, RA Rosati. Clinical judgment and statistics: lessons from a simulated randomized trial in coronary artery disease. Circulation 61(3):508–515, 1980.
8. MD Hale, AJ Nicholls, RE Bullingham, R Hene, A Hoitsma, JP Squifflet, W Weimar, Y Vanrenterghem, FJ Van de Woude, GA Verpooten. The pharmacokinetic-pharmacodynamic relationship for mycophenolate mofetil in renal transplantation. Clin Pharmacol Ther 64:672–683, 1998.
9. Y Hashimoto, LB Sheiner. Designs for population pharmacodynamics: value of pharmacokinetic data and population analysis. J Pharmacokinet Biopharm 19:333–353, 1991.
10. MK Al-Banna, AW Kelman, B Whiting. Experimental design and efficient parameter estimation in population pharmacokinetics. J Pharmacokinet Biopharm 18:347–360, 1990.
11. EI Ette, AW Kelman, CA Howie, B Whiting. Interpretation of simulation studies for efficient estimation of population pharmacokinetic parameters. Ann Pharmacother 27:1034–1039 and Correction 27:1548, 1993.
12. EI Ette, AW Kelman, CA Howie, B Whiting. Inter-animal variability and parameter estimation in preclinical animal pharmacokinetic studies. Clin Research Regul Affairs 11(2):121–139, 1994.
13. EI Ette, H Sun, TM Ludden. Design of population pharmacokinetic studies. Proc Am Stat Assoc (Biopharmaceutics Section), 487–489, 1994.
14. CD Jones, H Sun, EI Ette. Designing cross-sectional pharmacokinetic studies: implications for pediatric and animal studies. Clin Res Regul Affairs 13(3,4):133–165, 1996.
15. NE Johnson, JR Wade, MO Karlsson. Comparison of some practical sampling strategies for population pharmacokinetic studies. J Pharmacokinet Biopharm 24(6):245–272, 1996.
16. H Sun, EI Ette, TM Ludden. On error in the recording of sampling times and parameter estimation from repeated measures pharmacokinetic data. J Pharmacokinet Biopharm 24(6):635–648, 1996.
17. C Veyrat-Follet, R Bruno, R Olivares, GR Rhodes, P Chaikin. Clinical trial simulation of docetaxel in patients with cancer as a tool for dosage optimization. Clin Pharmacol Ther 68(6):677–687, 2000.

18. P Williams, JR Lane, CC Turkel, EV Capparelli, Z Dziewanowska, AW Fox. Dichloroacetate: population pharmacokinetics with a pharmacodynamic sequential link model. J Clin Pharmacol 41:259–267, 2001.
19. GV Glass. Primary, secondary and meta-analysis of research. Educ Res 5:3–8, 1976.
20. K Dickersin, JA Berlin. Meta-analysis: state-of-science. Epidemiol Rev, 14:154–176, 1992.
21. EI Ette. Statistical graphics in pharmacokinetics and pharmacodynamics: a tutorial. Ann Pharmacother 32:818–828, 1998.
22. KG Kowalski, MM Hutmacher. Design evaluation for a population pharmacokinetic study using clinical trial simulations: a case study. Stat Med 20:75–91, 2001.
23. SL Beal, LB Sheiner. NONMEM Users Guide, parts I–VI. San Francisco, CA: Division on Clinical Pharmacology, University of San Francisco, 1979–1992.
24. NHG Holford, M Hale, HC Ko, J-L Steimer, LB Sheiner, CC Peck. Simulation in Drug Development: Good Practices. Washington, DC: CDDS, Georgetown University, 1999. http://cdds.georgetown.edu
25. Pinheiro JC, Bates DM, eds. Mixed-Effects Models in S and S-PLUS. New York: Springer, 2000.
26. DO Stram, JW Lee. Variance components testing in the longitudinal mixed effects model. Biometrics 50:1171–1177, 1994.
27. U Wählby, EN Jonsson, MO Karlsson. Assessment of actual significance levels for covariate effects in NONMEM. J Pharmacokinet Pharmacodyn 28:231–252, 2001.
28. Gelman A, Carlin JB, Stern HS, Rubin DB, eds. Bayesian Data Analysis. London: Chapman and Hall, 1995.
29. P Bauer. Multiple testing in clinical trials. Stat Med 10:871–890, 1991.
30. G Hommel. Tests of overall hypothesis for arbitrary dependent structures. Biometrical J 25(5):423–430, 1983.
31. G Hommel. Multiple test procedures for arbitrary dependence structures. Metrika 33(6):321–336, 1986.
32. W Lehmacher, G Wassmer, P Reimer. Procedures for two-sample comparisons with multiple endpoints controlling the experimentwise error rate. Biometrika 67:655–660, 1991.
33. PC O'Brien. Procedures for comparing samples with multiple endpoints. Biometrics 40:1079–1089, 1984.
34. SJ Pocock, NL Geller, A Tsiatis. The analysis of multiple endpoints in clinical trials. Biometrics 43:487–498, 1987.
35. AJ Sankoh, MF Huque. A note on O'Briens OLS and GLS tests. Biometrics 51: 1580–1581, 1995.
36. DI Tang, C Gnecco, NL Geller. Design of group sequential trials with multiple endpoints. J Am Stat Assoc 84:776–779, 1989.
37. DI Tang, C Gnecco, NL Geller. An approximate likelihood ratio test for a normal mean vector with nonnegative components with application to clinical trials. Biometrika 76:751–754, 1989.
38. DI Tang, NL Geller, SJ Pocock. On the design and analysis of randomized clinical trials with multiple endpoints. Biometrics 49:23–30, 1993.
39. S Holm. A simple sequentially rejective Bonferroni Test Procedure. Scandinavian Journal of Statistics 6:65–70, 1979.

40. RJ Simes. An improved Bonferroni procedure for multiple tests of significance. Biometrika 67:655–660, 1976.
41. Y Hochberg. A sharper Bonferroni procedure for multiple significance testing. Biometrika 75:800–803, 1988.
42. G Hommel. A comparison of two modified Bonferroni procedures. Biometrika 76: 624–625, 1989.
43. GA Hommel. A stagewise rejective multiple test procedure based on a modified Bonferroni test. Biometrika 75:383–386, 1988.
44. CW Dunnet, AC Tamhanne. A step-up multiple test procedure. J Am Stat Assoc 87:162–170, 1993.
45. JW Tukey, JL Ciminera, JF Heyse. Testing the statistical certainty of a response to increasing doses of a drug. Biometrics 41:295–301, 1985.
46. SD Dubey. Adjustment of p values for multiplicities of intercorrelating symptoms. Proceedings of the VIth International Society of Clinical Biostatisticians, Germany, 1985.
47. P Armitage, M Parmar. Some approaches to the problem of multiplicity in clinical trials. Proceedings of the XIIth International Biometrics Conference, Seattle, Washington, 1986.
48. EI Ette, P Williams, H Sun, FO Ajayi, LC Onyiah. The process of knowledge discovery from large pharmacokinetic data sets. J Clin Pharmacol 41:25–54, 2001.
49. WS Cleveland, R McGill. Graphical perception and graphical methods of analyzing scientific data. Science 229:828–833, 1985.
50. GE Legge, Y Gu, A Luebker. Efficiency of graphical perception. Percep Psychophys 46:365–374, 1989.
51. JS Hollands, I Spence. Judgement of change and proportion in graphical perception. Human Factors 34:313–334, 1992.
52. EI Ette, H Sun, TM Ludden. Balanced designs in longitudinal population pharmacokinetic studies. J Clin Pharmacol 38:417–423, 1998.
53. EO Fadiran, CD Jones, EI Ette. Designing population pharmacokinetic studies: performance of mixed designs. Eur J Drug Metab Pharmacokinet 25:231–239, 2000.
54. WS Cleveland. Visualizing Data. Summit, NJ: Hobart Press, 1993.
55. MD Hale. Using population pharmacokinetics for planning a randomized concentration–controlled trial with a binary response. In: L Aarons, LP Balant, M Danhof, M Gex-Fabry, UA Gundert-Remy, MO Karlsson, F Mentre, PL Morselli, F Rombout, M Rowland, JL Steimer, S Vozeh, eds. The Population Approach: Measuring and Managing Variability in Response, Concentration and Dose. Commission of the European Communities, European Cooperation in the Field of Scientific and Technical Research; Brussels 1997, pp 228–235.

# 8

# Sensitivity Analysis of Pharmacokinetic and Pharmacodynamic Models in Clinical Trial Simulation and Design

**Ivan A. Nestorov***
University of Manchester, Manchester, England

## 8.1 INTRODUCTION

The concept of sensitivity analysis (SA) has been widely used in the fields of system theory and statistical modeling (1–4). It has been mentioned in connection with the analysis of compartmental systems (5, 6) in a theoretical setting, and applied to a number of chemical (combustion kinetics) and biochemical (metabolism kinetic) studies (7, 8). SA has received inadequate attention in pharmacokinetic (PK) and pharmacodynamic (PD) modeling and its potential has not been fully utilized. Several papers on sensitivity analysis of PK models have appeared recently (9–16), most of them related to physiologically based pharmacokinetic (PBPK) models for risk assessment and analysis of toxic drugs and chemicals. A very limited number of SA studies of the commonly used one- and two-compartment PK models (17) and PK-PD models (18, 19) has been published. The existing research tends to lack formality and comprehensiveness; most of the publications are confined to simple cases of perturbing one or more model parameters, simulating the system and registering the output concentration variations.

* *Current affiliation*: Amgen Corporation, Seattle, Washington, U.S.A.

The importance and benefits of SA in PK/PD modeling seem to be hardly disputed by anybody. There has been a revived interest in PK/PD-related SA recently and it is gradually becoming an integral part of all stages of the modeling process, providing an important insight into the system studied and its model performance. This revival can be attributed to the generally wider implementation of computer modeling and simulation in all phases of the drug development process, and especially in clinical trial design.

The purpose of this chapter is to present a systematic approach to SA of PK/PD systems in the context of the current clinical trial simulation and design (CTS&D) practices. The possible implications of SA are discussed and illustrated with examples wherever appropriate.

## 8.2 DEFINITIONS AND SENSITIVITY MEASURES

Sensitivity analysis in pharmacokinetic and/or pharmacodynamic modeling can be defined as the systematic investigation of the PK and/or PD model responses to perturbations and variations of the model factors. The model factors of interest can be either quantitative (e.g., inputs and/or parameters) or qualitative (e.g., structure and connectivity, model specifications, modeling assumptions, scenarios, etc.) (19). The model responses of interest to pharmacokinetics and pharmacodynamics are usually concentration and effect processes.

The role and significance of SA becomes evident if we understand some of the possible sources of perturbations and variations in the PK and PD systems (20). Following the constructivist view, any model is an approximate description of the real world. As such, models incorporate vagueness and imprecision; i.e., uncertainty is inherent to the model. Hence, there is no such thing as a "true" model structure and/or parameters; what one uses are just approximations and estimates of the latter, usually drawn from small experimental samples. On the other hand, variability is an inherent property of the real world and needs to be accounted for during the modeling process. Therefore, it is always necessary to study how the system will behave if its structure is not exactly as assumed and/or when the parameters vary. The uncertainty and variability involved can be viewed upon as perturbations occurring both in the system structure and/or parameters.

Almost all PK and PD systems are inherently nonstationary—their parameters change with time. At the same time the common PK/PD models assume that parameters such as clearance and bioavailability do not change in time. Therefore, one can consider the parameter drift as a perturbation, introduced to the system, and the question of how this drift influences the outputs (such as concentration or effect profiles) is of primary importance. Another possible perturbation in the system can be caused by the presence of nonlinearities. A typical example of model misspecification uncertainty is a system with a monocompartment behavior at low doses and a multicompartment behavior at higher doses. This effect can be considered as a perturbation to the system structure and how the latter affects the system responses should be considered.

The descriptive definition of SA, given above, shows the reason why it is often referred to as a "what-if" analysis. The main question that SA answers is: "*What* will happen with the system responses *if* a model factor (quantitative or qualitative) is perturbed?" Obviously, the results of SA go further beyond the answer of this basic question, but it leads quite naturally to the definition of a quantitative measure of sensitivity, the sensitivity index (SI), which relates the changes observed in the system states or outputs to the change in the factors causing them (20):

$$\text{Sensitivity index} = \frac{\text{change in response}}{\text{change in factor}} \tag{1}$$

This definition shows that if the model response is a function of time and/or other independent variables (e.g., dose for a PK/PD model), the SI itself is also a function of time and the same variables. In order to acknowledge this feature, the sensitivity indices are sometimes referred to also as sensitivity functions (20). Often, the SI are used in their normalized form, quantifying the relative (or percentage) change in the model response at a relative (or percentage) change of the factor:

$$\text{Normalized sensitivity index} = \frac{\text{change in response}}{\text{response}} \cdot \frac{\text{factor}}{\text{change in factor}} \tag{2}$$

In the vast majority of applications (7–13, 15–18), the change in the model factor is assumed to be (theoretically—infinitesimally) small and the SI is used in its derivative form. Using SI in the continuous (derivative) form is often referred to as "local" SA. For example, in the widespread pharmacokinetic notation, the index $dC_p/dCL$ is the measure of the sensitivity of the plasma concentration to a (small) perturbation of the total clearance; the derivative $\partial^2 C_p/(\partial F\ \partial CL)$ measures the sensitivity of the plasma concentration to simultaneous (small) perturbations of the bioavailability and the total clearance. The respective normalized SI are $dC_p/C_p \cdot CL/d\,CL = d\ln C_p/d\ln CL$ and $\partial^2 C_p/C_p^2 \cdot ((F \cdot CL)/(\partial F \cdot \partial CL))$.

When the SI are used in their finite difference form (e.g., $\Delta C_p/\Delta CL$, $\Delta^2 C_p/(\Delta F\ \Delta CL)$, with a physical meaning defined in the previous paragraph and corresponding normalized forms $\Delta C_p/C_p \cdot CL/\Delta CL$, $\Delta^2 C_p/C_p^2 \cdot ((F \cdot CL)/(\Delta F \cdot \Delta CL))$, the approach is referred to as "global" SA, as the model factor changes and the corresponding response variations can be arbitrarily large.

Consequently, at least two approaches for system SA are possible. The direct analytical approach is usually implemented for local SA when the SI are used in their derivative form. The respective derivatives are obtained either analytically by differentiation of the model output function expressions (given in an open form), or by numerical integration of a (set of) sensitivity differential equation(s) as in Refs. 5, 9, 20. An empirical approach, which can be applied to both local and global SA, uses model simulations and statistical techniques to calculate the finite difference form of the SI. Typical examples of the latter are the Monte Carlo and design of experiment (DOE) techniques (4, 20–22). It should be noted, however, that in both cases, the SI are calculated for a set of predefined "nominal" values of the factors of interest—i.e., both the analytical derivatives and/or the finite difference expressions are evaluated at (or around in the second case) the nominal values. There is empirical evidence (17) that local SA results are approximately valid for parameter changes of up to 30% of their nominal values.

An alternative approach to SA, based on matrix perturbation theory, can be used when the PK/PD model is in the form of a linear system of ordinary differential equations (LSODE). In this case, the dynamics of the system, i.e., the time course of the model states (e.g., the concentrations in various compartments) is determined by the eigenvalues of the LSODE matrix (compartmental matrix). Matrix perturbation theory studies the behavior of matrix eigenvalues under "small" perturbations of the matrix elements (9, 23, 24). This theory can be applied to the compartmental matrix of linear PK/PD models, the entries of which depend on the physiological and drug-related parameters. Therefore, any perturbations in the entries of the compartmental matrix are introduced by variations and uncertainties in the model parameters. Consequently, the sensitivity of the model states to parameter variations can be examined by studying the sensitivity of the eigenvalues to perturbations of the matrix elements. As this is an indirect SA method, it gives only a qualitative picture of the system sensitivity and hence its application in the PK/PD field is limited, as shown in Ref. 9.

A recent development in global SA is the introduction of the so-called statistical "variance-based" SA techniques which are intended to quantify how much of the existing output variability can be attributed to the action of the factors (parameters, inputs) of interest, taken at a time and/or in a combination (4, 22, 25). The variance-based sensitivity is obviously strongly related to the classical sensitivity concept; an example of this is the well-known approximate expansion (17):

$$\mathrm{Var}[C(t)] = \left[\frac{\partial C(t)}{\partial V}\right]^2 \mathrm{Var}[V] + \left[\frac{\partial C(t)}{\partial CL}\right]^2 \mathrm{Var}[CL] \tag{3}$$

where $V$ and CL are the volume of distribution and clearance, $C(t)$ is the plasma concentration and $\mathrm{Var}[\cdot]$ is the estimated variance of volume and clearance.

A naïve variance-based SA can be carried out by varying the variances (or equivalently, the coefficients of variation, CVs) of the factors of interest and exploring the resulting output variances (CVs), e.g., through a classical Monte

Carlo simulation exercise (12). This approach is directly related to the simulation-based power calculations often performed during CTS&D (32).

Recently, several new and more sophisticated methods for variance-based SA have been published, based on the decomposition of the model output variance into terms with increasing dimensionality (22, 25–28). The most important of those are the Fourier Amplitude Sensitivity Test (FAST) (25) and the Sobol's sensitivity indices (22, 28).

## 8.3 SENSITIVITY ANALYSIS OF PHARMACOKINETIC AND PHARMACODYNAMIC MODELS

The majority of the commonly used models in PK and PD share a similar structure. For example, the most popular whole body PK model structure used is the compartmental mammilary structure (Figure 1). This follows from the fundamental assumption that drugs are transported throughout the body by the blood circulation system. With a small modification, this structure also applies to the relatively complex whole-body PBPK models (Figure 2), which can be converted to a purely mammilary structure by simple structural transformations (19, 29). The

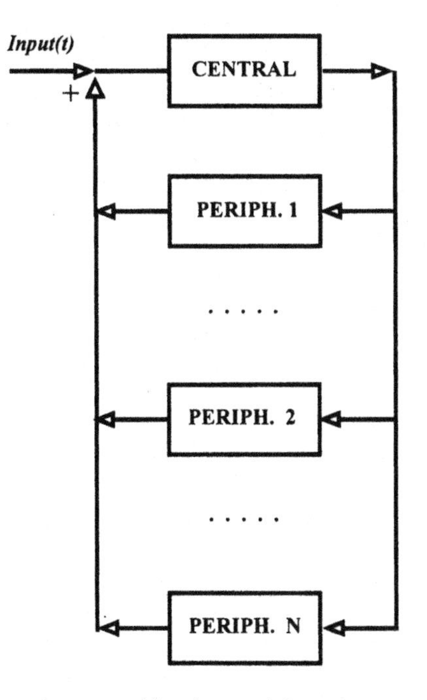


**Figure 1** Mammilary pharmacokinetic model structure.

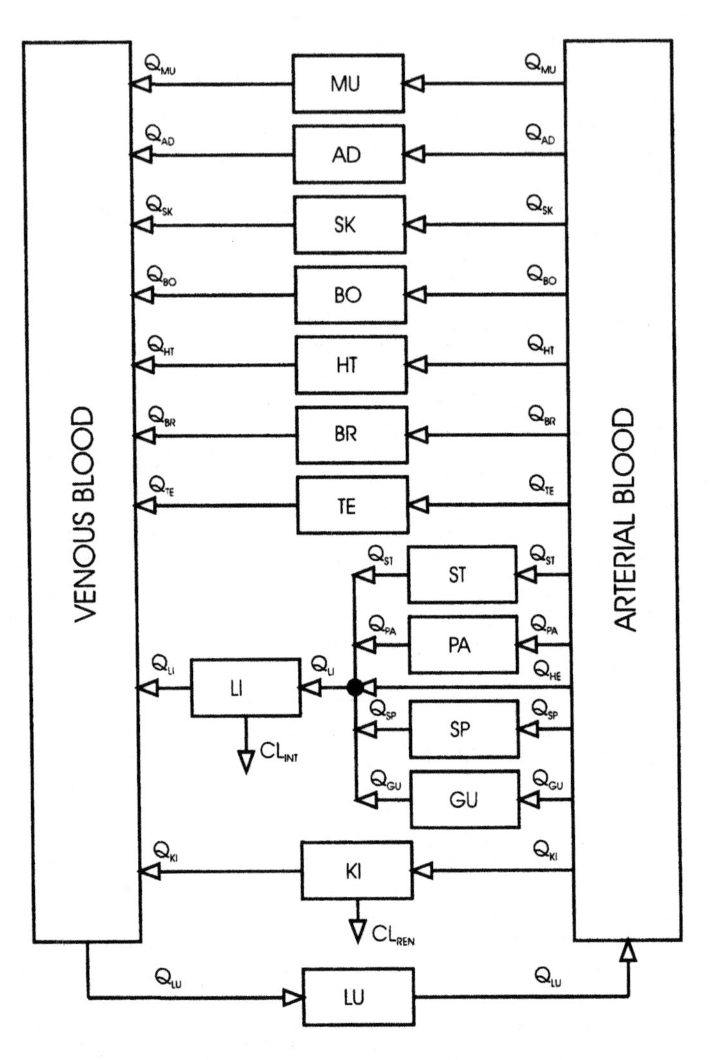


**FIGURE 2** A typical whole body PBPK model structure. ART, arterial compartment; VEN, venous compartment; LU, lungs; LI, liver; KI, kidneys; ST, stomach, PA, pancreas; SP, spleen; GU, gut; MU, muscle; AD, adipose tissue; SK, skin; BO, bone; HT, heart; BR, brain; TE, testes; $Q_i$, blood flow to tissue *i*; $CL_i$, clearance from tissue *i*.

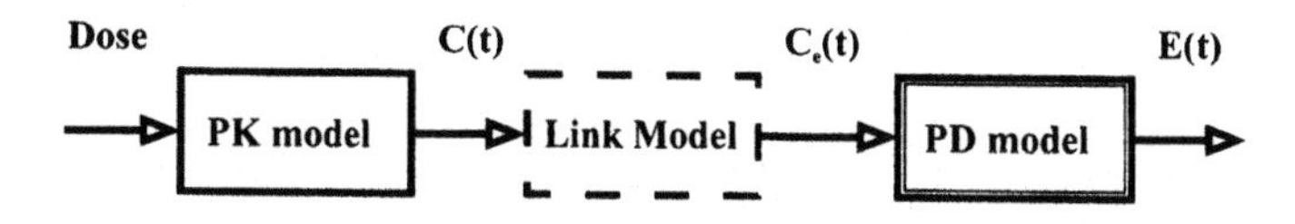


FIGURE 3 Sequential PK/PD model with a link compartment.

most important fact, relating to the structure of the whole-body PK models, which has an impact on the system sensitivity, is that there is a significant parallelism in the system as all the peripheral compartments are situated in parallel to the central compartment.

On the other hand, one of the most frequently used simultaneous PK/PD models has a clearly serial structure, as shown in Figure 3. If there is a hysteresis loop between the concentration and the observed effect, a serial link compartment is attached (18, 19, 30, 31) to collapse the hysteresis. If the effect is linked directly to the concentration (in plasma or at any other site), the link compartment may be missing.

The structural similarity of common PK and PK/PD models has specific sensitivity implications. Our previous findings with a fairly comprehensive whole body PBPK model (9) suggest that a significant part of the PK model sensitivity to perturbations can be explained by the model structure alone. This lead to the concept of structural sensitivity of PK/PD models, introduced recently by us (20)—a property determined solely by the structure and connectivity of the system and independent of the drug administered or the factors perturbed. The core of this concept is the factorization of the system SI into two multiplying components. The first one, called structural SI, has an analytical form, which depends solely on the structure and connectivity of the system and does not depend on the drug administered or the factor perturbed. The second multiplier, the parameter SI, depends on the drug properties, the tissue of interest and the parameter perturbed, but is largely independent of the structure of the system. The structural and parametric sensitivity indices can be evaluated and analyzed separately.

The most important benefit from the proposed structural approach to SA is that the conclusions drawn from the analysis of the structural SI of whole-body PBPK models are valid across all mammalian species, since the latter share a common anatomical and physiological structure. The general conclusion is that the sensitivity of mammilary PK models is low to moderate, which is explained by the large degree of parallelism in the structure of the mammilary system: Any perturbation occurring within the system is distributed by the closed circulation loop to the inputs of all parallel peripheral tissues and is, therefore, damped.

The impact of the serial structure of the PK/PD "link" model sensitivity will be illustrated with the following example. Let us consider the PK/PD model,

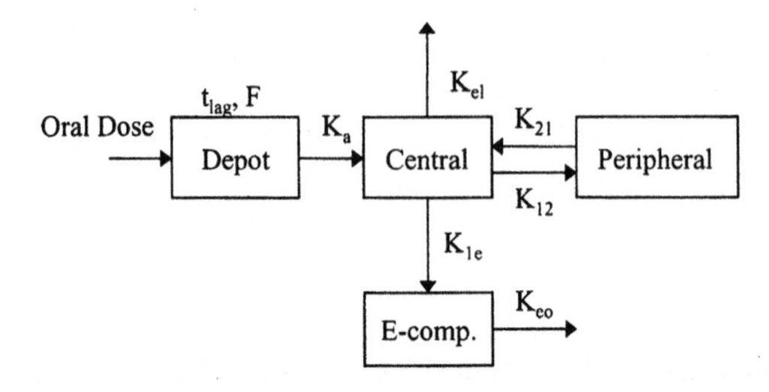


**FIGURE 4** Two-compartment PK model with "effect" $E$ compartment (32). Parameters: $F = 0.48\ h^{-1}$; $K_a = 0.65\ h^{-1}$; $K_{e0} = 1.2\ h^{-1}$; $K_{12} = 4.39\ h^{-1}$; $K_{21} = 1.03\ h^{-1}$; $K_{el} = 0.78\ h^{-1}$; $K_{1e} = 0.0012\ h^{-1}$; $V_1 = 23$ liters; $V_2 = 98.4$ liters; $V_e = 0.023$ liters; $b_1 = -2.06$; $b_2 = 1.93$; $b_3 = 3.96$; $b_4 = 2.34$.

given in Figure 4. The PK part is a common two-compartment model with first-order absorption and an effect compartment (30, 31). The PD part is a variant of the $E_{max}$ model. The mathematical description of the model is given by:

$$C(t) = \frac{k_a FD}{V_1}\left[\frac{k_{21} - \alpha}{(k_a - \alpha)(\beta - \alpha)} e^{-\alpha t} + \frac{k_{21} - \beta}{(k_a - \beta)(\alpha - \beta)} e^{-\beta t} + \frac{k_{21} - k_a}{(k_a - \alpha)(k_a - \beta)} e^{-k_a t}\right]$$

$$= \frac{k_a FD}{V_1}[A_1 e^{-\alpha t} + A_2 e^{-\beta t} + A_3 e^{-k_a t}]$$

$$C_e(t) = k_{e0}\frac{k_a FD}{V_1}\left[\frac{A_1(e^{-\alpha t} - e^{-k_{e0} t})}{(k_{e0} - \alpha)} + \frac{A_2(e^{-\beta t} - e^{-k_{e0} t})}{(k_{e0} - \beta)} + \frac{A_3(e^{-k_a t} - e^{-k_{e0} t})}{(k_{e0} - k_a)}\right] \quad (4)$$

$$E(t) = \frac{E_{max} C_e(t)}{C_e 50 + C_e(t)}$$

$$k_{12} = \frac{Q}{V_1},\ k_{21} = \frac{Q}{V_2},\ k_{10} = \frac{\mathrm{CL}}{V_1},\ b = k_{10} + k_{21} + k_{12},\ c = k_{10}k_{21},$$

$$\alpha = \frac{b + \sqrt{b^2 - 4c}}{2},\ \beta = \frac{b - \sqrt{b^2 - 4c}}{2}$$

where $V_1$ and $V_2$ are the volumes of the central and peripheral compartment, CL is the clearance, $Q$ is the intercompartmental clearance, $k_a$ is the absorption rate, $F$ is the bioavailability, $D$ is the dose administered, $k_{e0}$ is the effect compartment rate constant, $E_{max}$ is the maximal effect, $C_e50$ is the effect compartment concentration giving 50% of the maximum effect.

Equation (4) can be rearranged in the form:

$$\begin{aligned} C(t) &= C[t, D, \Theta] & \Theta &= [k_a, F, V_1, V_2, Q, \mathrm{CL}] \\ C_e(t) &= C_e[t, D, \Theta, k_{e0}] & & \\ E(t) &= E[t, C_e,(t, D, \Theta, k_{e0}), \Phi] & \Phi &= [E_{max}, C_e50] \end{aligned} \tag{5}$$

The normalized SI of the effect with respect to a pharmacokinetic parameter $\theta \in \Theta$ is given by:

$$\begin{aligned} \frac{\partial E(t)}{E(t)} \cdot \frac{\theta}{\partial \theta} &= \frac{\partial E(t)}{E(t)} \cdot \frac{C_e(t)}{\partial C_e(t)} \cdot \frac{\partial C_e(t)}{C_e(t)} \cdot \frac{\theta}{\partial \theta} \\ &= \frac{C_e50}{[C_e50 + C_e(t)]} \cdot \frac{\partial C_e(t)}{C_e(t)} \cdot \frac{\theta}{\partial \theta} \end{aligned} \tag{6}$$

The right-hand side multiplier in Eq. (6) is the normalized SI of the effect site concentration with respect to $\theta$, which is usually small to medium, for reasons stated above. The left-hand side multiplier is an expression, which is always smaller than unity for this PD model. Therefore, Eq. (6) shows that the sensitivity of the effect with respect to any of the pharmacokinetic parameters is even smaller than the sensitivity of the effect site concentration to the same parameter; i.e., the serial structure decreases the sensitivity further. This fact explains some of the parameter estimation problems, related to slow convergence, experienced when fitting simultaneous PK/PD models to experimental data.

The analysis of the structural sensitivity shows that the PK and/or PD model parameters can be rank-ordered with respect to their impact on the model outputs. The blood flows and other flow-related parameters (such as clearances) are generally more influential than the compartmental volumes and tissue distribution parameters (as partition coefficients). Physiologically, this relationship underlines the importance of the blood flows as material carriers of the drug. What is more, perturbations in clearance values influence the concentration-time profiles at later time points, whereas variations in the central volume model terms have the greatest impact in the initial times. All the above conclusions, based on the theoretical structure SA have been supported by results from empirical and experimental sensitivity studies both with comprehensive PBPK models (9–15) and with a simple one-compartment model (17).

The ranking between various types of parameters with respect to their impact on the model outputs leads to ranking between compartments or subsystems of the model. From the latter, direct recommendations regarding the design of experiments for whole body PK models can be derived. The central compartment and the richly perfused compartments (if more than one, e.g., in a PBPK model) have been found to be the most influential parts of the model, so special care has to be taken for their sampling and parametric identification.

It should always be remembered that the SI are functions of time. As a consequence, the ranking of the parameters with respect to sensitivity may change in time. This is a fact, observed by many researchers (9, 17, 20). Therefore, all conclusions drawn from the SA should be accompanied by the respective time frame of interest they relate to.

The compartmentization of the PK/PD systems leads to the introduction of the terms *auto-* and *cross-sensitivity*, by analogy to the terms auto- and cross-correlation in stochastic modeling (9, 20). Autosensitivity is defined as the sensitivity of a system state (concentration or effect in a compartment) to a parameter, belonging to the compartment the state is representing. Cross-sensitivity is defined as the sensitivity of a system state (concentration or effect in a compartment) to a parameter, belonging to a compartment, represented by another state. For example, the sensitivity of the central compartment (plasma) concentration to the clearance of the peripheral compartment of a two-compartment pharmacokinetic model is cross-sensitivity, while the sensitivity of the central compartment (plasma) concentration to the volume of the same compartment is autosensitivity.

It has been proved mathematically (20), that autosensitivity in most cases is greater than cross-sensitivity. What is more, due to the significant parallelism, a weak cross-sensitivity has been found to be typical for the pharmacokinetic mammilary systems (9, 20). This finding has very interesting implications to pharmacokinetic and pharmacodynamic modeling. Based on it, one can estimate the parameters of the peripheral compartments of very large mammilary pharmacokinetic models (e.g., PBPK models) by opening the closed circulation system and using fits of the central compartment profile as a forcing function. In this way each of the peripheral compartments concentration profiles is fitted independently, ignoring the interrelationship between them in the closed circulation loop. The latter is a technique often used by researchers in physiologically based modeling due to its simplicity, but without any discussion or reference to the underlying assumptions.

## 8.4 SENSITIVITY ANALYSIS AND CLINICAL TRIAL SIMULATION AND DESIGN

There are at least two ways to look at the relationship between SA and CTS&D. According to the more limited view, one can regard SA as a methodology for development and validation of the system of models used in CTS&D, including the PK/PD models. The potential of SA in model building in general are undisputed and have been thoroughly analyzed (1–4, 21, 22), but still remain to be widely utilized by the PK/PD modeling community.

The power and capabilities of SA for model building in CTS&D can be illustrated by an example from our experiences. In a recent exercise (Chapter 16), we needed to simulate an early phase II clinical trial with a second in line orally administered (in a tablet form) antimigraine drug. The pharmacokinetics of the drug was modeled by a classical two-compartment model with first-order absorption and an effect compartment (30–32), as described by Eq. (4) and Figure 4. The parameters of the PK model were estimated from phase I information with i.v., s.c., and p.o. (oral solution) data after administration of the drug to healthy volunteers.

The headache severity experienced by the patients was coded in a categorical scale with four categories: 0—no headache, 1—mild headache, 2—moderate headache, and 3—severe headache. The effect of the drug is "pain relief" and is measured on a binary scale. It is assumed that there is a therapeutic effect if a transformation from headache categories 2 or 3 to categories 0 or 1 occurs as a result of the drug administration. Consequently, the pharmacodynamics were modeled by a logistic regression function, giving the probability of success as

$$E(t) = \Pr(t) = \frac{e^{\chi}}{1 + e^{\chi}} \tag{7}$$

where

$$\chi = b_1 + b_2 \cdot \log(t) + \frac{b_3 C_e(t)}{b_4 + C_e(t)}$$

The parameters of the PD model, Eq. (7), were scaled up from the respective PD model of the first in line predecessor of the drug of interest. A detailed description of the scaling and the simulation results is given in Chapter 16 and Ref. 32.

The problems during the model development phase of the exercise were related to the absorption parameters of the PK model. There was no information regarding the absorption of the tablet form as the oral data from the phase I study were generated after administration of a solution. The healthy volunteer data also contained no information regarding the migraine-related changes of absorption parameters. Although we were able to identify some general information regarding the impact of migraine on the GI absorption, there was still a considerable uncertainty regarding the bioavailability $F$ and absorption rate $K_a$ to be used for simulations in the PK/PD model. Therefore, we decided to carry out a simple SA to quantify the impact of the uncertain absorption parameters on the model outputs. The output of primary interest was the drug effect.

In the context of Eq. (6), the normalized SI of the effect with respect to the effect compartment concentration is given by

$$\frac{\partial E}{E} \cdot \frac{C_e}{\partial C_e} = \frac{1}{1 + e^{\chi}} \cdot \frac{b_3 b_4 C_e}{(b_4 + C_e)^2} \tag{8}$$

The normalized SI of the effect compartment concentration with respect to the bioavailability is

$$\frac{\partial C_e}{\partial F} \cdot \frac{F}{C_e} = 1 \tag{9}$$

The normalized SI of the effect compartment concentration with respect to the absorption rate constant can also be computed by analytical differentiation of Eq. (4). A more practical way is using numerical integration of the sensitivity system of the PK model. For this purpose, the state space form of the PK model is specified as

$$\dot{\mathbf{x}} = \mathbf{A}\mathbf{x} + \mathbf{b} \tag{10}$$

where

$\mathbf{x}$ = $[\mathbf{x}_1, \mathbf{x}_2, \mathbf{x}_3]'$ = vector of drug amounts in the central compartment $\mathbf{x}_1$, the peripheral compartment $\mathbf{x}_2$, and the effect compartment $\mathbf{x}_3$

$$\mathbf{A} = \begin{pmatrix} -(K_{12} + K_{el} + K_{1e}) & K_{21} & 0 \\ K_{12} & -K_{21} & 0 \\ K_{1e} & 0 & -K_{e0} \end{pmatrix}$$

= compartmental matrix of the system

$$\mathbf{b} = \begin{pmatrix} K_a \cdot \mathrm{FD} \cdot e^{-K_a t} \\ 0 \\ 0 \end{pmatrix}$$

= absorption input vector

The sensitivity system of the PK model with respect to the absorption rate constant is formed by taking the derivative of Eq. (10) to $K_a$:

$$\frac{\partial \dot{\mathbf{x}}}{\partial K_a} = \frac{\partial \mathbf{A}}{\partial K_a}\mathbf{x} + \mathbf{A}\frac{\partial \mathbf{x}}{\partial K_a} + \frac{\partial \mathbf{b}}{\partial K_a} \tag{11}$$

where $\mathbf{u}(t) = \partial \mathbf{x}/\partial K_a$ is the vector of (nonnormalized) sensitivity indices of the drug amounts in the central compartment $u_1$, the peripheral compartment $u_2$, and the effect compartment $u_3$.

As the compartmental matrix $\mathbf{A}$ does not have entries, depending on the absorption rate constant, Eq. (11) can be rewritten in the form:

$$\dot{\mathbf{u}} = \mathbf{A}\mathbf{u} + \mathbf{v} \tag{12}$$

where

$$\mathbf{v}(t) = \frac{\partial \mathbf{b}}{\partial K_a} = \begin{pmatrix} (1 - K_a t) \cdot \mathrm{FD} \cdot e^{-K_a t} \\ 0 \\ 0 \end{pmatrix}$$

is the input vector of the sensitivity system.

If the linear sensitivity system, Eq. (12), is solved numerically together with the original system, Eq. (10), the normalized SI of the effect time concentration with respect to the absorption rate constant can be calculated by definition from the third component of the sensitivity vector $\mathbf{u} = [u_1, u_2, u_3]'$ and the third component of the state vector $\mathbf{x} = [x_1, x_2, x_3]'$.

The nominal values of the PK/PD model parameters (see also Chapter 16), around which the SA was done were as follows: $F = 0.48$ h$^{-1}$; $K_a = 0.65$ h$^{-1}$; $K_{e0} = 1.2$ h$^{-1}$; $K_{12} = 4.39$ h$^{-1}$; $K_{21} = 1.03$ h$^{-1}$; $K_{el} = 0.78$ h$^{-1}$; $K_{1e} = 0.0012$ h$^{-1}$; $V_1 = 23$ liters; $V_2 = 98.4$ liters; $V_e = 0.023$ liter; $b_1 = -2.06$; $b_2 = 1.93$; $b_3 = 3.96$; $b_4 = 2.34$.

As seen from Eqs. (6), (8), and (9), the normalized SI of the effect with respect to the bioavailability (Figure 5) coincides with the normalized SI of the

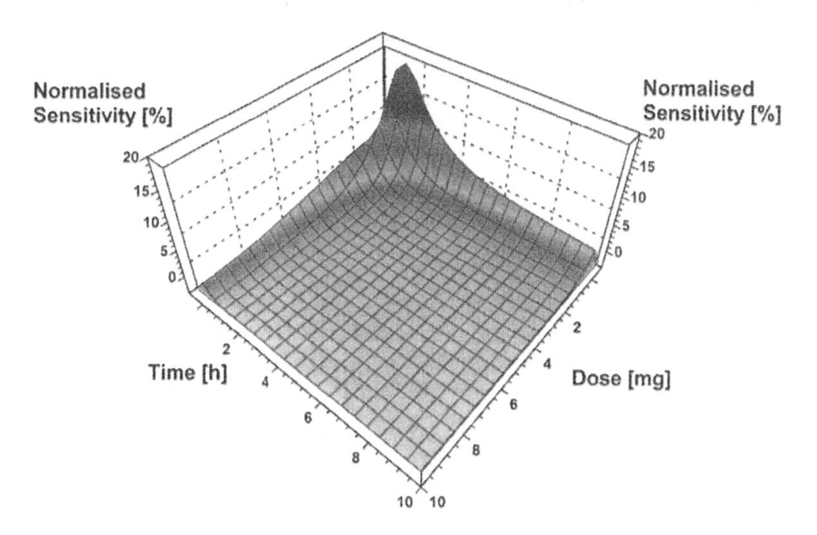


**Figure 5** Normalized SI of the effect with respect to the bioavailability coincides with the normalized SI of the effect with respect to the effect compartment concentration equation (8).

effect with respect to the effect compartment concentration, Eq. (8). Bioavailability is also bound to be the most influential parameter in the system. Yet, Figure 5 shows that the maximum value of the SI with respect to the bioavailability, attained at the lowest dose of 0.5 mg at 1.5 h after administration is around 20%. For the doses with expected therapeutic application (between 1 and 5 mg) at 2 h postadministration (the standard effect sampling time for headache relief) the SI with respect to bioavailability is between 1 and 8%, which confirms the expected low sensitivity of the system.

The SI of the effect with respect to the absorption rate constant is computed using equations (6), (8), and (12) and shown in Figure 6 (on the same scale as Figure 5) as a function of time and dose. The comparison with Figure 5 shows that this sensitivity is even lower than the one with respect to bioavailability. The highest sensitivity is again attained for the lower dose of 0.5 mg at time between 1 and 1.5 h and is less than 14%. For the doses with expected therapeutic application (between 1 and 5 mg) at 2 h post administration (the standard effect sampling time for headache relief) the sensitivity with respect to the absorption rate constant is between 0.5 and 5%.

The overall conclusion from this simple, but very useful SA exercise was that the nominal values of the bioavailability and the absorption rate constant that we intended to use in the PK model for clinical trial simulations did not have a critical impact on the effect modeled. Therefore, the uncertainty resulting from the incomplete information those values were derived from was not likely

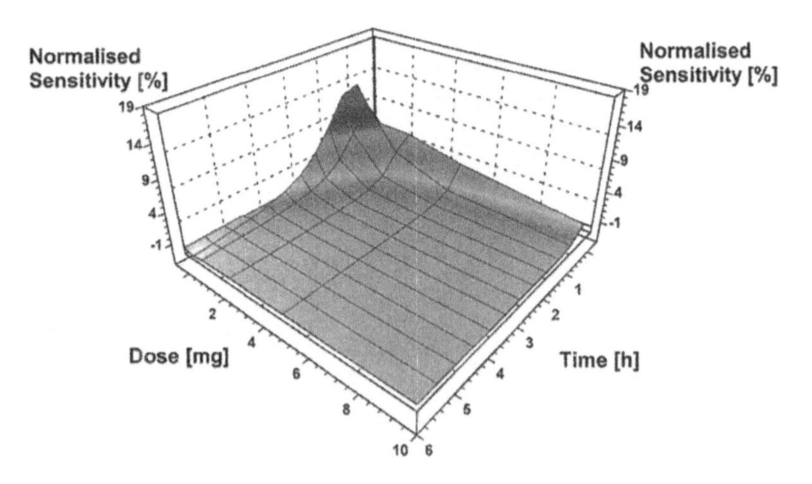


**FIGURE 6** Normalized SI of the effect with respect to the absorption rate constant.

to have a major impact on the model outputs and hence on the results of the simulations.

From a broader point of view, clinical trial simulations for trial design are aimed at evaluation of the therapeutic and adverse effect responses under different assumptions, limitations, and scenarios. Although this evaluation does not necessarily include estimation of explicit sensitivity indices of the outputs with respect to the factors of interest, it generates all the information necessary to do the latter. This circumstance is not often recognized by researchers, but shows the close relationship between CTS and SA and the potential of cross-fertilization between the two areas.

## 8.5 CONCLUSION

Clinical trial simulation and design (CTS&D) uses a combination of SA and uncertainty analysis techniques with the aim of designing rational and informative clinical trials, given the existing variability and all the uncertainties related to the assumptions, scenarios, and models involved. The comments and examples shown in this chapter confirm the undoubted potential of performing more rigorous SA as part of the CTS&D exercise.

The concept of structural sensitivity arises from the finding that a significant part of the PK/PD model sensitivity to perturbations can be explained by the model structure alone. The most important benefit from the proposed structural approach is that the conclusions drawn from the analysis of the structural sensitivity are largely drug and species independent.

The introduction and implementation of more sophisticated SA methodology (e.g., global variance based methods or Bayesian SA) combined with good modeling and simulation practices are likely to save considerable time and effort in CTS&D activities and will ultimately result in more informative clinical trials.

## 8.6 APPENDIX

This section is intended to give practical guidance to the most commonly used procedures for SA.

*Local SA.* The two most frequently used methods for computation of the local SI are as follows:

1. Direct differentiation of the concentration and/or effect equations with respect to the parameter of interest. This is possible when the equations are available in an open (analytical) form. As seen from the case of Eq. (4), for models with two or more compartments, this could be a cumbersome and impractical task.

2. Numerical solution of the combined PK/PD and sensitivity ordinary differential equations model. For the example shown in Section 8.4, the latter is given by Eqs. (10) and (11) as follows:

$$\begin{aligned}
\dot{X}_1(t) &= -(K_{12} + K_{el} + K_{1e}) \cdot X_1(t) + K_{21} \cdot X_2(t) + K_a \cdot \mathrm{FD} \cdot e^{-K_a t} \\
\dot{X}_2(t) &= K_{12} \cdot X_1(t) - K_{21} \cdot X_2(t) \\
\dot{X}_e(t) &= K_{1e} \cdot X_1(t) - K_{e0} \cdot X_e(t) \\
\frac{\partial \dot{X}_1(t)}{\partial K_a} &= -(K_{12} + K_{el} + K_{1e}) \cdot \frac{\partial X_1(t)}{\partial K_a} + K_{21} \cdot \frac{\partial X_2(t)}{\partial K_a} \\
&\quad + (1 - K_a \cdot t) \cdot \mathrm{FD} \cdot e^{-K_a t} \\
\frac{\partial \dot{X}_2(t)}{\partial K_a} &= K_{12} \cdot \frac{\partial X_1(t)}{\partial K_a} - K_{21} \cdot \frac{\partial X_2(t)}{\partial K_a} \\
\frac{\partial \dot{X}_e(t)}{\partial K_a} &= K_{1e} \cdot \frac{\partial X_1(t)}{\partial K_a} - K_{e0} \cdot \frac{\partial X_e(t)}{\partial K_a}
\end{aligned} \tag{13}$$

The notation above is the same as defined in Section 8.4 and the derivation of the equation is based on the fact that $\partial \mathbf{A}/\partial Ka = 0$. Assuming the sensitivity indices are $u_1(t) = \partial X_1(t)/\partial Ka$, $u_2(t) = \partial X_2(t)/\partial Ka$, $u_e(t) = \partial X_e(t)/\partial Ka$, any numerical ordinary differential equations integration software can be used to solve the system (13), including those routinely used for PK/PD modeling.

*Global SA.* The procedure to combine MC simulation with global SA is outlined below. For illustration purposes, the steps, specific to SA, have been italicized.

As seen, the modifications to the standard MC procedure are minimal and the added time/memory requirements are insignificant compared to the potential benefit from the SA results generated. If the MC software used does not allow the execution of user-generated code, there is no problem taking the SA out of the MC loop, i.e., carrying out SA after the MC process has been completed and the results of the simulations saved.

---

*Calculate nominal model outputs at the prespecified nominal values of the model factors/parameters of interest to SA.*
**Start MC simulation.**
Cycle on the number of simulations required.

- Sample the predefined probability distributions of the model factors/parameters.
- Calculate model outputs with the sampled factor/parameter values.
- Save results.
- Calculate global sensitivity indices (see the finite difference form of

the SI given in Section 8.2) using the nominal and the current model output values. Save results.

End cycle on the number of simulations.

**End MC simulation.**

---

*SA Software Considerations and Other References.* An excellent recent reference for those interested in the more technical aspects of SA methodology, including the advanced variance based approach, is Ref. 33. It also gives an overview of the available SA software, ranging from general-purpose commercial applications (like SPSS, Excel spreadsheet add-ins) to specialized programs, developed specifically for implementing various SA techniques.

The commonly used PK/PD modeling software does not offer ready to use solutions for SA. The only possible exemption to this rule is ACSL Optimize (34), which has a sensitivity analysis tool. This software, however, is no longer supported.

## REFERENCES

1. M Eslami. Theory of Sensitivity in Dynamic Systems: An Introduction. Heidelberg: Springer Verlag, 1994, p 600.
2. R Tomovic. Sensitivity Analysis of Dynamic Systems. New York: McGraw-Hill, 1964, p 115.
3. JB Cruz, ed. System Sensitivity Analysis. Benchmark papers in electrical engineering and computer science series. Stroudsburg, PA: Dowen, Hutchinson and Ross Inc., 1973, p 428.
4. JPC Kleijnen. Sensitivity analysis and related analyses: a review of some statistical techniques. J Statist Comput Simul 57:111–142, 1997.
5. K Godfrey. Compartmental Models and Their Application. Cambridge: Academic Press, 1983, p 293.
6. DA Anderson. Compartmental Modeling and Tracer Kinetics. Lecture Notes in Biomathematics, vol. 50, Heidelberg: Springer-Verlag, 1983, p 302.
7. RA Yetter, FL Dryer, H Rabitz. Some interpretative aspects of elementary sensitivity gradients in combustion kinetics modeling. Combustion and Flame 59:107–133, 1985.
8. PM Schlosser, T Holcomb, JE Bailey. Determining metabolic sensitivity coefficient directly from experimental data. Biotechn Bioeng 41:1027–1038, 1993.
9. I Nestorov, L Aarons, M Rowland. Physiologically based pharmacokinetic modelling of a homologous series of barbiturates in the rat: a sensitivity analysis. J Pharmacokin Biopharm 25:413–447, 1997.
10. MV Evans, WD Crank, HM Yang, JE Simmons. Applications of sensitivity analysis to a physiologically based pharmacokinetic model for carbon tetrachloride in rats. Toxicol Appl Pharmacol 128:36–44, 1994.
11. DM Hetrick, AM Jarabek, CC Travis. Sensitivity analysis for physiologically based pharmacokinetic models. J Pharmacokin Biopharm 19:1–20, 1991.

12. P Varkonyi, JV Bruckner, JM Gallo. Effect of parameter variability on physiologically-based pharmacokinetic model predicted drug concentrations. J Pharm Sci 84: 381–384, 1995.
13. RC Spear, FY Bois, T Woodruff, D Auslander, J Parker, S Selvin. Modeling benzene pharmacokinetics across three sets of animal data: parametric sensitivity and risk implications. Risk Analysis 11:641–654, 1991.
14. JL Gabrielsson, T Groth. An extended physiological pharmacokinetic model of methadone disposition in the rat: Validation and sensitivity analysis. J Pharmacokin Biopharm 16:183–201, 1988.
15. D Krewski, Y Wang, S Bartlett, K Krishnan. Uncertainty, variability, and sensitivity analysis in physiological pharmacokinetic models. J Biopharm Stat 5:245–271, 1995.
16. JH Clewell, BM Jarnot. Incorporation of pharmacokinetics in noncancer risk assessment: Example with chloropentafluorobenzene. Risk Analysis 14:265–276, 1994.
17. G Wu. Sensitivity analysis of pharmacokinetic parameters in one-compartment models. Pharm Res 41:445–453, 2000.
18. WL van Meurs, E Nikkelen, ML Good. Pharmacokinetic-pharmacodynamic model for educational simulations. IEEE Trans Biomed Eng 45:582–589, 1998.
19. I Nestorov. System Sensitivity Analysis in Pharmacokinetic and Pharmacodynamic Modelling. Proceedings of the International Conference on Health Sciences Simulation, Western MultiConference, San Diego, CA, January 2000, pp 117–122.
20. I Nestorov. Sensitivity analysis of pharmacokinetic and pharmacodynamic systems: I. A Structural approach to sensitivity analysis of physiologically based pharmacokinetic models. J Pharmacokin Biopharm 27:577–596, 1999.
21. RCH Cheng, W Holland. Sensitivity of computer simulation experiments to errors in input data. J Statist Comput Simul 57:219–241, 1997.
22. GEB Archer, A Saltelli, IM Sobol. Sensitivity measures, ANOVA-like techniques and the use of bootstrap. J Statist Comput Simul 58:99–120, 1997.
23. DE Watkins. Fundamentals of Matrix Computations. New York: John Wiley, 1991.
24. P Lancaster. Theory of Matrices. Cambridge: Academic Press, 1969.
25. D Draper, A Saltelli, S Tarantola. Scenario and parametric sensitivity and uncertainty analyses in nuclear waste disposal risk assessment: The case of GESAMAC. In: A Saltelli, K Chan, M Scott, eds. Mathematical and statistical methods for sensitivity analysis. New York: John Wiley, 1999.
26. A Slatelli, TH Andres, T Homma. Sensitivity analysis of model output: an investigation of new techniques. Computational Statistics and Data Analysis 15:211–238, 1993.
27. A Saltelli, J Hjorth. Sensitivity analysis for model output. Performance of blackbox techniques on three international benchmark exercises. Computational Statistics and Data Analysis 14:211–238, 1992.
28. IM Sobol. Sensitivity estimates for nonlinear mathematical models. Mathematical Modelling and Computational Experiments. 1:407–414, 1993.
29. IA Nestorov, LJ Aarons, PA Arundel, M Rowland. Lumping of whole-body physiologically based pharmacokinetic models. J Pharmacokin Biopharm 26:21–46, 1998.
30. E Fuseau, LB Sheiner. Simultameous modelling of pharmacokinetics and pharma-

codynamics with a nonparametric pharmacodynamic model. Clin Pharm Ther 45: 733–741, 1984.

31. LB Sheiner, DR Stanski, S Vozeh, RD Miller, J Ham. Simultaneous modeling of pharmacokinetics and pharmacodynamics: Application to *d*-tubocurarine. Clin Pharm Ther 40:358–371, 1979.
32. I Nestorov, G Graham, S Duffull, L Aarons, E Fuseau, P Coates. Modelling and simulation for clinical trial design involving a categorical response: a phase II case study with naratriptan. Pharm Res 18:1210–1219, 2001.
33. A Saltelli, K Chan, EM Scott, eds. Sensitivity Analysis. Wiley series in probability and statistics, Chichester, UK: John Wiley, 2000.
34. ACSL Optimize. User's Guide. Concord, MA: MGA Software, 1997.

# 9

# Choice of Best Design

**Jonathan P. R. Monteleone**
University of Auckland, Auckland, New Zealand

**Stephen B. Duffull**
University of Queensland, Brisbane, Queensland, Australia

## 9.1 INTRODUCTION

The pharmaceutical literature is full of theories and methods for arriving at a "best" design (1–17), but ultimately the choice of which method to use depends on the goal of the clinical trial. Sheiner (18) provides a useful framework for establishing the question an experiment should address. Experiments can be performed to quantitate some feature of the system being investigated (LEARN) or confirm an observation about the system (CONFIRM). "What is the bias in an estimated parameter?" is an example of a learning question. "Does the drug have a significant effect compared to placebo?" is an example of a confirming question. With this framework in mind, design properties can be identified for optimization using analytical or simulation techniques.

For example, learning trials may be redefined as pharmacokinetic (PK, dose concentration) and/or pharmacodynamic (PD, concentration effect) model parameter estimation studies. Thus, the PK/PD parameter estimation design problem can be redefined as selecting the optimal combination of design properties to precisely estimate model parameters. This optimization problem can sometimes be solved analytically, using design of experiments (DOE) techniques (19–21), which can be either analytical approaches, which solve the model directly, for independent variables (design properties such as sampling times) or numerical approaches to approximate the minimum/maximum of some scalar measure (design metric) representing the parameter estimation precision resulting from the design (20, 21), e.g., the D-optimality criterion (22).

The design problem can also be approached empirically using clinical trial simulation (CTS) methods (23–25). CTS often uses a Monte Carlo approach to manually explore typical system behavior and the influence of controllable trial variables (design properties) such as dosing and sampling schedules, which are examined in the presence of changes to uncontrollable trial variables (simulation factors) such as model structure and parameter variability. Simulation results, from different combinations of design properties and simulation factors, are analyzed using various statistical techniques to determine the design properties of the clinical trial that produce the best outcomes of the trial with respect to the prespecified goals (26). The specifics of CTS are discussed in detail elsewhere in this book, and other sources (24, 27, 28).

The future of prospectively selecting successful, informative, and cost-effective trial designs will require the integration of DOE and CTS. The well-established theory and methodology for DOE (20, 29) provides an ideal backdrop for incorporating the robustness and versitility that CTS offers in designing a clinical trial. The primary focus, in this chapter, is to show how CTS can be integrated into a DOE framework. This is something that has been alluded to previously (16), but no clear proposal for such a union has been presented to date.

### 9.1.1 Motivation

Simulation of clinical trials has recently gained attention as an emerging technique for knowledge synthesis and exploration of possible trial results based upon a mathematical/stochastic model of the trial, including PK/PD and disease process submodels (30–33). The scientific motivation is obvious; the more informative and reliable the information from a clinical trial, the better the PK/PD characteristics of a drug can be known, resulting in efficient and appropriate use by prescribers (34).

The business motivation for choosing the best possible clinical trial design is to reduce the cost involved to bring a medicine to market. In 1997, research and development costs were estimated to be greater than $359 million US dollars (35), the majority of which was associated with conducting clinical trials. There are many reasons for the sizeable investment, but one clear reason is the high percentage of failed clinical trials due to ambiguous trial results that are attributed to design flaws such as inappropriate dose selection and poorly timed observations (33, 36, 37). Fewer and better designed trials would produce more informative results leading to an overall smaller investment for the knowledge aquired about the potential new medicine.

### 9.1.2 Optimal Design of Experiments in PK/PD

In the design of PK/PD experiments, it has been established that the precision of estimated parameters is correlated with the sampling design properties, i.e., the number, the range, and the spacing of sampling times (1, 5, 38). This statistical property–design property relationship has been used to develop design metrics, which provide an optimal design yielding the most precisely estimated parameters. An often used design metric is the D-optimality criterion (4, 22, 39, 40), which selects sampling designs minimizing the determinant of the variance-covariance matrix [the inverse of this matrix is also known as the Fisher information matrix (FIM)]. D-optimality as well as other design measures derived from manipulation of the FIM have basis in the asymptotic property of maximum likelihood estimators (41). Advantages to this approach include nondegenerate designs (all parameters are identifiable), the fact that optimal design is not sensitive to parameter transformation, at least for single subject designs, and a nice geometric interpretation. Geometrically, maximizing the determinant of FIM is equivalent to minimizing the determinant of $\text{FIM}^{-1}$; therefore, it corresponds to minimizing the volume of the asymptotic joint confidence ellipsoid on the parameters (22).

To construct an optimal design using D-optimality requires two important assumptions. First, the form of the model must be known. It is a difficult task to establish the appropriateness of a model, and only thorough model building and validation methods can provide evidence of the model's utility (42–45). Second, information is available about the magnitude and distribution of the parameter values. Construction of the FIM requires initial information about the model parameter values. If these values are not well known, the determinant of the FIM will also be in question (4, 21, 39). If either of these assumptions are poorly known, the resulting design will be suboptimal, often producing imprecisely estimated parameters.

For PK/PD parameter estimation experiments, often the model and parameter values are not well known because of unknown physiological mechanisms and large variabilities in the population parameter estimates (46). In these situations, D-optimality often fails to provide useful designs. To overcome the parameter variability problem, Draper and Hunter (47) proposed addition of a covariance matrix to FIM, thus allowing the determinant of the new matrix to account for the parameter variability. Unfortunately, the resulting design measure becomes suboptimal if the information in the added parameter covariance matrix becomes dominant in the overall information content of the FIM (41).

Pronzato and Walter (41) offered design measures to account for the variability in parameter values. Like D-optimality, the measures resulted from manipulation of the FIM. For an ED-optimal design, the idea is to maximize the expected value (over a given parameter distribution) of the variance-covariance matrix, for low-information parameters. This produces a constraint based on the largest values possible from the least-known parameters, and allows the range of the parameter values to be known, thus reducing the susceptibility of the optimality calculation to the parameter magnitude and distribution limitation. The second design metric proposed minimizes the expected value of the determinant of the $FIM^{-1}$, which has the advantage of generating designs, on average, minimizing the asymptotic uncertainty of the estimated parameters. D'Argenio (6) also derived a design metric to incorporate parameter variability. The approximation to the preposterior information (API) criterion uses an asymptotic interpretation of the expected posterior information provided by an experiment. Comparison of the parameter estimation precision resulting from designs proposed by the ED-optimal, D-optimal, and API design metrics showed API produced more precisely estimated parameters compared to the ED-optimal results, and an average estimator precision approaching those of local D-optimal designs (6).

Bayesian design optimization can sometimes overcome parameter value uncertainty by using prior information about a parameter's variability (46). Wakefield (48) used a Bayesian approach to derive an optimal dosing regimen in a population PK/PD problem. Merle et al. (49), considered optimal design with discrete, nonparametric priors based on historical data. Merle and Mentré (13) compared three Bayesian design criteria: the determinant of the Bayesian information matrix, the determinant of the preposterior covariance matrix, and the expected information provided by an experiment. The preposterior covariance matrix and the inverse of the Bayesian information matrix resulted in different designs for some examples, specifically, using an $E_{max}$ model and multiplicative error. The expected information provided by an experiment and the determinant of the pre-posterior covariance matrix generally lead to the same designs. Duffull et al. (17) optimized sampling times, doses, and patient groupings for a PD experiment using the drug Ivabradine. The optimization was performed using the method of Mentré et al. (46) and involved calculation of the population Fisher information matrix, for nonlinear mixed-effect models. This design metric incorporates population parameter variability into the variance-covariance matrix.

### 9.1.3 Simulation in PK/PD and Drug Development

The increasing use of simulation in drug development (30) has led to the creation of a good practices guide (27) detailing the necessary components of a clinical trial simulation and issues to consider when organizing and performing a clinical trial simulation. The PK/PD and drug development literature is full of examples where computer simulation has been used for one reason or another, but in general, three categories of simulation projects have been identified (24).

Simulation is used to evaluate statistical properties and analytical methods, which are used for model building and parameter estimation. To explore the impact of estimation method on parameter estimation efficiency, Sheiner and Beal compared naïve pooled (NP), standard two-stage (STS), and NONMEM (extended least squares) approaches to estimating population PK parameters, from simulated data, using monoexponential (50), biexponential (51), and Michaelis-Mentin (52) models. The NP approach failed to estimate between subject variability (BSV) and imprecisely estimated population parameters compared to the STS and NONMEM approaches. The STS approach produced good estimates for mean kinetic parameters and residual error, but poor estimates for BSV, and the NONMEM approach produced the best parameter estimation efficiency for mean kinetic, BSV, and residual error parameters. These results revealed that different estimation methods produce different final estimates and care should be taken when selecting an estimation method.

Simulation is used to evaluate how well an experiment has been designed. Typically, a design is determined using a nonsimulation strategy, and then compared to parameter estimation results generated using simulated data. D'Argenio (5) compared the precision and bias of estimated PK parameters from optimal sampling designs and standard intensive sampling designs using simulation. Ette et al. (3) explored how parameter variability affected D-optimal PK sampling design as well as the influence of sample size on the estimation of population PK parameters (53). As a method for selecting best designs, Al-Banna et al. (25) manually altered design properties, including sampling times, number of subjects, and number of samples per subject, using a PK one-compartment, first-order input model to illustrate the influence of design on population parameter estimation efficiency (parameter estimation bias and precision). Jonsson et al. (54) manually altered the same properties as Al-Banna, using a one-compartment, first-order input, steady-state model including occasion to occasion variability (55). Both studies demonstrated the utility of simulation as a method, in its own right, for determining best designs. An extensive categorization of simulation literature can be found in Holford et al. (24).

Perhaps the most useful application of simulation, in drug development, is for the prospective investigation of full clinical trial designs. This is ultimately what trial designers would want from clinical trial simulation. The ability to look at a range of possible trial designs and, using simulation, determine which designs will best answer the questions of the trial. In a clinical trial design problem, a simulation model was developed characterizing the pharmacokinetics, tumor tissue pharmacodynamics, and healthy tissue pharmacodynamics, for an oral anticancer drug (23). The simulation model was used to answer a dosing schedule–

related design question. Comparison of intermittent and continuous treatments and the effects on tumor growth and skin toxicity revealed comparable effectiveness for two intermittent regimens, whereas continuous treatment caused greater adverse effects in some individuals. This schedule dependence issue could be answered a priori, using nonstochastic simulation. A stochastic simulation experiment was required to create a dose adaptation rule based on the occurrence of adverse events. If a 5% loss of healthy tissue (adverse event) led to a 50% dose reduction (adaptation), the adverse event was reduced in 7 out of 10 patients. Using results from these simulations, a design was determined, which would test the balance between safety and effectiveness predicted by the simulation.

## 9.2 OPTIMAL DESIGN OF EXPERIMENTS

Optimal design of experiments has been an active area of research (20) since the 1920s, when Fisher established the role of statistics in experimental design. He proposed that statistical analysis was only informative if the data itself was informative (56), thus suggesting that statistical methods be employed to prospectively identify informative experimental designs. The utility of optimal design theory has been applied across many different areas of science offering improved designs for everything from block design construction to the precise estimation of model parameters (29). Extensive literature exists on the theory and application of optimal design theory, and much of the early work is covered in Fedorov's text (19), and an excellent review of developments in the field is given by Atkinson (20). A good general starting guide can be found on the Internet (29).

Figure 1 presents one possible organization of the various types of design optimization methods. In most texts on the subject, simulation is not part of the methodology discussed, but is included here (shaded region) to demonstrate how it may be incorpated as another method in a design of experiments framework. The following sections discuss each part of Figure 1 in more detail.

### 9.2.1 Information and Goals

Any prospective design of experiments (simulation or numerical) requires a mathematical model reflecting the system under investigation, parameters defining the model, prior information about the system, e.g., idea of parameter magnitudes, and a range of design properties to be explored. For clinical trial simulation, the full simulation model typically requires submodels for relating dose-concentration-effect (PK/PD model), disease effects on the PK/PD model, accounting for patient specific features, e.g., weight (covariate distribution model), and a model for incorporating failures in following the specified trial protocol (trial execution model) (see Chapters 2 and 4). A stochastic model is also required to account for population parameter variability and experimental measurement error.

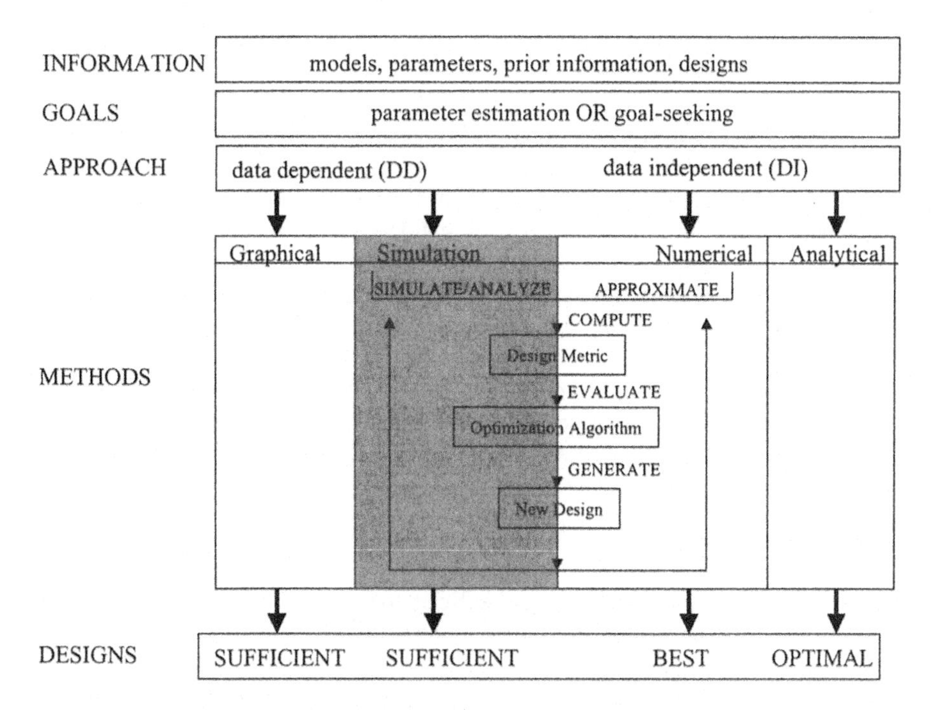


**FIGURE 1** Design of experiments (DOE) framework. The information and goals of designing experiments are linked to methods of data dependence or data independence. The gray area is the integration of simulation into the DOE framework.

Whether the goal is to answer a practical question (goal seeking), or the precise estimation of the model's parameters (optimization), the fundamental question is always the same, i.e., "How to choose a design providing maximally informative data?"

### 9.2.2 Data Dependence Versus Data Independence

The output from an experiment is data. This data takes the form of observational measurements of responses generated following input into the system. The responses are modeled to describe relationships between the input, observations, and parameters of a mathematical model developed to represent the system being investigated. In PK, consider a one-compartment model with i.v. bolus drug input. Drug concentration can be determined from this model, in Eq. (1), with two parameters, CL and $V$.

$$C = \frac{D}{V} \cdot e^{-(\mathrm{CL}/V)\cdot \mathrm{time}} \tag{1}$$

A response, i.e., concentration $C$, is predicted by Eq. (1), the model. The responses for one individual or multiple individuals create data used to describe the change in response, when inputs, e.g., dose $D$ or time, are altered. Likewise, information about the parameters is different depending on the inputs. This approach for using responses to relate inputs, observations, and model parameters is often referred to as a data-dependent (DD) technique.

It is also possible to solve Eq. (1) using mathematical solution to relate the inputs and model parameters. This technique would be considered data independent (DI) because it was not necessary to generate a response to relate inputs to the model parameters. It is these two concepts about relating inputs to parameter information that allow simulation to be incorporated into analytical methods for designing experiments.

### 9.2.3 A Graphical Basis

Selecting optimal sampling times can be illustrated with a simple graphical example. If we want to determine the best sampling times for estimating these parameters, we choose sample times that provide the largest fluctuation in drug concentration for a given change in expected parameter value. For example, if dose = 200 mg, CL = 10 liters/h, and $V$ = 100 liters, the middle curve in Figures 2a,b would be plotted as the concentration-time profile. The upper and lower curves in each figure represent the drug concentrations expected, when the $V$ or CL is increased or decreased by 25%. The times when the three curves are furthest apart would be times when the parameter values influence the concentration measurement the most. For $V$, this can be seen early in the concentration-time profile, while for CL it is late in the concentration-time profile.

Another way to view this concentration-parameter relationship is using a plot of change in drug concentration versus change in parameter value as a function of time (Figure 3)—sensitivity analysis. This partial derivative-time profile shows the time of maximum change as 0 h for $V$ and 10 h for CL demonstrating a two-sample optimal design to be 0 and 10 h. These methods assume the data has equal weighting. With data weighted according to the variance in each data point, different optimal sampling times may be expected. Other graphical techniques for determining optimal experiment designs can be found in Gabrielsson and Weiner (57).

### 9.2.4 Analytical Optimization

The optimal sampling times in Section 9.2.3 may also be determined analytically. Taking the first derivative of each parameter with respect to Eq. (2) produces

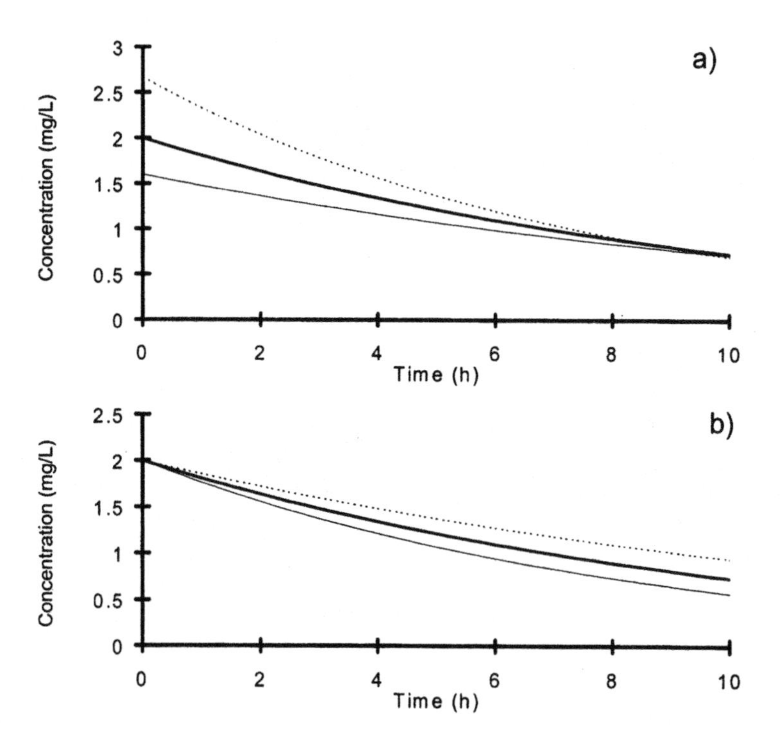


**FIGURE 2** Graphical basis for optimal design. Concentration-time profiles (solid line) are presented for experiments where volume of distribution (a) and clearance (b) are perturbed by +25% (dotted line above solid line) and −25% (dotted line below solid line).

expressions for $dC/dV$ and $dC/d\text{CL}$, which can be solved for a maximum/minimum value. For $V$,

$$\frac{\partial C}{\partial V} = -\frac{D}{V}\left(e^{-(\text{CL}/V)\cdot\text{time}} \cdot \frac{\text{CL}}{V} \cdot \text{time}\right) - \frac{D}{V^2}\left(e^{-(\text{CL}/V)\cdot\text{time}}\right) \tag{2}$$

it can be seen from Eq. (2) that the expression $dC/dV$ increases continuously as time decreases with a maximum value at time equal to zero. The maximum value of Eq. (2) is at time zero since the dose $D$ is zero at earlier times and, therefore, is the value of $dC/dV$. The $dC/dV$ profile, in Figure 3, agrees with this finding. So, the optimal sampling time for precisely estimating $V$ would be at zero time. This is not practical, so a time as soon after administering the dose as possible should be used. For CL, the answer is different.

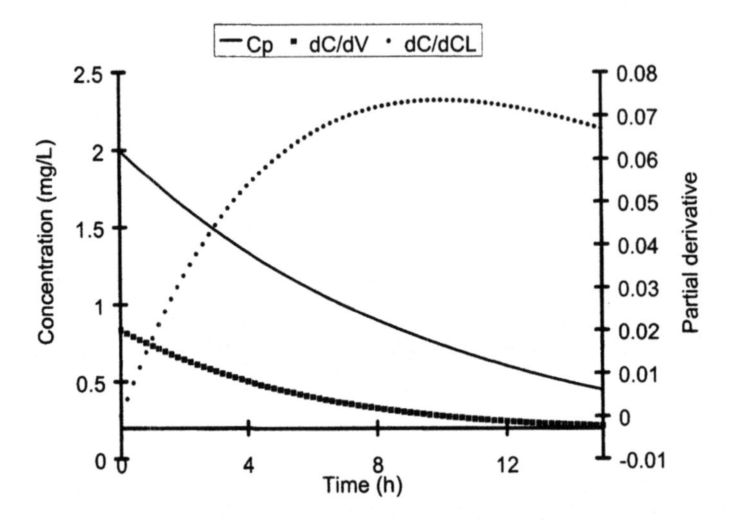


**FIGURE 3** Partial derivative-time profile. The concentration-time profile (solid line) is shown on the left *y* axis. The partial derivatives profiles for volume (dashed line) and clearance (dotted line) are show on the right *y* axis. The maximum/minimum points on partial derivatives profiles indicated optimal times obtaining parameter information from the response (concentration).

$$\frac{\partial C}{\partial \mathrm{CL}} = -\frac{D}{V}\left(e^{-(\mathrm{CL}/V)\cdot \mathrm{time}} \cdot \frac{\mathrm{time}}{V}\right) \tag{3}$$

Taking the derivative of Eq. (3) with respect to time gives

$$\frac{\partial^2 C}{\partial \mathrm{CL}\, \partial_{\mathrm{time}}} = -\frac{D}{V} \cdot e^{-(\mathrm{CL}/V)\cdot \mathrm{time}}\left(-\frac{\mathrm{CL}\cdot \mathrm{time}}{V^2} + \frac{1}{V}\right) \tag{4}$$

Setting this result equal to zero and solving for time produces

$$-\frac{\mathrm{CL}\cdot \mathrm{time}}{V^2} + \frac{1}{V} = 0 \tag{5}$$

or

$$\mathrm{Time} = \frac{V^2}{\mathrm{CL}\cdot V} = \frac{V}{\mathrm{CL}} = \frac{100}{10} = 10\ \mathrm{h} \tag{6}$$

which agrees with the optimal sampling time for CL, from the graphical technique.

Notice the generation of responses was not required to solve for the optimal sampling times, using the analytical technique. The mathematical model was solved directly and parameter values were used to solve for the optimal sampling time. This approach is appealing because it can be done quickly and provides the mathematically optimal time for precisely estimating the model's parameters. Unfortunately, real-world design problems do not often provide examples so analytically tractable.

### 9.2.5 Numerical Optimization

When analytical techniques cannot solve the optimization problem, numerical methods are used. Numerical methods apply approximation techniques to compute the statistical criterion being used (design metric) to evaluate the design's ability to answer the desired questions asked of the experiment. Approximations are required because no mathematically exact solution exists for the solution of the problem. The design metric value is used by an optimization algorithm to minimize/maximize the value. The algorithm requires a range of design properties to vary in the search for the minimum/maximum design metric value.

The process is repeated for a specified number of iterations, and results in location of a best design. The quality of the design is dependent upon the optimization algorithm and number of iterations. The optimization algorithm is a black box requiring input and producing the next design as output. Once it has a design measure value, how it produces the next design depends upon the algorithm.

### 9.2.6 Simulation Methods

Simulation can be used in the same way, the difference being instead of approximating a design metric value, it is calculated directly from the results, e.g., parameter estimates, of analyzing a simulated dataset. What would be the approximation step, for numerical methods, then becomes the simulation/analysis step using simulation methods. So optimal design of experiments using simulation requires data simulation/analysis, design metric calculation, and an optimization algorithm to select the next design. Again, the process is repeated for a specified number of iterations, with the quality of the design dependent upon the number of iterations performed and the optimization algorithm.

### 9.2.7 Sufficient, Best, and Optimal Designs

There is a terminology issue that needs to be resolved to understand the kinds of designs produced from the different methods. The term *optimal* is used quite often, when talking about a maximally informative design determined by any method appropriate. This term should only be applied to designs determined using an analytical method because the only way to truly know the optimal design is for the model to be solved directly without approximation or simulation, i.e., in a data-independent manner. For certain design problems this is possible, but most problems can not be solved so cleanly.

The term *best design* (maximally informative design is used interchangeably) is used to describe designs that were determined using numerical approximations because an exact solution was not possible. Often, numerical methods are referred to as producing optimal designs, but really they are producing the best design given the approximations used. This is not so much an issue for linear design optimization problems, where approximations are well understood and in most cases can approach an optimal design, but for nonlinear design problems the meaning of *best* does not always contain the implicitness of optimality. This is especially true when population experiment designs are required adding an extra level of complexity, which must be incorporated into the nonlinear design problem.

The term *sufficient* is used to describe any design or designs that provide answers to the questions the experiment addresses. It is important to note that optimal and best designs typically produce a single design, whereas sufficient designs can be one or more different designs. The only criteria of a sufficient design is that it provide information to answer the experiment questions. Clinical trial simulation is one method of design optimization that produces sufficient designs to address the clinical trial questions.

## 9.3 DESIGN MEASURES

A measure of design quality is needed to allow the comparison of different designs. The design metric is derived, using statistical considerations, based on questions the experiment is to answer. Typically, since accurate and precise parameter estimation is a desired goal from PK/PD experiments, these estimation properties become the design measures for design optimization.

### 9.3.1 Parameter Estimation Precision

Any statistical measure reflecting precision can be used as a design metric. The standard error, $SE(\theta)$, for a given parameter estimate can be determined empirically or asymptotically estimated. Empirical determination involves estimating the parameter $N$ times (replications) and then calculating the $SE(\theta)$ using a summary statistic, e.g., standard deviation (SD). Alternatively, some software, e.g., NONMEM, can be asked to provide an asymptotic estimate of $SE(\theta)$ from only a single replication.

Another representation of parameter precision is the coefficient of variation,

$$CV = \frac{SE(\theta)}{\theta} \cdot 100 \tag{7}$$

where $\theta$ is the parameter estimate, and CV is a unitless measure of precision which allows comparison of parameter estimates on different scales. A small CV value represents a more precisely estimated parameter. As a design metric, the CV is easy to calculate and useful when precise estimation of a single parameter is desired.

### 9.3.2 Fisher Information Matrix

When the experiment objective is the precise estimation of multiple parameters, a more complicated design metric is required. Given sampling times $T = \{t_i: i \in 1, \ldots, N\}$, a model $y(t_i) = n(t_i, \theta)$, structural parameters $\theta^* = [\theta_1, \ldots, \theta_P]$ and measurement error $\sigma^2(t_i) = \text{VAR}(y(t_i))$ the Fisher information matrix (FIM) can be calculated by

1. Creating an array of partial derivatives for each parameter with respect to the model

$$f(t_i, \theta) = \left[\frac{\partial y(t_i)}{\partial \theta_1}, \ldots, \frac{\partial y(t_i)}{\partial \theta_P}\right] \tag{8}$$

2. Building a matrix $X$ [Eq. (9)], using Eq. (8) and each sampling time, i.e., $t_1, \ldots, t_N$ and an error matrix $V$ [Eq. (10)], using the reciprocal of the measurement errors.

$$\text{X} = \begin{bmatrix} f(t_1, \theta) \\ f(t_\text{N}, \theta) \end{bmatrix} \tag{9}$$

$$\text{V} = \begin{bmatrix} \frac{1}{\sigma^2(t_i)} & 0 \\ 0 & \frac{1}{\sigma^2(t_i)} \end{bmatrix} \tag{10}$$

3. Finally, multiply the transpose of $X$ with $V$ and $X$ to produce the final matrix $M$.

$$\text{M} = \text{X}^\text{T} \cdot \text{V} \cdot \text{X} \tag{11}$$

This matrix $M$ is also known as the moment matrix (hence the symbol $M$) and reflects changes in parameter information with time. Scalar measures derived from manipulation of $M$, e.g., D-optimality, can be used to reflect the overall precision of the parameter estimation procedure. D-optimality (22) is the most popular design metric used in PK/PD optimization. It is defined as the $\det(M^{-1})$ and is useful because it reflects the volume of the joint asymptotic confidence region for a model's parameters. Minimizing this volume results in precisely estimated parameters.

It is possible to calculate other useful metrics from $M$. The C-optimality criterion (58), is the trace of a $WM^{-1}$ matrix, where $W$ is a diagonal matrix with $k$th diagonal equal to $1/\theta_k^2$ and is interpreted as the average squared coefficient of variation of the parameter estimates. Both D- and C-optimality offer measures for estimating the precision of multiple parameters as a scalar value that can be used to compare different designs. There are other "alphabetic" design measures (59), but D and C design measures are the most commonly used in PK/PD. A good review of optimal design, for PK parameter estimation experiments, can be found in Rostami (40).

### 9.3.3 Parameter Estimation Accuracy

Design measures for precision, e.g., C- or D-optimality, do not account for the accuracy of a parameter estimate. Estimation accuracy is an important part of parameter estimation efficiency that reflects the deviation of the parameter estimate from the known or true parameter value. Frequently, true values are not known for a parameter, and so, the accuracy of a parameter estimate cannot be determined exactly. Often, a priori values for parameters are known and may be used as the "true" parameter values for calculating parameter estimation accuracy. Computer simulation allows exploration of parameter estimation accuracy because the true values are the parameter values used by the simulation model, and can be compared to the parameter estimates produced by the estimation procedure.

Accuracy is typically expressed as prediction error (pe) by $\theta - \theta_T$, where $\theta$ is the estimated parameter value and $\theta_T$ is the true value. This difference results in the deviation of the parameter estimate from the true value. For multiple replications, a mean prediction error (me) is used,

$$\text{me} = \frac{1}{N}\sum_{i=1}^{N} \text{pe}_i \tag{12}$$

to estimate the magnitude of the systematic error component, where $N$ is the number of replications. The pe can also be expressed as a percent prediction error (PE%),

$$\text{PE\%} = \frac{\text{pe}}{\theta_T} \cdot 100 \tag{13}$$

producing a unitless value, which allows comparison of parameter accuracy for parameter estimates on different scales. A small PE% value represents a more accurately estimated parameter. As a design metric, the PE% is easy to calculate and useful when the accurate estimation of a single parameter is required of the design.

### 9.3.4 Parameter Estimation Efficiency

Measures for precision and accuracy each provide half the available information about parameter estimation efficiency, but an ideal measure would combine both accuracy and precision information into a single value. A design measure combining accuracy and precision is mean-squared error (MSE) (60). This is a common statistical measure and represents the sum of the squared bias and squared precision

$$\mathrm{MSE} = \mathrm{me}^2 + \mathrm{pev} \tag{14}$$

where prediction error variance is

$$\mathrm{pev} = \frac{1}{N}\sum_{i=1}^{N} (\mathrm{pe}_i - \mathrm{me})^2 \tag{15}$$

Equation (14) works well with a single parameter, but scaling is required to use the MSE measure for multiple parameters. Two steps are involved in modifying MSE to reflect the parameter estimation efficiency for multiple parameters. In the first step, the MSE for each parameter $\theta$ requires scaling, so they can be summed into a single scalar value. The scaled MSE is

$$\mathrm{MSE}_s = \frac{\mathrm{MSE}}{\theta_T^2} \tag{16}$$

The MSE value for $\theta$ is scaled to the square of the parameter's true value, $\theta_T^2$. This creates a dimensionless value by keeping the scale of the denominator and numerator the same. In the second step, the scaled MSE values for each parameter are averaged together producing a single statistic reflecting the overall parameter estimation efficiency. The smaller the $\mathrm{MSE}_s$, the better the resulting parameter estimation efficiency from a given design.

The $\mathrm{MSE}_s$ is an absolute measure of parameter estimation efficiency, for a given design, produced under a specified set of simulation factors. Changing simulation factors potentially changes the magnitude of the $\mathrm{MSE}_s$ value, making comparison of a design under different simulation factors difficult. To compare designs between simulation experiments, further normalization is required.

$$n\mathrm{MSE}_s = \frac{\mathrm{MSE}_s}{\mathrm{MIN}(\mathrm{MSE}_s)} \tag{17}$$

The denominator is the minimum $MSE_s$ from all designs being compared. This includes results under different conditions. The resulting $nMSE_s$ value reflects how a given design compares to the best design from all the simulation experiments examined. These simple mathematical manipulations of the basic statistical metric, MSE, allow designs to be standardized for comparison in the parameter estimation efficiency they provide.

## 9.4 OPTIMIZATION ALGORITHMS

Optimization algorithms are primarily function minimization/maximization methods. The goal is to optimize an objective function value (design metric) with respect to a set of continuous and/or discrete controllable parameters subject to specified constraints.

Assuming an appropriate model and initial parameter values, the general sequence of events for an optimization algorithm include:

1. Choose a starting design.
2. Calculate the design metric.
3. Evaluate the design metric, and select a new design.
4. Repeat steps 2 and 3 until the design metric value does not significantly change, or until a user-defined stopping value has been reached.

The following methods of optimization all build upon the basic steps 1 through 4, with different techniques applied to overcome the common obstacle of local minima.

### 9.4.1 Global Methods

Global methods refer to those algorithms that search over the entire design space. Examples include the grid search methods and non-adaptive random searches.

#### 9.4.1.1 Grid Methods

A design space is a continuous surface where points on the surface are different designs related by varying scalar values of design properties (e.g. sampling times, dose, number of subjects). Each design property represents a spatial dimension and contains a finite set of elements. One classical approach of covering a region of space is to lay a grid of points (designs) over the space where replacement occurs and ordering is important (61). The grid has $N^n$ possible points, where $n$ denotes the number of dimensions containing $N$ elements. As an example, a design space of possible sampling times is required for a two-sample experimental design. Each sample is a dimension, i.e., two dimensions, containing let us say a set of 5 possible times (elements) to use in the design. The total number of possible designs in the grid space is then $5^2$ or 25 different designs.

The grid sampling approach is the "gold standard" for defining a space because it represents an exhaustive means for examining every point surrounding every other point in a specified spatial region. For most problems, the grid approach is not practical because it is impossible to evaluate every location in a realistic design space.

#### 9.4.1.2 Nonadaptive Random Searches

The nonadaptive random search is a method similar in spirit to the grid methods; however, the design locations selected from within the design space are done so at random. Its benefits are due to its simplicity. All that is required is the ability to generate sets of design properties at random from the prior design space and test each different design to determine which provides the best design according to the design metric. The downfall of this method is that it is computationally intensive especially with high dimensional problems; the search will not in all likelihood locate the exact maximum/minimum of the design space.

### 9.4.2 Local Methods

Local methods refer to those that search a limited area of the design space in accordance with some search criteria. These methods are considerably more efficient than global methods but the trade-off for this efficiency is the possibility of the method returning a local minimum/maximum.

#### 9.4.2.1 Gradient Methods

The steepest ascent (descent) technique uses a fundamental result from calculus (gradient points in the direction of the maximum increase of a function) to determine how the initial settings of the parameters should be changed to yield an optimal value of the response variable. The direction of movement is made proportional to the estimated sensitivity of the performance of each variable.

Although quadratic functions are sometimes used, one assumes that performance is linearly related to the change in the controllable variables for small changes. Assume for the sake of argument that a good approximation is a linear form. The basis of the linear steepest ascent is that each controllable variable is changed in proportion to the magnitude of its slope. When each controllable variable is changed by a small amount, it is analogous to determining the gradient at a point. For a surface containing $N$ controllable variables, this requires $N$ points around the point of interest.

#### 9.4.2.2 Simplex Method

This technique runs first at the vertices of the initial simplex; i.e., a polyhedron having $N + 1$ vertices. A subsequent simplex (moving toward the optimum) is formed by three operations performed on the current simplex: reflection, contraction, and expansion. At each stage of the search process, the point with the highest design metric value is replaced with a new point found via reflection through the centroid of the simplex. Depending on the design metric value at this new point, the simplex is either expanded, contracted, or unchanged. The simplex technique starts with a set of $N + 1$ factor settings. These $N + 1$ points are all the same distance from the current point. Moreover, the distance between any two points of these $N + 1$ points is the same. Comparing the design metric values, the technique eliminates the factor setting with the worst value and replaces it with a new factor setting, determined by the centroid of the $N$ remaining factor settings and the eliminated factor setting. The resulting simplex either grows or shrinks, depending on the response value at the new factor settings. The procedure repeats until no improvement can be made by eliminating a point, and the resulting final simplex is small. This technique generally performs well for unconstrained problems, but for constrained problems it may collapse to a point on a boundary of a feasible region, causing the search to come to a premature halt. This technique is effective if the response surface is generally bowl shaped even with some local optimal points.

#### 9.4.2.3 Response Surfaces

Response surface search attempts to fit a polynomial to the surface. If the design space is suitably small, the performance function may be approximated by a response surface, typically first order or perhaps second order. The response surface method requires running the simulation in a first-order experimental design to determine the path of steepest descent. Simulation runs made along this path continue, until one notes no improvement in the performance function. The analyst then runs a new first-order experimental design around the new optimal point reached, and finds a new path of steepest descent. The process continues, until there is a lack of fit in the fitted first-order surface. Then, one runs a second-order design, and takes the optimum of the fittest second-order surface as the estimated optimum.

#### 9.4.2.4 Simulated Annealing

This technique resembles a nonadaptive random search except that it directs its random search towards the minimum/maximum. It was originally used in thermodynamics to determine the rate of cooling (annealing) of solids in order that they achieve their most stable state. The algorithm allows for searches away from the minimum/maximum during the searching process in order to reduce the possibility of locating a local minimum/maximum and as the search progresses the movement away from the minimum/maximum is reduced. This procedure reduces the number of runs required to yield a result; however, there is no guarantee that the design found is actually the maximum/minimum. It is merely the best design of the ones explored. Of course, the more designs selected, the more likely the true optimum is to be found. Replications can be made on the designs selected, to increase the confidence in the optimal design. However, a price is paid in terms of a large increase in the computational time required. It can be proven that the technique will find an approximated optimum. The annealing schedule might require a long time to reach a true optimum.

Various pure adaptive search techniques have been suggested for optimization in simulation. Essentially, these techniques move from the current solution to the next solution that is sampled uniformly from the set of all better feasible solutions.

Simulated annealing as an optimization technique was first introduced to solve problems in discrete optimization, mainly combinatorial optimization. Subsequently, this technique has been successfully applied to solve optimization problems over the space of continuous decision variables.

#### 9.4.2.5 Exchange Algorithms

Exchange algorithms are a collection of techniques for optimizing experimental design. They have become an important part of numerical optimization techniques for experimental design since they are more efficient when the number of design variables is large (40). The earliest version was described by Fedorov (Fedorov's algorithm), and although not generally used in its current form, it remains an appropriate place to introduce these methods. The method involves the univariate optimization of each design variable in turn and repeating this process until the change in the optimization criteria (e.g., the change in the determinant of the information matrix) is sufficiently small.

Consider a problem with two design variables, e.g., two sampling times. The first step involves the choice of an initial design for which the criteria can be evaluated, i.e., nondegenerate. In the second step, the optimal sampling time for one of the sampling times is located (while the other is fixed at its first estimate). In the third step, the change in criterion is compared to the convergence criterion and the algorithm terminated if appropriate; otherwise return to the second step to repeat the process using the other sampling time. The algorithm is termed "exchange" since each design variable is being exchanged for a "more optimal" one at each step.

Difficulties with Fedorov's algorithm may occur since for some problems the univariate exchange might not be sufficient to avoid local minima. Later algorithms, such as the Fedorov-Wynn algorithm, do not have this problem.

## 9.5 THE SIMULATION EXPERIMENT

It is useful to think of a computer simulation as an experiment, with questions to be answered. An overview of this "in silico" experiment is presented in Figure 4.

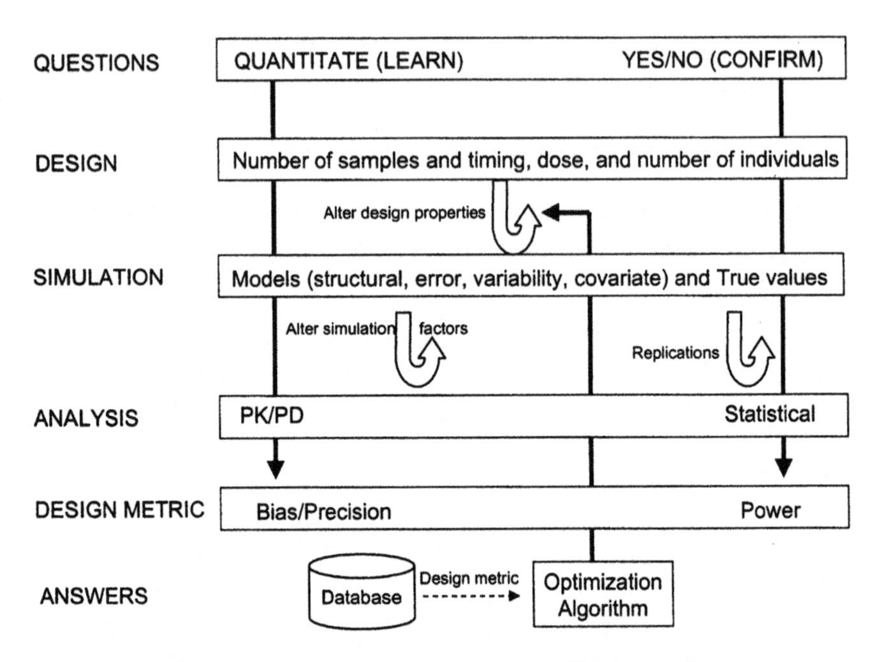


**Figure 4** Overview of a simulation experiment. Trial questions to answers are presented as an experiment, where designs, simulation, and analysis components can be altered to compare different designs. Design metrics are stored in a results database or can be sent to an optimization algorithm for automatic generation of a different design. In this manner, a sufficient design can be selected.

### 9.5.1 Questions

Before optimizing an experimental design, the objective of the experiment must be defined so the methods for optimizing can be stated clearly. Sheiner (18) provides a useful framework for establishing the questions an experiment is to address. Experiments can be performed to quantitate some feature of the system being investigated (LEARN) or confirm an observation about the system (CONFIRM). By recognizing the type of question, the design properties and simulation factors to be explored can be identified.

### 9.5.2 Design and Simulation

These components of a simulation experiment are the same as mentioned above.

### 9.5.3 Analysis and Design Metrics

Analysis of the simulated data tests how well the simulated data, and its assumptions, resulted in the desired outcome. Quantitative analyses can be performed using PK/PD modeling or statistical analysis of the simulated data. The choice of analysis method or methods will depend on the questions being addressed by the simulation experiment.

A simulation experiment may contain replications. A single trial replication is an experiment, where the unit of replication is each individual in the simulated data set. A simulation experiment replication is an experiment, where the unit of replication is the clinical trial (24). This analysis of a simulation experiment is a meta-analysis of multiple replication analysis results.

Any design measure from Section 9.3 can be used as the design metric for a simulation experiment. Once again, the metric used depends on the goal of the experiment.

### 9.5.4 Databases and Optimization

Using the database of simulation experiment outcomes, learning and confirming questions can be answered. Learning questions about the parameter estimation efficiency, i.e., bias and precision, under the conditions of the simulation experiment can be answered by calculating a design metric using the final parameter estimates from the data analysis and true parameter values from the data simulation. Comparing the design metric when design properties and/or simulation factors are altered reveals the conditions resulting in the best parameter estimation efficiency.

Confirming questions such as, "Does the drug work?" can be answered by conducting simulation experiments with a placebo and active treatment group. Using the means and standard deviations of the two groups, statistics can reveal if a significant difference in drug effect exists between the two groups. The question can also be answered by analyzing concentration-response data. Using pharmacodynamic parameters representing the maximum effect $E_{max}$ and drug concentration necessary to produce half the maximum effect ($EC_{50}$), a relationship can be described between the concentration and response. If the $E_{max}$ is significantly different from zero, then the drug worked. By testing this hypothesis repeatedly, i.e., multiple replications, the power can be determined.

Optimization can occur during the simulation experiment. The design component of a traditional simulation experiment is a list of designs to be explored, using the simulation. Instead of a finite list of designs, if a range of values for each design property is used, an optimization algorithm can be applied to select the next design. The design metric serves as the criterion being optimized because it expresses the goals of the experiment as a measurable value. The optimization algorithm can select the next design by selecting a design that seeks to minimize/

maximize the design metric value. The database can still be saved for future analysis if other questions arise; however, the extra steps of sending the design metric to the optimization algorithm and having the algorithm select a new design are required, as compared with the database generation approach.

## 9.6 LIMITATIONS OF SIMULATION

### 9.6.1 Computationally Expensive

Nonlinear regression is often required to estimate parameters of the mathematical model used in a PK/PD simulation experiment. The computational expense associated with nonlinear regression is related to the numerically intensive, iterative parameter estimation process. Increasing the number of parameters to be estimated increases the amount of computing resource required.

Replication requires the simulation and analysis of data multiple times. This compounds the computational expense by having to estimate a model's parameters for each replication.

### 9.6.2 Manually Tedious and Prone to Human Error

Initially, the simulation/analysis model building process may require many changes to achieve a final model adequately reflective of the system. Once a simulation/analysis model is developed, the simulation experiment must be designed and the appropriate files configured, by the experiment designer (user), to execute the simulation experiment. This may require the user to manually modify the simulation experiment factors. Often only small changes are required, e.g., altering the simulated amount of parameter variability, but it is easy for a wrong value to be entered when there are many factors to keep track of and they may vary over the experiment.

The steps involved in performing a simulation experiment are similar, and often, work on one step reveals problems with a previous step requiring it to be revisited. This iterative process aids in understanding how underlying assumptions affect the quality of information obtained from an experiment. Unfortunately, this iterative process requires additional interaction of the user with the relevant simulation experiment factors, opening the experiment up to additional human error.

### 9.6.3 Time Requirements

The time to complete a simulation experiment depends on the complexity of the simulation/analysis models, size of the data sets being simulated, and number of different design properties and simulation factors being altered to answer the experimental questions. The more complex the model and data set, the longer the simulation and/or analysis time. The larger the space of designs and simulation factors being altered, the longer the time requirement to complete the simulation experiment. The generation of a simulation experiment database may take days or weeks compared with the numerical approaches which commonly only take minutes to hours to locate a best design.

## 9.7 SOFTWARE

Any software that allows for the modeling of data can be used to perform a simulation, but since 1996, a few products have been released which focus on simulation of clinical trials. A teaching motivation led to the educationally sponsored development of the RIDO (RIght DOse first time) software (62). The RIDO program was developed to educate pharmaceutical scientists about the basic principles of clinical pharmacology and the role of variability in understanding why clinical trials are difficult to design. A commercially motivated clinical trial simulation software package is the Pharsight Trial Simulator (63). This package is designed to help the trial designer build simulation models to evaluate different designs and their impact on information content for generated data. A popular PK/PD nonlinear regression package, NONMEM, also has the ability to perform data simulation as well as analysis.

## 9.8 SOURCES OF INFORMATION

The internet is a powerful means of allowing people to locate information on almost any topic they desire. It is also easy to access summarized information on specific issues compared to reading an article or book to answer a simple question. The clinical pharmacokinetics LISTSERV (64) (an email discussion group on topics in clinical pharmacokinetics) is an excellent example of an internet source of useful summaries of important topics related to clinical pharmacokinetics as well as links to many other websites.

StatSoft is a statistical software company with a website (29) introducing many key concepts on optimal design of experiments in a simple and organized fashion. Professor Hossein Arsham (65) has a site discussing, in detail, the use of computer simulation in design experiments to answer engineering oriented questions. This site also outlines how various optimization algorithms can be used with computer simulation, giving a guide to how clinical trial simulation can integrate optimization procedures.

## 9.9 FUTURE CHALLENGES

Integration of simulation and optimization methods is the future of clinical trial simulation. The components of simulation based optimization methods are being expanded and refined with each new article on the subject, but much work is still required to make the use of such methods commonplace in the pharmaceutical industry.

Computational studies of techniques for problems with a large number of controllable parameters and constraints are required. Simulation's strength is its generality and ability to strictly specify all assumptions about a problem. With the potential for large numbers of parameters, each one with its own constraints, the impact on computing resources, design metric derivation, and appropriate optimization procedures are unknown factors at this time.

The development of parallel and distributed resource allocation for conducting large simulation-based trial design optimization still requires research. The outcome for meeting these challenges will be an expert system software package that requires the trial designer to input trial questions to be answered, specifications for design constraints, relevant models, and initial parameter estimates. The program then derives the appropriate design measure to answer the trial questions and selects the best optimization method. The program runs, and outputs a list of sufficient designs to answer the questions being asked.

## 9.10 SUMMARY

While trial designers and mathematicians agree that the quality of the model developed depends heavily on the quality of the data collected, it must be admitted that they often disagree on the best way for designing a trial. At the risk of being provocative, one might say that most trial designers think of design as an art owing much to intuition and very little to mathematics, while mathematicians view it as an essentially statistical problem. The importance of the designer's intuition and prior knowledge is obvious, but a mathematical treatment can also be helpful in designing appropriate trials. Both approaches have strengths, but clearly a good trial designer can use elements from either approach to achieve the goal of determining a sufficient design. Clinical trial simulation serves as a tool to achieve this integration by allowing the explicit stating of assumptions and specifying mathematical models describing relationships, such as dose-concentration-effect.

"What is a best design?" A best design is a rationally determined specification of design properties, which maximizes the ability of the data to provide the best answers to the goals of the trial. Such a design of a clinical trial can result in less failed trials and improve information about a drug's PK/PD in humans. "How to choose a best design?" Analytical methods exist for the optimization of experiment designs, but often fall short of adequately providing optimal designs for clinical trials. Simulation offers an approach for choosing a best design, with its primary advantage being its generality. Unfortunately, simulation is underutilized as a means of locating best designs.

Simulation offers a data dependent method that can be incorporated into a design of experiments optimization framework increasing the range of solvable design problems compared to use of data independent (analytical) methods alone. Using either analytical or simulation methods, a design measure and optimization algorithm is required to optimize the design. Design measures include parameter estimation accuracy, precision, or a combination of both (parameter estimation efficiency) metrics. Optimization algorithms include global and local search methods.

Addition of an optimization algorithm to a simulation experiment can allow a manual comparison of different designs to be conducted automatically, in an organized and logical fashion. The future of designing informative clinical trials will assuredly utilize methods that offer simulation and optimization methods.

## REFERENCES


1. JJ DiStefano III. Optimized blood sampling protocols and sequential design of kinetic experiments. Am J Physiol 240 (Regulatory Integrative Comparative Physiology, 9):R259–265, 1981.
2. EI Ette, AW Kelman, CA Howie, B Whiting. Interpretation of simulation studies for efficient estimation of population pharmacokinetic parameters. Ann Pharmacother 27(9):1034–1039, 1993.
3. EI Ette, H Sun, TM Ludden. Design of Population Pharmacokinetic Studies. Proceedings of American Statistical Association (Biopharmaceutics Section):487–492, 1994.
4. EM Landau. Optimal design for individual parameter estimation in Pharmacokinetics. In: Rowland M, Sheiner LB, Steimer J, eds. Variability in Drug Therapy—Description, Estimation, and Control: A Sandoz Workshop. New York: Raven Press, 1985, pp 187–200.
5. DZ D'Argenio. Optimal sampling times for pharmacokinetic experiments. Journal Pharmacokinetics & Biopharmaceutics 9(6):739–756, 1981.
6. DZ D'Argenio. Incorporating prior parameter uncertainty in the design of sampling schedules for pharmacokinetic parameter estimation experiments. Math Biosci 99: 105–118, 1990.
7. GL Drusano, A Forrest, KI Plaisance, JC Wade. A prospective evaluation of optimal sampling theory in the determination of the steady-state pharmacokinetics of piperacillin in febrile neutropenic cancer patients. Clin Pharmacol Ther 45(6):635–641, 1989.
8. GL Drusano, A Forrest, MJ Snyder, MD Reed, JL Blumer. An evaluation of optimal sampling strategy and adaptive study design. Clin Pharmacol Ther 44(2):232–238, 1988.
9. PHL Chen. A two-stage design for comparing clinical trials. Biometrical J 34(1): 29–35, 1992.
10. JG Fadel. Simulation of traditional and optimal sampling schedule designs for digestion and passage models. J Theor Biol 157(1):109–126, 1992.

11. LG Ensign, EA Gehan, DS Kamen, PF Thall. An optimal three-stage design for phase II clinical trials. Statistics in Medicine 13(17):1727–1736, 1994.
12. KC Carriere. Crossover designs for clinical trials. Statistics in Medicine 13(10): 1063–1069, 1994.
13. Y Merle, F Mentre. Bayesian design criteria: computation, comparison, and application to a pharmacokinetic and a pharmacodynamic model. J Pharmacokinetics & Biopharmaceutics 23(1):101–125, 1995.
14. RA Markus, J Frank, S Groshen, SP Azen. An alternative approach to the optimal design of an LD50 bioassay. Statistics in Medicine 14(8):841–852, 1995.
15. FY Bois, TJ Smith, A Gelman, HY Chang, AE Smith. Optimal design for a study of butadiene toxicokinetics in humans. Toxicol Sci 49(2):213–224, 1999.
16. P Muller. Simulation Based Optimal Design. In: Berger JO, Bernardo JM, Dawid AP, Smith AFM, eds. Bayesian Statistics. Oxford University Press, 1999.
17. S Duffull, F Mentre, L Aarons. Optimal design of a population pharmacodynamic experiment for ivabradine. Pharm Res 18(1):83–89, 2001.
18. LB Sheiner. Learning versus confirming in clinical drug development. Clin Pharmacol Ther 61(3):275–291, 1997.
19. VV Fedorov. Theory of Optimal Experiments. New York: Academic Press, 1972.
20. AC Atkinson. Recent developments in the methods of optimum and related experimental designs. International Statistical Review 56(2):99–115, 1988.
21. E Walter, L Pronzato. Qualitative and quantitative experiment design for phenomenological models—a survey. Automatica 26(2):195–213, 1990.
22. GEP Box, HL Lucas. Design of experiments in non-linear situations. Biometrika 46:77–90, 1959.
23. R Gieschke, BG Reigner, J-L Steimer. Exploring clinical study design by computer simulation based on pharmacokinetic/pharmacodynamic modelling. Int J Clin Pharmacol Ther 35(10):469–474, 1997.
24. N Holford, H Kimko, J Monteleone, C Peck. Simulation of Clinical Trials. Annu Rev Pharmacol Toxicol 40:209–234, 2000.
25. MK Al-Banna, AW Kelman, B Whiting. Experimental design and efficient parameter estimation in population pharmacokinetics. J Pharmacokinet Biopharm 18(4): 347–360, 1990.
26. PID Lee. Design and power of a population pharmacokinetic study. Pharm Res 18(1):75–82, 2001.
27. N Holford, M Hale, H Ko, J-L Steimer, L Sheiner, C Peck. Simulation in drug development: good practices 1999. http://cdds.georgetown.edu/SDDGP.html
28. DW Taylor, EG Bosch. CTS: a clinical trials simulator. Statistics in Medicine 9(7): 787–801, 1990.
29. I StatSoft. Experimental Design (Industrial DOE), 1997. http://www.ic.polyu.edu.hk/posh97/general/Download/Statistics_textbook/stexdes.html
30. M Hale, WR Gillespie, SK Gupta, B Tuk, NH Holford. Clinical trial simulation—streamlining your drug development process. Applied Clinical Trials 5(8):35–40, 1996.
31. CC Peck, RE Desjardins. Simulation of clinical trials: encouragement and cautions. Applied Clinical Trials 5:31–32, 1996.

32. R Krall, K Engleman, H Ko, C Peck. Clinical trial modeling and simulation—work in progress. Drug Inf J 32(4):971–976, 1998.
33. CC Peck. Drug development: improving the process. Food Drug Law J 52(2):163–167, 1997.
34. DD Breimer, M Danhof. Relevance of the application of pharmacokinetic-pharmacodynamic modelling concepts in drug development—the "wooden shoe" paradigm. Clin Pharmacokinet 32(4):259–267, 1997.
35. KJ Watling, RA Milius, M Williams. Recent advances in drug discovery. RBI neurotransmissions. Newsletter for the Neuroscientist, Nov.:1–5, 1997.
36. Pharsight Corporation. Knowledge-Accelerated Drug Development. San Francisco: Pharsight Corporation, 1999.
37. L Lesko, M Rowland, C Peck, T Blaschke. Optimizing the science of drug development: opportunities for better candidate selection and accelerated evaluation in humans. Pharm Res 17(11):1335–1344, 2000.
38. DZ D'Argenio, K Khakmahd. Adaptive control of theophylline therapy: importance of blood sampling times. J Pharmacokin Biopharm 11(5):547–559, 1983.
39. L Endrenyi. Design of Experiments for Estimating Enzyme and Pharmacokinetic Parameters. New York: Plenum Press, 1981.
40. HJ Rostami. Experimental design for pharmacokinetic modeling. Drug Inf J 24:299–313, 1990.
41. E Walter, L Pronzato. Optimal experimental design for nonlinear models subject to large uncertainties. American Journal of Physiology (Regulatory Integrative Comparative Physiology) 253(22):R530–R534, 1987.
42. L Lacey, A Dunne. The design of pharmacokinetic experiments for model discrimination. J Pharmacokinet Biopharm 12(3):351–365, 1984.
43. L Sheiner, J-L Steimer. Pharmacokinetic/pharmacodynamic modeling in drug development. Annu Rev Pharmacol Toxicol 40:67–95, 2000.
44. JW Mandema, D Verotta, LB Sheiner. Building population pharmacokinetic—pharmacodynamic models. I. Models for covariate effects. J Pharmacokin Biopharm 20(5):511–528, 1992.
45. M Karlsson, E Jonsson, C Wilste, J Wade. Assumption testing in population pharmacokinetic models: illustrated with an analysis of moxonidine data from congestive heart failure patients. J Pharmacokin Biopharm 26:207–246, 1998.
46. F Mentre, A Mallet, D Baccar. Optimal design in random-effects regression models. Biometrika 84(2):429–442, 1997.
47. NR Draper, WG Hunter. The use of prior distributions in the design of experiments for parameter estimation in non-linear situations. Biometrika 54:147–153, 1967.
48. J Wakefield, A Racine-Poon. An application of Bayesian population pharmacokinetic-pharmacodynamic models to dose recommendation. Statistics in Medicine 14(9-10):971–986, 1995.
49. Y Merle, F Mentre, A Mallet, AH Aurengo. Designing an optimal experiment for Bayesian estimation: application to the kinetics of iodine thyroid uptake. Statistics in Medicine 30 13(2):185–196, 1994.
50. LB Sheiner, SL Beal. Evaluation of methods for estimating population pharmacokinetic parameters. III. Monoexponential model: routine clinical pharmacokinetic data. J Pharmacokin Biopharm 11(3):303–319, 1983.

51. BL Sheiner, SL Beal. Evaluation of methods for estimating population pharmacokinetic parameters. II. Biexponential model and experimental pharmacokinetic data. J Pharmacokin Biopharm 9(5):635–651, 1981.
52. LB Sheiner, SL Beal. Evaluation of methods for estimating population pharmacokinetics parameters. I. Michaelis-Menten model: routine clinical pharmacokinetic data. J Pharmacokin Biopharm 8(6):553–571, 1980.
53. EI Ette, H Sun. Sample size and population pharmacokinetic parameter estimation. Clin Pharmacol Ther 57(2):188, 1995.
54. EN Jonsson, JR Wade, MO Karlsson. Comparison of some practical sampling strategies for population pharmacokinetic studies. J Pharmacokin Biopharm 24(2):245–263, 1996.
55. MO Karlsson, LB Sheiner. The importance of modeling interoccasion variability in population pharmacokinetic analyses. J Pharmacokin Biopharm 21(6):735–750, 1993.
56. DM Steinberg, WG Hunter. Experimental design: review and comment. Technometrics 26(2):71–97, 1984.
57. J Gabrielsson, D Weiner. Pharmarcokinetic and Pharmacodynamic Data Analysis: Concepts and Applications, 2nd ed. Stockholm: The Swedish Pharmaceutical Press, 1997.
58. EM Landau. Optimal experimental design for biologic compartmental systems with application to pharmacokinetics. (Ph.D. dissertation), University of California at Los Angeles, 1980.
59. F Pukelsheim. Optimal Design of Experiments. New York: Wiley-Interscience, 1993.
60. LB Sheiner, SL Beal. Some suggestions for measuring predictive performance. J Pharmacokin Biopharm 9(4):503–512, 1981.
61. GL Gaile, CJ Willmott. Spatial Statistics and Models. Boston: D. Reidel Publishing Company, 1984.
62. R Amstein, J Steimer, N Holford, T Guentert, A Racine, D Gasser, C Peck, S Kutzera, H Rohr, F Buehler. RIDO: Multimedia CD-ROM software for training in drug development via PK/PD principles and simulation of clinical trials. Pharm Res 13: S452, 1996.
63. Pharsight Corporation. Pharsight Trial Designer User's Guide. In. 1.0.0 ed. Palo Alto, California: Pharsight Corporation; 1997.
64. D Bourne. PharmPK LISTSERV, 2001. http://www.boomer.org/pkin
65. H Arsham. Modeling and Simulation, 1999. http://ubmail.ubalt.edu/~harsham/simulation/sim.htm

# 10

# Clinical Trial Simulation (CTS)

## A Regulatory Clinical Pharmacology Perspective

**Peter I. D. Lee and Lawrence J. Lesko**
Center for Drug Evaluation and Research, Food and Drug Administration, Rockville, Maryland, U.S.A.

### 10.1 INTRODUCTION

A number of published reports have addressed the fundamental problems that affect the pharmaceutical and biotechnology industries today (1, 2). The cost of drug development continues to increase, the time from drug discovery to peak sales is thought to be too long, and the amount of data and information arising from clinical trials is increasing exponentially, thereby complicating its interpretation. The solution to these problems is not always clear or simple. Increasingly, the pharmaceutical industry is looking toward new paradigms of drug development to deal with these issues, and in almost all cases, new technology is viewed as one way to overcome the challenges and the obstacles to achieving a more affordable, efficient, and informative drug development process. Among the technology-enabling solutions for which there are high expectations in the next 5–10 years include such platforms as pharmacogenomics, biomarkers, bioinformatics, and modeling and simulation (M&S). This chapter will focus only on the regulatory perspectives of M&S, or more specifically clinical trial simulation (CTS).

In the past 40 years, M&S, generally speaking, has increasingly permeated our society in many different ways, from forecasting the weather to virtual prototyping of automobiles and airplanes. M&S of pharmacokinetic data has become routine practice over the past 10 years and the FDA has published a guidance for industry that incorporates principles and recommendations related to M&S in the analysis of sparse PK samples (3). The FDA is in the process of developing a draft guidance for industry on study design, data analysis, and regulatory applications of exposure-response relationships that includes specific recommendations related to the M&S process. Several recent publications provide an excellent background on good practices of M&S in the simulation of clinical trials (4).

In contrast to traditional M&S, stochastic CTS has the potential to revolutionize the way that late-phase clinical trials in drug development are planned and conducted. CTS has the potential to transform the drug development process by making better use of prior data and information and to explore the importance and impact of clinical trial design variables. We feel that the use of CTS will result in better decision making, not only in drug development, but also in regulatory assessment of clinical data submitted in investigational new drug applications (INDs) and new drug applications (NDAs). Sometimes pharmacometricians and other reviewers in the Office of Clinical Pharmacology and Biopharmaceutics (OCPB) in the Center for Drug Evaluation and Research (CDER) at the Food and Drug Administration (FDA) utilize CTS as a review tool to assess the results of clinical studies or to test different "what if" scenarios related to drug dosing and/or dosing regimen adjustments in special populations. In doing so, they have found that CTS is an effective way to share clinical pharmacology knowledge and insights with biostatisticians and medical officers. From our experience with CTS in NDA reviews, however, there are still only limited instances where CTS has had a major impact on drug development or on regulatory decisions related to efficacy or safety (5).

Some scientists in industry have expressed concern that regulatory authorities may not be receptive to M&S and perceive that this may inhibit more widespread use of CTS. From our experience, we do not believe that this is the case. In fact, M&S is predicted to be incorporated into as many as 75% of drug development programs of major pharmaceutical companies in the next 2 years (6), and the FDA looks forward to seeing cases in which sponsors have used CTS for designing potentially better studies and selecting more optimal doses or dosing regimens. We also expect that more extensive use of CTS will be used in regulatory reviews and feel confident that the CDER has the expertise to deal with CTS discussions and submissions as part of the IND and NDA review process.

Unfortunately, there is scant evidence of the impact and value of CTS with few published, real-life examples of use in drug development or regulatory decision making. This suggests a need for more prospective integration of CTS into actual drug development programs to evaluate and determine the impact that CTS technology will have on the "return on investment" as measured by the efficiency, quality, cost-effectiveness, and informativeness of clinical trials. As a new technology, use of CTS by the pharmaceutical industry or by regulatory authorities may not always be straightforward because it depends on having an understanding of, and developing, a model of drug action, clinical response, and disease progression that can integrate prior knowledge and certain assumptions, for which there may not be consensus, into the decision-making process. CTS also requires the ability to rapidly integrate new data, sometimes across disciplines, as it becomes available in real time in order to improve the quality and usefulness of the model, and this will require a shift to a more collaborative working environment between disciplines. Further, we anticipate that there will be a period of time needed to uncover unexpected scientific complexities of CTS and to build communication links and confidence in the use of CTS for decisions in drug development and regulatory assessment.

In this chapter, we will provide a regulatory perspective on the use of CTS. We will present some general principles of M&S, and important regulatory policy initiatives in relation to M&S.

## 10.2 BACKGROUND

Based on an internal survey of a cohort of 779 NDAs submitted to the Center of Drug Evaluation and Research between 1992 and 2000, we found that 30% of these submissions failed to show either adequate efficacy or safety (7). The average drug development process takes 12 years and in the order of $350–$600 million (2). With the increasing cost of drug development, the pharmaceutical industry is looking at M&S as one way to reduce the failure rate of NDAs. CTS may provide the opportunity to integrate early-phase clinical data (Dose-PK, PK/PD relationships) in order to simulate late-phase clinical trials, leading to decisions by a sponsor that the FDA will be interested in, e.g., the design features of phase III studies, the selection of the number and range of doses to be tested, the number of subjects to enroll, the number of blood samples to obtain, and the number and type of clinical endpoints to evaluate. This trend of applying M&S in drug development and the availability of appropriate software points toward increasing numbers of INDs and NDAs involving M&S over the next 5 years.

At this time, there are no formal standards in industry or in the FDA that assure consistent criteria for assessment of the quality of M&S projects. Evaluation of in silico drug development will benefit from standards for M&S. Some CDER guidances, e.g., population PK/PD and renal impairment, recommend the use of M&S for regulatory decisions, but do not provide detail on the "best practices."

M&S exercises may frequently take place and provide valuable information over the course of drug discovery and development process. The purpose of these exercises may vary from exploring the first dose in humans, to determining dosages and dosing regimens in clinical studies, optimizing the design of efficacy and safety studies, and supporting labeling claims for dosage selection in special populations. The typical M&S methodologies include population pharmacokinetic modeling, pharmacokinetic and pharmacodynamic modeling, and CTS. These processes have been primarily used in a retrospective manner to reanalyze the data obtained from clinical trials. More recently emphasis has been placed on the prospective use of such models in designing clinical trials.

## 10.3 REGULATORY REVIEWS OF M&S

To consider the need for standards of M&S studies and analysis in regulatory reviews, the Office of Clinical Pharmacology and Biopharmaceutics (OCPB) is taking a systematic approach to:

1. Assess the current state of the art of M&S
2. Determine attributes to assess M&S suitability
3. Develop standard formats and evaluation criteria for M&S outputs
4. Build an interdisciplinary infrastructure for evaluation of M&S projects
5. Evaluate the need for a guidance for industry on "best M&S practices"
6. Maintain a training program for CDER reviewers

### 10.3.1 Current Focus

The current activities related to M&S within the FDA include the following areas.

#### 10.3.1.1 Study Design and Conduct

We are discussing the advantages and limitations of different study designs (e.g., crossover, parallel, titration) from the M&S (data analysis) point of view. Appropriate M&S methodology for each type of study is being evaluated. Methods for handling drawbacks in study designs will also be investigated for the purpose of identifying desirable approaches.

#### 10.3.1.2 Database Management

The quality and timeliness of available PK and PD data are major obstacles in the industry and in the regulatory agency in applying M&S. Several issues need to be addressed to facilitate the data management procedure, including such issues as data collection and auditing procedures, database integrity (storage), database merge, data format manipulation, and data transfer among software input/output files.

#### 10.3.1.3 Modeling Procedure

The procedure on how to implement a modeling project will be detailed, with an emphasis on consistency between other current guidances that touch on M&S. Presenting clear statements of the objectives for the M&S project with a prospective analysis protocol is recommended as most desirable. In this protocol, sources of data and methods of collection should be detailed, along with any critical assumptions that have been made. Some recommendations on model building should also be given, again attempting to maintain consistency between existing guidances for industry. It is important to establish basic methods of evaluating structural components of models, as well as assessing the addition of covariates. Finally, the process of model validation should be refined.

#### 10.3.1.4 Simulation Procedure

While the choice of simulation methods is case specific, certain common practices pertaining to the following characteristics of models are important: a problem statement of the simulation, a protocol of simulation, assumptions, simulation model building (sources of structural and covariate models, the links between model components, the use of less relevant data (establishing prediction error), and the definition of model validation.

#### 10.3.1.5 Software/Hardware Selection

Currently thorough and universally accepted methods of software/hardware verification are not readily available and have not been implemented. Standards regarding the verification of software/hardware selection should include computer verification, software verification protocol and associated report, cross-verification among software, and numerical method selection.

#### 10.3.1.6 Interpretation of Modeling and Simulation Results

The primary objectives of M&S are to describe data and make predictions. Different levels of detail of interpretation of M&S results should be discussed. These include (1) the interpretation of model parameters (e.g., $EC_{50}$), covariate effects (e.g., population PK), and hypothetical dependent variables (e.g., peripheral compartment and effect compartment concentrations), (2) the confirmation of a mechanism of action [e.g., existence of tolerance and active metabolite(s)], (3) the prediction (interpolate or extrapolate) of alternative scenarios (e.g., application of PK/PD in labeling of special population), and (4) the estimation of study power [e.g., using CTS (7)].

#### 10.3.1.7 Technical Issues

Despite the advancement in the technology of M&S, there are still operational issues requiring further discussions. The following are some of the topics that may be relevant to setting M&S standards: selection of numerical methods, objective function and convergence criterion, local vs. global minimum, number of model parameters, outlier handling, numerical instability, and mechanism-based vs. empirical models.

#### 10.3.1.8 Audit Trail of M&S Procedure

One way to ensure repeatability of an M&S exercise is to keep a computerized record of the procedure. The audit trail also allows future applications of the same method to analyze similar or even different studies. Several issues related to the audit trail of M&S projects are important: record keeping, summary of procedure, and auditing methods.

*FDA ACPS Recommendations.* The regulatory applications of M&S were discussed at a FDA Advisory Committee for Pharmaceutical Science (ACPS) meeting on Nov. 16, 2000 (9). The ACPS members and a group of invited industry and academic experts were assembled for the discussions. A series of questions related to regulatory applications of M&S were posted to the committee and the experts. These questions and the ACPS recommendations are listed below.

1. How does industry use simulation to facilitate the drug development process? Simulation techniques are being used in the industry from preclinical development to phase IV clinical studies, and with applications from simulating drug action at the molecular level to CTS. The ACPS recommended a survey on companies' internal practice of simulation projects. Subsequently, PhRMA conducted such a survey in 2001 (5) and the results showed that the applications of simulation in drug development will likely increase from 41 to 49% in the next 2 years, and all companies responding to the survey will conduct simulations in their drug development programs.
2. Is M&S appropriate for drug development and regulatory decisions? M&S is already an integral part of drug development process, and it is appropriate for drug development and regulatory decisions. Some of the examples include using CTS to reduce the time to market, designing studies to extrapolate adult efficacy data to pediatrics, identifying clinically relevant biomarkers and verifying them in clinical trials, and using M&S in supporting product line extensions.
3. What are the important attributes for a meaningful simulation practice? The important attributes for a meaningful simulation practice may include but are not limited to the following: use of a prospective simulation analysis plan, a clearly defined scope, clear assumptions, a statement about the implementation of the project, a team based approach to model building, evaluation of M&S results as compared to observed data, a full team participation in the interpretation of results, specifying

predetermined acceptance criteria of the modeling results, and prospective study/analysis plan.

4. Do we need a regulatory guidance for industry regarding the best practices of M&S? With the current review situation, case-by-case reviews of simulation may be prone to inconsistency, and M&S guidance will be useful. Such guidance should probably be broad and cover general framework for the best practice, as the field evolves.
5. If yes to question 4, what is the important information that should be involved in the guidance? An M&S guidance should include the following information: current best practices, advice on extrapolation, examples of regulatory applications, the systematic process of model building and simulation, components of the analysis plan, necessary information to assess the M&S exercise, model validation, a research plan, the basis of assumptions, impact of (sensitivity to) assumptions, methodology for parameter estimation, and disease progression and placebo response models. The ACPS further suggested that this should be a broad guidance and that the guidance should discuss joint involvement of CDER and industry to work together to develop best practice guidelines for M&S practice. The ACPS also suggested that the FDA could try to encourage sharing information on disease progression models between sponsors and FDA

## 10.4 GENERAL POINTS TO CONSIDER

The good review practice of simulation for the purpose of regulatory assessments and decisions should include the following steps: (1) formulate relevant questions, (2) prepare a decision tree, (3) establish study selection criteria, (4) define process for data review, (5) define process for model building, (6) present key results, and (7) justify interpretations. Clinically relevant questions are defined first; these address the potential labeling claims based on the simulation outcomes. The decision tree of applying a simulation approach is then determined; this lays out the regulatory conclusions corresponding to different simulation outcomes. The appropriate studies are then selected as the source of input information for model building. The data from these studies are reviewed for their quality and quantity. The model-building process should be reviewed carefully for its assumptions, model selection, and validation. Finally, the simulation outcomes should be reviewed and interpreted according to the predetermined decision tree.

The applications of simulation can generally be divided into two categories: (1) predicting the outcome of a clinical scenario, and (2) assuming the outcome of a scenario, estimate the power of a given study design to confirm the desired outcome. Examples of the first application include predicting multiple-dose plasma concentration profiles with single-dose pharmacokinetic data and models, and estimating the clinical consequence (change in effectiveness and toxicity) of changes in pharmacokinetics in special populations or due to drug-drug interactions and other intrinsic or extrinsic factors based on pharmacokinetic and pharmacodynamic relationships. Examples of the second application include estimating the power of population pharmacokinetic/pharmacodynamic studies, and efficacy/safety conclusions.

The acceptance of simulation results for regulatory decisions is largely dependent on the significance of the regulatory decision, the quality and quantity of evidence to support claims, and the soundness of the simulation method. Thus, applications of simulation to predict efficacy/safety require more stringent standards than applications of simulation for designing studies which will provide actual data to confirm efficacy/safety. The quality and quantity of source data used to build the model for simulation should be examined according to the simulation objectives. It is important to evaluate the study design and study conduct to ensure the proper data set is used for the simulation exercise. The critical study design and conduct factors influencing simulation outcomes may include the study power, the number of subjects and samples, whether the exposure and response endpoints are relevant to the simulation objectives, use of placebo or active control, sufficient baseline data, study record keeping, missing data, handling of outliers, and range of doses studies.

The reliability of simulation results is dependent on the amount of data as well as the methods of simulation. Simulation methods can be categorized based on the relationship between the simulated scenario and the original scenarios where the source data come from:

1. Resampling observed data is one of the simplest simulation exercises. This may involve bootstrapping the observed data or regenerating the distribution of a variable (e.g., based on observed mean and standard deviation). This simulation result is generally acceptable for regulatory decision.
2. Interpolation based on either empirical or mechanism-based models to describe the smoothness of data is another straightforward application of simulation. A data point within the range of the source data can be reasonably interpolated based on the established model.
3. Simulations based on well-validated models are generally acceptable. A validated model is one that has been consistently observed in different studies or data sets. For example, if a one compartment model is observed to be fitted to data observed in several pharmacokinetic studies with extensive blood samples, the model would be considered "well" validated.
4. When a simulation is used to predict a scenario that is different from

the original scenario of the source data for building the model, the acceptance of the simulation outcomes will then depend on the validity of the assumptions made to link the two scenarios. Examples may include using simulations to predict pharmacokinetic or pharmacodynamic outcomes in different populations with different PK (e.g., from adults to pediatrics), different formulations (e.g., from immediate release to modified release) with varying bioavailability, different clinical effect endpoints (e.g., from biomarkers to clinical outcomes), and different time points in a PK curve. The assumptions that the same model is applicable to the new scenario must then be validated.

5. Simulations based on models from multiple sources should be interpreted with extreme caution. The assumptions that models from different sources are applicable in conjunction with each other under the new scenario should be well justified with evidence to support such a claim.

## 10.5 SUMMARY

In the current health care environment dominated by the growth of managed care organization, generic competition, and therapeutic substitution, there is increased pressure on pharmaceutical companies to compete in the marketplace through implementation of optimal drug development strategies. M&S has gradually become an integrated part of drug development and regulatory assessment processes, since it has provided valuable information for decision making and has the potential for improving these processes. The recently emerging CTS technique presents tremendous promise in facilitating informative and efficient study designs. The CDER is currently evaluating the CTS methodology and the best practices for conducting and reviewing M&S applications.

## REFERENCES

1. Pharma 2005. An Industrial Revolution in Research and Development, Price Water House Coopers, 1999.
2. W Wierenga. Strategic alliances and the changing drug discovery process, Pharmaceutical News 3(3):13–16, 1996.
3. FDA Guidance: Population Pharmacokinetics, FDA, 1999.
4. Simulation in Drug Development: Good Practices. Washington, DC: Center for Drug Development Sciences, Georgetown University, 1999. http://cdds.georgetown.edu
5. NHG Holford, HC Kimko, JPR Monteleone, CC Peck. Simulation of clinical trials. Annu Rev Pharmacol Toxicol 40:209–234, 2000.
6. M Sale. "Pop PK/PD and simulation in the pharmaceutical industry—survey results, May 2001. Bethesda, MD: FDA/PhRMA Clinical Pharmacology and Drug Metabolism Committee Meeting, May 30, 2001.

7. P Lee. Applications of pharmacokinetics-safety information in regulatory decision making in dose regimen selection for intrinsic/extrinsic factors. Washington, DC: FDA/DIA Sponsored Workshop on Applications of Pharmacokinetics/Safety Information in Drug Development and Regulatory Decisions, April 26, 2001.
8. P Lee. Design and power of a population pharmacokinetic study. Pharmaceutical Research, 18(1):75–82, 2001.
9. FDA website. http://www.fda.gov

# 11

# Academic Perspective

## Modeling and Simulation as a Teaching Tool

**Nicholas H.G. Holford**
University of Auckland, Auckland, New Zealand
**Mats O. Karlsson**
University of Uppsala, Uppsala, Sweden

### 11.1 INTRODUCTION

This chapter briefly reviews the history of clinical trial simulation and applications of how it might be used for teaching in an academic setting.

### 11.2 HISTORY OF CLINICAL TRIAL SIMULATION

The history of clinical trial simulation and recent applications have been reviewed by Holford et al. (1).

#### 11.2.1 Stochastic Simulations

Early clinical trial simulators were based solely on predicting the stochastic aspects of how observations might arise but did not use models for the underlying pharmacological processes. Maxwell (2) produced a game for learning the principles of clinical trial design. The use of a computer program to simulate patient outcomes was a by-product of the actual game process. The application of sequential analysis to adaptive clinical trials (3), the influence of dropouts (4) and premature termination of trials (5) were all explored using simulation.

### 11.2.2 Mechanistic Simulations

The application of pharmacokinetic and pharmacodynamic models as the basis for simulation of clinical trial observations took off in the middle of the 1980s. Important precursors were the growing knowledge linking drug concentration to effect (6, 7) and the application of mixed effect models in pharmacology (8, 9). Simulation of the physiological response to circulating tissue-type plasminogen activator (t-PA) was used to show that fibrinogenolysis was not sensitive to a variety of concentration profiles arising from alternative dosing patterns. (10, 11). The properties of the randomized concentration-controlled trial were explored by simulation and define the kinds of clinical trials which might benefit most from prospective adjustment of doses based on drug concentrations. (12, 13).

In 1993 the European Course in Pharmaceutical Medicine (now the European Center for Pharmaceutical Medicine) received a grant from the COMETT program of the European Union to develop a computer-based multimedia CD-ROM teaching tool. The goal was to teach the principles of clinical pharmacology with a particular emphasis on their application to clinical drug development. This project was called RIDO (RIght DOse first time) (14).

Simulations of concentration time profiles were included in an interactive clinical pharmacology trainer to illustrate properties of the basic input-output model. Because the goal was to teach about drug development, a more advanced simulation system was developed to teach clinical trial design and analysis.

The RIDO clinical trial simulator includes covariate distribution, input-output, and execution model features. The output of each simulated clinical trial can be analyzed using a set of prebuilt NONMEM models. Students can use RIDO to explore alternative clinical trial designs by changing the number of subjects, type, and number of observations, and varying subject drop out rates. Each design can then be analyzed using NONMEM to see if it is capable of answering specific PK/PD model-based hypotheses. A simple pharmacoeconomic model is also incorporated so that the cost consequence of different clinical trial choices can be appreciated. The RIDO clinical trial simulation engine was licensed to Pharsight Corporation and used as the basis for their Trial Designer clinical trial simulation product.

Around the same time that the Pharsight Trial Designer was released, another product, ACSL Biomed, based on the advanced continuous simulation language (ASCL), was developed. Pharsight subsequently bought ACSL Biomed and produced a new product (Trial Simulator) with features taken from Trial Designer and ACSL Biomed Trial Simulator, with other extensions, was released in 1999.

## 11.3 A SIMPLE CLINICAL TRIAL SIMULATION

Simulation can be used to teach basic principles of pharmacological experimental design and data analysis. These ideas are typically presented in basic statistics courses but often lack direct relevance to students studying clinical pharmacology. A 2 hour workshop has been used at the University of Auckland to introduce students to pharmacometric principles in their last year of a 3 year BSc degree course. The students use a Microsoft Excel workbook to simulate and analyse data.

### 11.3.1 Input-Output Model

Interpretation of pharmacological observations most commonly involves an experiment that is designed to show that a particular intervention has a detectable effect; e.g., does blood pressure fall when a new drug is given compared with giving a placebo?

#### 11.3.1.1 The Model

The population input-output (IO) model is simple (see Chapter 2). The blood pressure (BP) fall with active drug (drug) is predicted from the population parameter for active drug. $E_{drug}$. The placebo blood pressure fall is similarly predicted from the population parameter for placebo, $E_{placebo}$.

$$BP = E_{placebo} * (1 - TRT) + E_{drug} * TRT \quad (1)$$

The group IO model is also defined in Eq. (1) for the treatment assignment to active drug (TRT = 1) or placebo (TRT = 0) as a group covariate.

The individual IO model predicts the BP fall in each subject, $BP_i$:

$$BP_i = BP + \eta_i \quad (2)$$

where $\eta_i$ is the individual difference in blood pressure response predicted by a random effects model based on a normal distribution with variability defined by the parameter SD. This can be simulated using Microsoft Excel using the NORMINV ( ) function:

$$\eta_i = \text{NORMINV(RAND ( ), 0, SD)} \quad (3)$$

Equation (3) includes the observation IO model with random effects arising both from individual differences in response to treatment and to blood pressure measurement error. The execution model is implicit and assumes that all treatments are taken as assigned and that there are no missing observations.

### 11.3.2 Design

The only experimental design property is the number of subjects in each treatment group. The number of subjects can be varied to test how it influences the statistical

**Table 1** Example of Values Computed by Excel for the Simple Blood Pressure Lowering Clinical Trial Simulation[a]

| Parameter | True | Experiment | Property | Subjects | Placebo | Drug |
|---|---|---|---|---|---|---|
| Eplacebo | −5 | N | 10 | 1 | 15.6 | −11.6 |
| Edrug | −10 | | | 2 | 10.2 | −3.4 |
| SD | 10 | | | 3 | −11.4 | 2.2 |
| | | | | 4 | 15.7 | −8.2 |
| Placebo | Estimate | Drug | Estimate | 5 | −21.2 | −17.4 |
| AVGplacebo | −1.31 | AVGdrug | −11.91 | 6 | −5.4 | −14.0 |
| SDplacebo | 12.76 | SDdrug | 9.34 | 7 | 5.1 | −8.5 |
| Nplacebo | 10 | Ndrug | 10 | 8 | 0.6 | −9.2 |
| SEplacebo | 4.03 | SEdrug | 2.95 | 9 | −9.9 | −32.5 |
| | | | | 10 | −12.4 | −16.5 |
| Hypothesis Testing | | | | | | |
| Alpha | 0.05 | | | | | |
| P_t_test | 0.05 | | | | | |
| Reject Null | Yes! | | | | | |

[a] The student has chosen to have 10 subjects in each group. In this particular case the *t*-test has a *P* value exactly equal to the alpha critical value for rejecting the null hypothesis.

analysis (see Table 1). The worksheet automatic recalculation feature has been turned off so that the student must press the F9 key to generate a new set of simulated observations based on the number of subjects.

### 11.3.3 Analysis

The average blood pressure can be calculated after giving each treatment and there will inevitably be some difference in the averages. A hypothesis test is used to determine how likely the observed difference arose by chance (the null hypothesis). If the probability is low enough (less than or equal to $\alpha$) that the observed difference arose by chance, then the null hypothesis is rejected.

#### 11.3.3.1 Student's *t*-test

The average and standard deviation of the observed responses in each group is computed and Student's *t*-test used to compute a *P* value for comparison with a nominal $\alpha$ value to test the null hypothesis.

### 11.3.4 Meta-Analysis

Each time the worksheet is recalculated, it is equivalent to performing a new clinical trial. Different designs based on the number of subjects can be examined.

Assumptions about the size of the active drug and placebo response and its variability can be included by varying these simulation parameters.

#### 11.3.4.1 Power

Students are asked to count whether the null hypothesis has been rejected after each time they press the F9 key. After 100 key presses they use the count to estimate the power of the design.

## 11.4 PARAMETER ESTIMATION

The concepts of more complex IO models and the ability to estimate their parameters from typical data is taught by graphical and numerical methods. Graphical display of the true model predictions, simulated observations, and the predictions from a set of parameter estimates are used to demonstrate how a model can be fitted to data by guesses at parameter values and evaluating visually how close the estimation model predictions resemble both the true model predictions and the simulated observations.

### 11.4.1 Student vs. Solver

A numerical approach to evaluating goodness of fit is demonstrated by computing an objective function value based on the current set of parameter estimates. Students are asked to try to minimize the objective function by changing the parameter estimates. They are then asked to use the Microsoft Excel Solver add-in. The Solver is set to minimize the same objective function by varying model parameters. The lowest objective function value obtained by the student can be compared with the value obtained by the Solver. The residual unknown variability (RUV) for the objective function and for the observation model is a mixed proportional and additive model (see Chapter 2). Details of the RUV are not discussed with students at this stage.

#### 11.4.1.1 The Straight Line

The first model examined is a linear model with slope and intercept parameters.

$$Y = \text{intercept} + \text{slope} \cdot X \tag{4}$$

Students can often guess parameters that are very close to the true model parameters. They see that their simulated line and the true model line match but their objective function is worse than that obtained by the Solver. This introduces them to the idea that the objective function is based on the actual observations which have error and that the true model cannot be known by the Solver.

| Estimate | Pest | X | Y | Ytrue | Yobs | OLSi | Virlsi | IRLSi | Velsi | ELSi |
|---|---|---|---|---|---|---|---|---|---|---|
| | | 0 | 100.0 | 100.0 | 101.3 | 1.74 | 10000.00 | 0.00 | 10000.00 | 9.21 |
| V | 10.00 | 0.3 | 86.1 | 91.4 | 38.7 | 2246.02 | 7408.18 | 0.30 | 7408.18 | 9.21 |
| cl | 5.00 | 0.5 | 77.9 | 86.1 | 55.5 | 503.01 | 6065.31 | 0.08 | 6065.31 | 8.79 |
| pwr | 2.00 | 1 | 60.7 | 74.1 | 55.7 | 24.60 | 3678.79 | 0.01 | 3678.79 | 8.22 |
| | | [illegible] | [illegible] | 54.9 | 49.2 | 153.98 | 1353.35 | 0.11 | 1353.35 | 7.32 |
| | | | | 40.7 | 49.5 | 737.61 | 497.87 | 1.48 | 497.87 | 7.69 |
| | | | | 30.1 | 34.8 | 450.69 | 183.16 | 2.46 | 183.16 | 7.67 |
| | | | | 16.5 | 10.8 | 34.21 | 24.79 | 1.38 | 24.79 | 4.59 |
| | | | | 9.1 | 12.2 | 106.83 | 3.35 | 31.85 | 3.35 | 33.06 |
| | | | | 5.0 | 3.5 | 8.03 | 0.45 | 17.70 | 0.45 | 16.91 |
| | | | | 2.7 | 2.3 | 4.15 | 0.06 | 67.47 | 0.06 | 64.68 |
| | | | | 1.5 | 0.6 | 0.22 | 0.01 | 26.47 | 0.01 | 21.68 |
| | | | | 0.8 | 0.6 | 0.30 | 0.00 | 267.25 | 0.00 | 260.46 |
| | | | | 0.5 | 0.6 | 0.31 | 0.00 | 1998.94 | 0.00 | 1990.16 |
| | | | | 0.1 | 0.2 | 0.04 | 0.00 | 30993.98 | 0.00 | 30980.48 |
| | | | | | | | | | | |
| | | | | | | 4271.75 | | 33409.48 | | 33430.14 |
| | | | | | | 4271.75 | | 33409.48 | | 33430.14 |
| | | | | | | 1 | | 1 | | 1 |
| | | | | | | 100 | | 100 | | 100 |
| | | | | | | | | | | |
| | | | | | | | | 33395.703 | | 33416.36 |

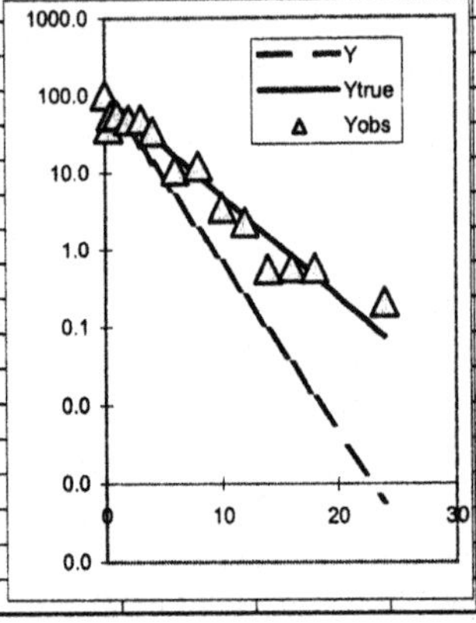

**FIGURE 1** One-compartment model objective function calculations.

### 11.4.1.2 One-Compartment Bolus Input

The second example is nonlinear and uses clearance and volume as parameters of a one-compartment disposition, bolus input pharmacokinetic model (see Figure 2).

$$C(t) = \frac{\text{dose}}{V} \cdot \exp\left(-\frac{\text{CL}}{V} \cdot t\right) \tag{5}$$

The students quickly appreciate how much harder it is to guess values of clearance and volume and learn that the Solver is a general-purpose technique for estimating parameters of arbitrary functions.

## 11.4.2 Objective Functions

A more advanced group of students taking graduate level courses use the same kind of Excel-based simulation models to learn about the properties of RUV models and different forms of least-squares objective function. Excel is used to calculate and display the contribution to the objective function from each observation so that the student can appreciate the consequences of different RUV model assumptions.

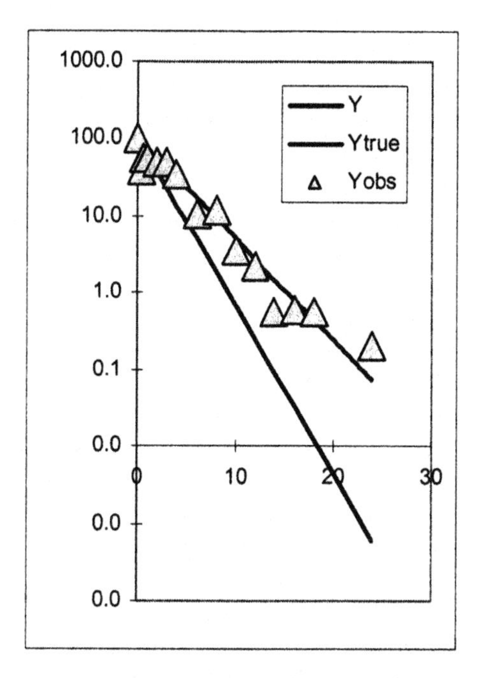

**FIGURE 2** One-compartment model true and estimated model predictions.

### 11.4.2.1 Ordinary Least Squares

The ordinary least squares objective function assumes a constant RUV for all observations. In the example in Eq. (6) the RUV is 1.

$$\text{OLS} = \sum_{i=1}^{i=\text{Nobs}} \frac{(Y_i - \hat{Y}_i)^2}{1} \tag{6}$$

### 11.4.2.2 Weighted Least Squares

The weighted-least-squares objective function relies on a fixed function of either the observed value (WLS) or the model predicted value (IRLS) to compute a weight which is inversely proportional to the RUV for that observation ($\text{Var}_i$).

$$\text{WLS} = \sum_{i=1}^{i=\text{Nobs}} \frac{(Y_i - \hat{Y}_i)^2}{\text{Var}_i}$$
$$\text{Var}_i = \frac{1}{W_i} = \frac{1}{f(Y_i)}\,(\text{WLS}) \qquad \text{or } \frac{1}{f(\hat{Y}_i)}\,(\text{IRLS}) \tag{7}$$

#### 11.4.2.3 Extended Least Squares

The extended least squares objective function uses a parametric model for RUV ($\mathrm{Var}_i$) which is a function of the IO model prediction and parameters of the RUV model such as SD.

$$\mathrm{ELS} = \sum_{i=1}^{i=\mathrm{Nobs}} \left[ \frac{(Y_i - \hat{Y}_i)^2}{\mathrm{Var}_i} + \ln(\mathrm{Var}_i) \right]$$
$$\mathrm{Var}_i = f(\hat{Y}_i, \mathrm{SD}, \ldots) \tag{8}$$

## 11.5 MODEL BUILDING

The graduate class is taught how to build models for simulation and parameter estimation using a range of software representative of the tools currently used in contemporary drug development (http://www.health.auckland.ac.nz/courses/Pharmacol716).

### 11.5.1 Excel

Microsoft Excel (http://www.microsoft.com) is the first simulation tool that is taught. Most students are familiar with basic spreadsheet operations and are quickly able to simulate IO models and display them as graphs. They are taught how to name cells and ranges so that formulas can be written in symbolic form rather than rely on cell references.

Observation IO models are simulated using the NORMINV( ) function. A proportional RUV model with a coefficient of variation approximately equal to the parameter CV is shown in Eq. (9).

$$\mathrm{Cobs}(t) = \mathrm{C}(t) \cdot \exp(\mathrm{NORMINV}(\mathrm{RAND}(\,), 0, \mathrm{CV})) \tag{9}$$

Poptools is an Excel add-in that includes (among many other statistical and modeling features) the ability to specify models as differential equations (http://sunsite.univie.ac.at/Spreadsite/poptools/).

### 11.5.2 ModelMaker

The ModelMaker program developed by Cherwell Scientific (http://www.modelkinetix.com) provides a simple graphical interface for building simulation models expressed as differential equations or closed-form solutions. It is particularly useful for demonstrating how to build complex models from simpler components, e.g., linking pharmacokinetic models to pharmacodynamic models.

### 11.5.3 WinNonLin

Parameter estimation and model building ideas are introduced using WinNonLin (http://www.pharsight.com). Although WinNonLin has an extensive library of common pharmacokinetic and pharmacodynamic models, students are taught to write their own user-defined models. They use observations simulated with Excel as the source of data and try different models to describe the observations. By learning how to write a model as an Excel formula, as a ModelMaker equation and a WinNonLin function they learn the general properties of a model, and are exposed to different dialects of expressing mathematical expressions.

### 11.5.4 NONMEM

Population approaches to parameter estimation and model building are taught using NONMEM (http://c255.ucsf.edu/nonmem0.html) and Wings for NONMEM (http://wfn.sourceforge.net). Students are shown how to construct a simple pharmacodynamic model and to discover influential covariates using a real set of concentration and effect data collected from a concentration-controlled trial of theophylline (15, 16). The main features of using NONMEM are learned after a 2 h workshop coupled with an assignment which students complete unsupervised taking an additional 6 h.

### 11.5.5 Trial Simulator

A brief introduction to a full clinical trial simulation project is taught by using Trial Simulator (http://www.pharsight.com). Students are asked to work through the Quick Tour introduction and then try to simulate a clinical trial based on the theophylline concentration-controlled trial they analyzed using NONMEM.

## 11.6 ANALOGUE PHARMACOKINETIC SIMULATION

The teaching materials described so far have all relied upon computer based simulation methods. One of the most rewarding and insightful clinical trial simulations that is undertaken by undergraduate and graduate classes involves an analogue simulation of a compartmental pharmacokinetic model.

### 11.6.1 Model

The one compartment model involves a single glass beaker with water flowing through it at constant rate. A red dye is added to the beaker and samples of water flowing out of the beaker used to measure the dye concentration. Dye concentration is measured by spectrophotometry. Graduate students use an extension to

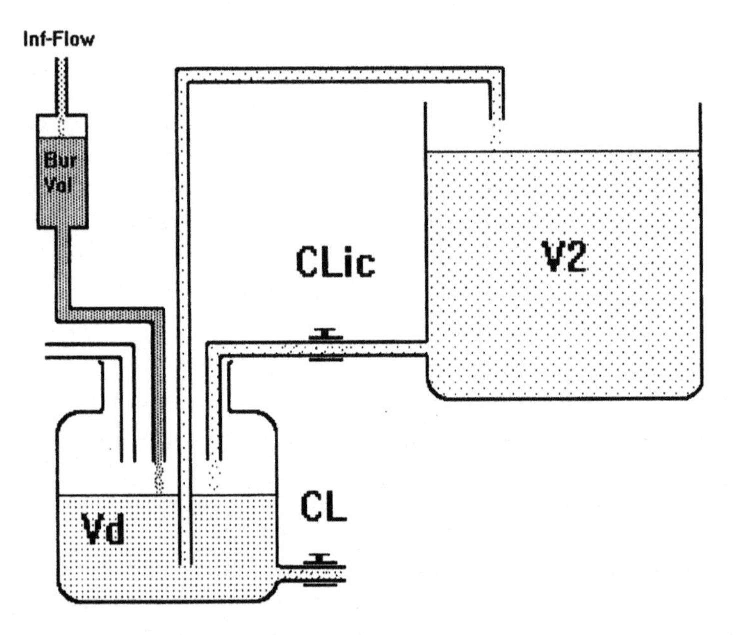


**Figure 3** Two-compartment analogue model.

the system with a bucket connected to the beaker by a recirculating pump system. The beaker and bucket system simulates a two-compartment pharmacokinetic model

## 11.6.2 Experiments

The input of dye to the beaker can mimic a bolus or a zero-order constant rate infusion or a combination of both. The time course of resulting concentrations is used to understand the relationship between the volume of water in the beaker (volume of distribution) and the flow rate of water out of the beaker (clearance). For the two-compartment model the bucket represents the tissue compartment and the recirculating pump flow is intercompartmental clearance (Figure 3).

## 11.6.3 Analysis

The calibration and measurement of dye samples using the spectrophotometer is used to teach the concepts of using a standard curve and to describe assay performance in terms of the bias and precision of replicate measurements of standard concentrations.

The time course of dye concentration in both the beaker and bucket can be observed and a pharmacokinetic model for central and tissue compartments used to fit the data. By combining a bolus dose of dye with a constant rate input from a standard hospital drip set the difficulty of maintaining a constant target concentration is demonstrated. It is then possible to show how the same apparatus can be use to administer an additional first-order input in order to achieve a constant concentration.

## 11.7 CLINICAL DRUG DEVELOPMENT WORKSHOP

A 4–5 day workshop using clinical trial simulation is used at Uppsala University to teach PK, PK/PD, and aspects of clinical drug development to student groups with varying backgrounds. In the following, the principal components of the workshop are presented, thereafter some variations depending on the level of the student group, and finally some general comments about the intentions and experiences with the workshop.

### 11.7.1 Workshop Outline

In the workshop, the students design and analyze clinical studies within a fictional drug development program. The aim is to bring the investigational drug from the first-in-human study until the end of phase III. The drug selected for this exercise is a thrombin inhibitor, where the students are provided with two articles (17, 18) describing the preclinical characteristics of the thrombin inhibitor inogatran, which was previously under development by AstraZeneca. The intended indication is prevention of thrombosis in patients with acute myocardial infarction. A full PK/PD model has been created for the drug and main metabolite pharmacokinetics in plasma and urine, a biomarker (activated partial thromboplastin time—APTT) and clinical endpoints, both the primary clinical endpoint (presence/absence of thrombosis during the first week following hospitalization) and the primary toxicity endpoint (presence/absence of bleeding incidents). In the model, APTT increases nonlinearly with drug plasma concentration, thrombosis hazard decreases with increasing APTT, whereas bleeding hazard increases with increasing APTT values. The model is loosely based on the clinical profile of inogatran and other thrombin inhibitors. It describes the typical subject behavior and variability components, including covariate relations, both for healthy volunteers and patients. The model is not disclosed to the students, but is used to generate study results based on the designs decided upon by the students. Initially, the model was implemented with NONMEM, later with the Pharsight Trial Designer, and presently the Pharsight Trial Simulator is used.

In addition to the preclinical information, the students are initially provided with a fictional project team discussion, where the team members (project leader, clinician, bioanalytical chemist, pharmaceutic pharmacist, etc.) are outlining the expectations with respect to benefit and toxicity, the relative seriousness of thrombosis versus bleeding, competitor profile with respect to thrombosis and bleeding frequencies, desired dosing schedule (24 h infusion followed by daily intramuscular doses), formulation characteristics (intramuscular release rate in vitro and in animals), bioassay characteristics, etc. The first task for each of the students is to summarize all the preclinical information relevant to the design of the first-in-human studies (one for the intravenous and one for the intramuscular formulation) in an abstract. Thereafter all work is carried out in groups of 6–8 students, starting with the design of the first-in-human studies. The students are expected to specify the design with respect to study population (sex, age range, etc.), number of subjects, doses, escalation aspects including stopping rules, what to measure (drug and/or metabolite concentration in plasma, plasma water, urine and/or faeces, and APTT), number and timing of samples, etc. They outline what information is sought and how the data will be analyzed to obtain this information. The design is finalized during a consensus discussion among the groups. The discussion is led by one of the groups, which provides the teacher with all the aspects necessary to generate the study data, based on the population PK/PD model. The teacher then uses the consensus design to simulate the outcome of a clinical trial. The following day each group analyzes the generated data using Excel, and the estimated PK (noncompartmental, standard two-stage analyses) and PD (various regression models using pooled data) characteristics are summarized after yet another consensus discussion.

The work then progresses sequentially in a similar manner with the design and analysis of a phase II dose finding study (active control optional) and the design and analysis of two phase III studies (active control necessary). The information generated should result in a suggested dosing regimen (individualized by, e.g., weight or renal function, or not) and knowledge about strengths and weaknesses relative to the active control (heparin). Based on the information generated, each group is finally assigned to make a presentation. Examples of presentations are: a suggested text for the PK and PD part of the product label, a summary of the PK characteristics for regulatory pharmacokineticists, or a summary of the pharmacodynamic characteristics for a group of clinicians working in the therapeutic area.

During the workshop a lecture is given on regulatory considerations when reviewing pharmacokinetic documentation. A lecture is also given by a pharmacokineticist with experience from the clinical development of thrombin inhibitors. In association with these lectures, the role of other clinical studies (e.g., drug interaction and bioequivalence studies) that could not be incorporated in this very condensed development is discussed. (Course material is available from mats.karlsson@farmbio.uu.se)

### 11.7.2 Workshop Variations Depending on Students' Previous Knowledge

The workshop was initially created as the PK/PD part of a 20 week full-time course in clinical drug development for students with a previous academic degree (usually in medicine, pharmacy, toxicology, biomedicine, or nursing). As these students, with exception of the pharmacists, have no or little previous training in PK and PK/PD, short lectures on these topics were included in the course. To make group discussions productive, teacher involvement in this course needs to be at a relatively high level, with approximately one teacher per three groups being continuously occupied.

The workshop has mainly been used for fourth-year pharmacy students who previously had a 6 week course in pharmacokinetics. In this setting, teacher involvement in group discussions is usually low. In this course, this workshop is followed by another one, where the students are trained in the principles of clinical trial simulation and the use of the Pharsight Trial Simulator. They are asked, in groups of two or three, to design a clinical PK or PK/PD study of their choice with the restriction that it has to involve real, rather than hypothetical, drugs and diseases and that the study should not have been performed and reported in the literature. Examples of studies that students have chosen to design are drug or food interaction studies, special population studies, and bioavailability/bioequivalence studies. Their choices of study and design are then presented and discussed during a seminar. Thereafter, the students collect information about PK and PK/PD characteristics that allow them to implement a best-guess PK or PK/PD model using the Pharsight Trial Simulator. The resulting data from a single simulation using the chosen design and the best-guess model are then analyzed as decided beforehand. Should the analysis indicate a severe shortcoming of the design, the students have the opportunity to improve the design. The model and result of the study is presented orally at a seminar and individually in writing. This extension to the workshop involves an additional 2 weeks of full-time studies during which students receive lectures on various aspects of clinical drug development.

The workshop has been used for civil engineering students specializing in drug industry and who have had courses in physiology, pharmacology, pharmaceutics, and pharmacokinetics. In the analysis of PK and PD data, these students have, in contrast to the pharmacy students, used nonlinear regression as can be implemented using the Solver function in Excel.

A course in clinical trial simulation and clinical drug development for postgraduate students used the workshop with the addition that the students also were trained in principles of clinical trial simulation and the use of clinical trial simulation software, where the structure of the underlying model was revealed to the students for them to implement and use in the optimization of their study designs.

### 11.7.3 Workshop Objectives and Experience

The workshop is designed to provide the students with training in several different aspects. These are:

1. *Basic PK principles.* In particular for the student groups with no or little previous training in this area, the concepts of clearance, volume of distribution, half-life, bioavailability, and absorption rate and how these are calculated and used are the main topics. For students at higher levels, this workshop serves as an update on these issues.
2. *Basic PD principles.* For all student groups, the concept and potential use of a biomarker, as well as basic pharmacological models, are discussed. The search for the appropriate balance between beneficial effect and toxicity, the utility, in drug development is illustrated.
3. *Study design.* The students are trained in designing informative studies and consider practical, economical, and ethical aspects in their designs. The necessity for them to prespecify how the data are to be analyzed provide insight in the interaction between design and method of analysis. The specific characteristics of first-in-human dose finding and phase III studies are covered.
4. *Clinical drug development principles.* Although the development program is rudimentary, it provides an appreciation of the sequential nature of a program and the nature of the PK and PD information sought in each phase and for what purpose. The students are required to contemplate how to use preclinical information in the design of clinical studies.
5. *Statistics.* Apart from basic descriptive statistics, the students are trained in sample size calculation in the design of patient studies and in exploratory data analysis. The latter in particular from the dose finding study where the relations between outcomes and covariates, including concentration and biomarker, are explored. Also, relations between clearance and covariates are investigated. Various techniques of performing exploratory data analysis are discussed.
6. *Clinical trial simulation.* Although in the main workshop students are not asked to optimize their design through the use of clinical trial simulation, discussions around the weaknesses of their selected designs and how they can be detected provide an insight into the usefulness of

clinical trial simulation. A short presentation on the principles of clinical trial simulation is also provided to all students. For fourth-year pharmacy students, a more comprehensive course in clinical trial simulation is given.

7. *General skills*. The students get experience applying their PK and PD knowledge in discussions and decision making. Also, the consensus discussions allow them training in leading a discussion. The final presentations make them familiar with presenting PK and PD information in a manner that suits the audience.

Teacher involvement in study design discussions is to make sure, if necessary, that unethical, entirely uninformative, or unpractical designs are avoided. Designs are generally suboptimal in one or several aspects, but allow sufficient information to be collected to proceed to the next phase. For the development to be considered successful, it is required that the investigational drug in phase III is significantly ($p < 0.05$) superior to heparin in either beneficial effect or toxicity, while not being inferior in either aspect. The drug characteristics in the simulation model are selected such that in a narrow dose range, the investigational drug is superior to heparin in both aspects. Thus, if the students have managed to find a dose within this range and perform phase III studies with sufficient sample sizes, superiority in both aspects may result. However, more commonly, significant superiority in only one of the aspects is found.

## REFERENCES

1. NHG Holford, HC Kimko, JPR Monteleone, CC Peck. Simulation of clinical trials. Annu Rev Pharmacol Toxicol 40:209–234, 2000.
2. C Maxwell, JG Domenet, CRR Joyce. Instant experience in clinical trials: a novel aid to teaching by simulation. J Clin Pharmacol 11(5):323–331, 1971.
3. SJ Pocock, R Simon. Sequential treatment assignment with balancing for prognostic factors in the controlled clinical trial. Biometrics 31(1):103–15, 1975.
4. DR Jones. Computer simulation as a tool for clinical trial design. International Journal of Bio-Medical Computing 10:145–150, 1979.
5. DL DeMets. Practical aspects of decision making in clinical trials: the coronary drug project as a case study. The Coronary Drug Project Research Group. Controlled Clin Trials 1981; 1(4):363–376.
6. NHG Holford, LB Sheiner. Understanding the dose-effect relationship: clinical application of pharmacokinetic-pharmacodynamic models. Clin Pharmacokin 6:429–453, 1981.
7. WA Colburn. Simultaneous pharmacokinetic and pharmacodynamic modeling. J Pharmacokin Biopharm 9(3):367–387, 1981.
8. LB Sheiner, TH Grasela. Experience with NONMEM: analysis of routine phenytoin clinical pharmacokinetic data. Drug Metabol Rev 15(1–2):293–303, 1984.

9. TH Grasela, LB Sheiner. Population pharmacokinetics of procainamide from routine clinical data. Clin Pharmacokin 9(6):545–554, 1984.
10. DA Noe, WR Bell. A kinetic analysis of fibrinogenolysis during plasminogen activator therapy. Clin Pharmacol Thera 41(3):297–303, 1987.
11. AJ Tiefenbrunn, RA Graor, AK Robison, FV Lucas, A Hotchkiss, BE Sobel. Pharmacodynamics of tissue-type plasminogen activator characterized by computer-assisted simulation. Circulation 73(6):1291–1299, 1986.
12. LP Sanathanan, C Peck, Temple R, Lieberman R, Pledger G. Randomization, PK-controlled dosing, and titration: An integrated approach for designing clinical trials. Drug Inf J 25:425–431, 1991.
13. L Endrenyi, J Zha. Comparative efficiencies of randomized concentration- and dose-controlled clinical trials. Clin Pharmacol and Thera 56:331–338, 1994.
14. R Amstein, JL Steimer, NHG Holford, TW Guentert, A Racine, D Gasser, et al. RIDO: Multimedia CD-Rom software for training in drug development via PK/PD principles and simulation of clinical trials. Pharm Res 13:S452, 1996.
15. NHG Holford, Y Hashimoto, LB Sheiner. Time and theophylline concentration help explain the recovery of peak flow following acute airways obstruction. Clin Pharmacokin 25(6):506–515, 1993.
16. NHG Holford, P Black, R Couch, J Kennedy, R Briant. Theophylline target concentration in severe airways obstruction—10 or 20 mg/L? Clin Pharmacokin 25(6): 495–505, 1993.
17. AC Teger-Nilsson, R Bylund, D Gustafsson, E Gyzander, U Eriksson. In vitro effects of inogatran, a selective low molecular weight thrombin inhibito. Thrombosis Res 85(2):133–145, 1997.
18. UG Eriksson, L Renberg, U Bredberg, AC Teger-Nilsson, CG Regardh. Animal pharmacokinetics of inogatran, a low-molecular-weight thrombin inhibitor with potential use as an antithrombotic drug. Biopharm Drug Dispos 19(1):55–64, 1998.

# 12

# Modeling and Simulation of Clinical Trials

## An Industry Perspective

**Timothy Goggin,* Ronald Gieschke, Goonaseelan (Colin) Pillai,† Bärbel Fotteler, Paul Jordan, and Jean-Louis Steimer†**
F. Hoffmann-La Roche, Basel, Switzerland

### 12.1 INTRODUCTION

Current trends within the pharmaceutical industry and within the offices of some regulatory authorities [in particular the U.S. Food and Drug Administration (FDA)] suggest a promising future for modeling and simulation (M&S)—incorporating pharmacokinetic and pharmacodynamic (PK/PD) modeling and clinical trial simulation (CTS) (1–5). The methodology has the potential to become an indispensable tool in pre-/nonclinical development, in early and late clinical development, and in the postmarketing phases. As such it will provide the scientific evidence for key decision points during the development of clinical entities.

Modeling and simulation has a strong mathematical foundation and has been used extensively and for a long time in areas other than the pharmaceutical industry (e.g., automobile and aerospace industries) to design and develop products more efficiently and safely (6, 7). Both modeling and simulation rely on the

* *Current affiliation*: Serono International S.A., Geneva, Switzerland.
† *Current affiliation*: Novartis Pharma AG, Basel, Switzerland.

use of mathematical and statistical models that are essentially simplified descriptions of complex systems under investigation. Models that are used in the context of PK/PD are named pharmacostatistical models.

Clinical trial simulation is a process for combining premises/assumptions to generate a large number of replications of virtual (representations of actual) clinical trials. Each replicate of a computer simulated trial is viewed as one realization from a probability distribution of all possible trial outcomes with the same design properties (system parameters). The emphasis is on forecasting expected trial outcomes, exploring the sensitivity of the outcome to system parameters and covariates, and understanding the impact of premise and model on the trial outcome.

In this chapter we present our (sometimes-biased) perspective on CTS—including the essential prerequisite to CTS viz. PK/PD modeling. We maintain a philosophical approach but provide appropriate references where applicable. In keeping with our clinical bias, we admit to paying less attention in this chapter to preclinical and postapproval applications of M&S.

## 12.2 CLINICAL TRIAL SIMULATION—A SWISS ANALOGY

Communication of the methods involved with new and emerging technologies is often difficult. In a recent internally published article, the Roche Modeling and Simulation team chose to draw an analogy between M&S and skiing in the Alps:

*The preparation*

- Finding the appropriate location with pleasant mountain scenery is like identifying the clinical project, developing the mechanism-based PK/PD hypothesis and identifying the data with which to build the model.
- Obtaining the approval of the local community to commence construction is like liaising with the clinical team and subject specialists.

*The hard work*

- Building the ski lifts and all the other necessary infrastructure is like building the pharmacostatistical model.

*The reward*

- Skiing down the primary slope is like using the model to simulate (many times) the "base design" study.

*The benefit*

- Finally skiing down all the other slopes to find the optimal way (according to one's personal objectives and abilities) is like changing the base design and performing other simulations in order to optimize the study design with regard to the likely information content and success as predefined.

*The risks*

- The busy orthopedic experts in Switzerland will verify the risks associated with skiing in the Alps. In this chapter, we attempt to identify and highlight the conditions necessary to reduce the risks associated with M&S.

It will be apparent from this analogy that the clinical trial simulations themselves are the "easy" bit, represented above by the reward and the benefit. The preparation and in particular the hard work are, however, key to the success of the entire undertaking, and will therefore also be considered in this chapter.

During discussions with our "nonmodeler" drug development colleagues, we also found it useful to present the following distinction between modeling and simulation:

- The *model* is a mathematical construct to mimic reality, built based on data, from the given drug and "similar" compounds and the current scientific understanding of the processes involved. *It looks backward in time!*
- The *simulation* involves use of the model (incorporating random variability) to play various "what-if scenarios" in order to determine the best one. *It looks forward in time!*

## 12.3 MODELING AND SIMULATION IN DRUG DEVELOPMENT

In an idealized future paradigm, M&S can play an increasingly important role in guiding drug development. In Figure 1, the role for M&S methodologies and the corresponding activities at key decision points in drug development are outlined.

### 12.3.1 Pre-/Nonclinical Development

Methods for predicting human PK characteristics based on preclinical data are important in identifying the best candidates for further development. In an environment of high throughput molecule identification, it is desirable to have methods that can rapidly and reliably predict respective human PK characteristics. Data including lipophilicity, ionization, solubility, ex vivo protein binding, metabolic stability in liver preparations, and membrane permeability are used as input data and prior knowledge to predict characteristics such as fraction of dose ab-

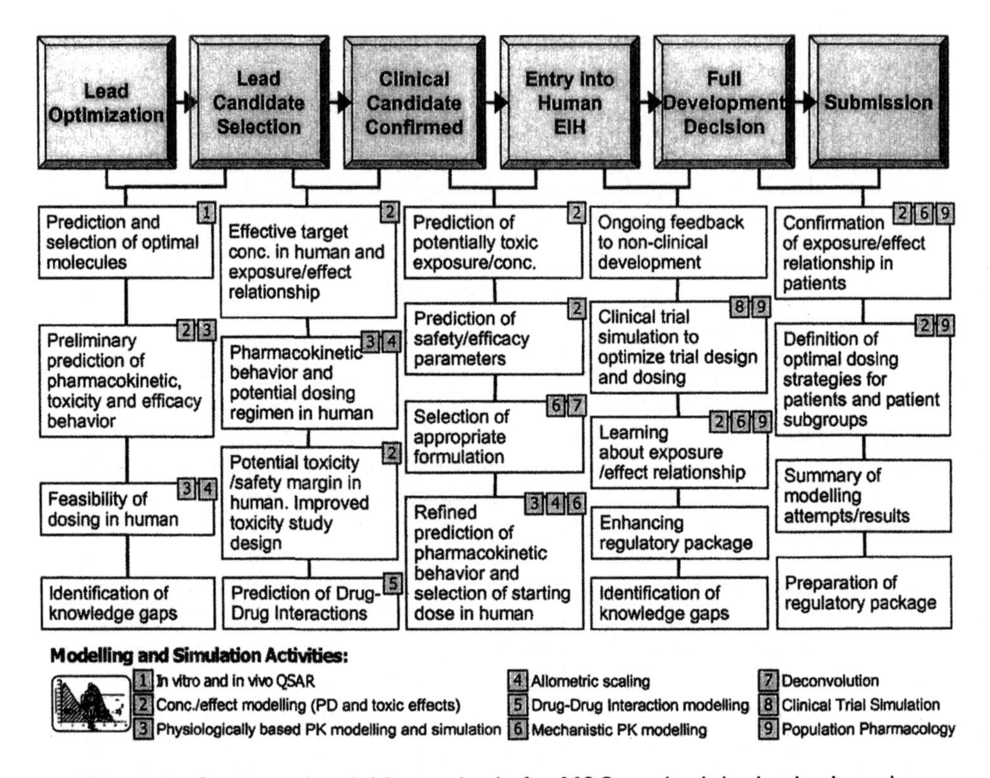


FIGURE 1 Suggested activities and role for M&S methodologies in drug development (with acknowledgment to ideas developed during an internal F. Hoffmann-La Roche workshop on M&S, 1–2 March 2001, Basel, Switzerland).

sorbed, fraction metabolized, and extent of distribution. These characteristics are subsequently used to predict the time course of the clinical candidate's plasma and tissue concentration profiles.

Physiologically based pharmacokinetic models (PBPK) are increasingly being used in preclinical departments of the pharmaceutical industry (8–15). Compared to (standard) compartmental models, PBPK models aim at a more physiologically oriented description of the system under investigation with obvious advantages regarding the interpretation of results. While the main aim of applying these modeling techniques is to guide and optimize clinical candidate selection, it can also provide the basis for further M&S work during the compound's clinical development. However, given the internal organizational structure of many pharmaceutical companies (i.e., the separation of pre-/nonclinical research from clinical development), this is currently an ideal rather than a reality.

### 12.3.2 Early Clinical Development

In the pharmaceutical industry the earliest efforts in learning and extracting knowledge using PK/PD modeling was probably conducted by clinical pharmacologists and/or pharmacokineticists using data from phase I studies. This activity has a long history, probably more than a quarter of a century, and the clinical pharmacology-pharmacokinetic literature provides many examples of its application in early clinical development (3, 16). Naïve pooled modeling or individual subject modeling, using linear and nonlinear regression techniques on extensive PK data, were the methods predominantly applied, and software such as SAAM and Metzler's NONLIN was employed (17–20). Unlike the situation with noncompartmental analysis, the activity was never institutionalized and often PK/PD modeling depended on the individual lonely efforts of a multitalented scientist with the interest to perform such work. Even when performed to a high standard, this work often had 'little' impact on the overall development process and was generally regarded as a "nice to have" rather than a "must have" part of the drug's development program. Fortunately, as a result of initiatives from within the industry, from academia and from the regulatory bodies, this situation has changed and continues to develop.

Applications of M&S techniques and a strategy for the development of drugs based on a target concentration strategy have been published (21). The benefits of PK/PD principles applied to drug development have been positively evaluated by industrial and academic scientists (1, 22). All too often, however, the industry as a whole viewed early phase I work as limited to the determination of the maximum tolerated dose, although there has been repeated evidence and encouragement (3, 23) to extract additional information from the data. There is a unique opportunity presented during phase I of exploring dose levels that one is unlikely to be allowed to test again. As a consequence, the pharmacological knowledge gained by the individual clinical pharmacologist-pharmacokineticist during phase I was often lost once therapeutic studies were started. In addition, in the experience of the authors, there was a tendency to move clinical candidates with suboptimal pharmacologic characteristics into phases II and III, due to a lack of understanding of the consequences of the M&S efforts undertaken using the phase I data. From a business perspective this tendency had an adverse effect and most pharmaceutical companies have their legacy of unnecessary and potentially avoidable late phase failures (24).

However, the picture is not entirely bleak. In early 2000, an expert meeting of drug development scientists and academics was convened in Europe to discuss the role, main advantages, and obstacles to rational use of M&S in early drug development (25). In the course of making suggestions for practical implementation—some of which are explored further in this chapter—the experts expressed optimism that the need for data driven modeling will increase in clinical development.

### 12.3.3 Late Clinical Development

With the advent of "therapeutic drug monitoring" in the early 1980s regulatory scientists in the FDA and other agencies became aware of the lack of PK and exposure information in the target patient population included in the registration packages of new chemical entities. At about the same time, during the 1970s, there were advances in the application of estimation methods including maximum likelihood and particularly in the area of nonlinear mixed effects modeling. The release of NONMEM Version 1 in 1980 (26) was a major milestone. Numerous applications brought the demonstration of its utility in the analysis of PK/PD data, especially sparse data as collected in patients. This has been repeatedly documented along the years in a variety of survey papers and chapters in textbooks (27–35). The area became known as "population pharmacokinetics" or "the population approach." There was an ongoing spread of knowledge on nonlinear mixed effects modeling in academia, the pharmaceutical industry, and the regulatory agencies. This resulted in a steady increase in NONMEM usage and publications on population pharmacokinetics, with an approximate doubling of the number every 5 years (Figure 2).

During the routine use of some low therapeutic index drugs such as the anticonvulsants, digoxin, warfarin, and theophylline, the clinical value of PK-based "therapeutic drug monitoring" could be established (36). However, in many cases it was clear that accounting for pharmacokinetic variability alone was insufficient to explain the absence or presence of a therapeutic or adverse response. Development of analytical methods to measure the concentrations of drugs in body fluids with a high sensitivity, precision, and accuracy, at relatively low cost and with a high throughput, lead to increased collection of PK data during clinical development, and also in patients. The attempt to formally approach dose and dosage regimen determination through proper linkage of exposure and effect(s) culminated in the concept of "a target concentration strategy" for drug development (21) with a "marketing" objective of PK/PD principles toward the medical community.

These developments led to a flurry of activity in the first half of the 1990s with the formation of scientific forums dedicated to the population approach for analyzing PK/PD data, such as PAGE (Population Approach Group Europe), ECPAG (East Coast Population Approach Group) and MUFPADA (Midwest

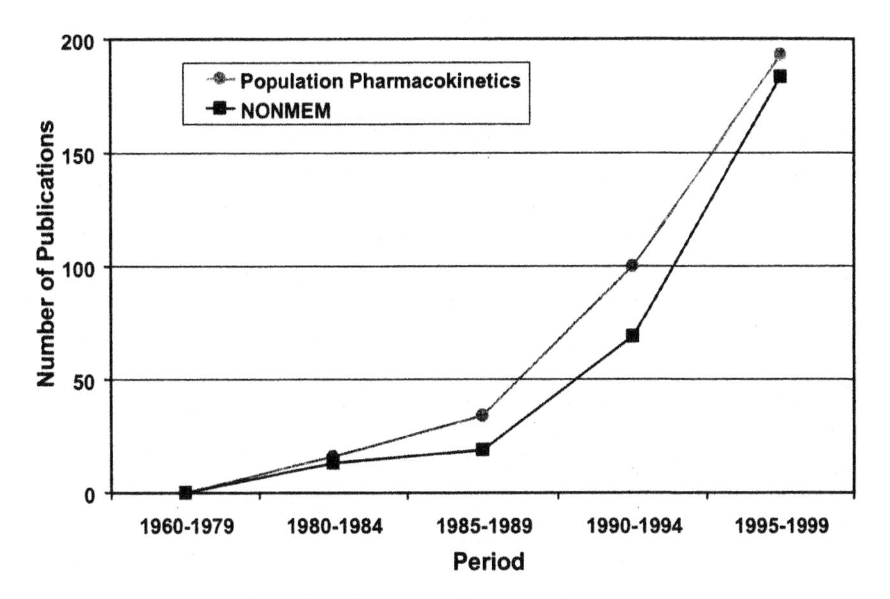


**FIGURE 2** Trends in population pharmacokinetics as reflected in the scientific literature: results of a Medline search using the keywords *Population Pharmacokinetics* and *NONMEM.*

Users Forum for Population Approaches in Data Analysis), with annual meetings attracting an increasing number of (pharmacological) scientists from around 50 to above 200 participants (37, 38). There was further "regulatory" interest and this led to a buildup of knowledge in the industry. However, there were still very few new drug applications (NDAs) supported by population PK data and their model-based analysis.

In the second half of the 1990s most large companies recognized the need to develop a core competence in population PK modeling. In general, it was regarded as having a supportive role rather than a leading role in the drug development process. Nonetheless, the use of population PK increased in phase II and many more NDAs included key labeling statements based on such analyses (39, 40). Recognition of the importance of this methodology to the labeling of new drugs was evident in the FDA's population PK guideline published in 1999 (41), and in addition there was increased regulatory demand for population PK, and increasingly, PK/PD information. Other regulatory agencies followed the FDA's lead, e.g., Australia (42) and Japan (43, 44).

However, global pharmaceutical companies face the vexing challenge of varying awareness and receptivity of regulators in the different countries, especially when "novel" methodologies are employed in support of the experimental evidence. While the International Conference on Harmonisation attempts to resolve these, discrepancies still exist in their own guidelines; e.g., ICH-E4 (45) mentions modeling as one analysis technique for establishing dose response (in its section 4.6), while ICH-E9 (46), although published later, does not mention modeling and simulation of clinical trials as one useful technique for clinical trial design and data analysis.

Currently, in the pharmaceutical industry, the application of M&S is still largely retrospective—often in an attempt to rescue failed trials or to respond to health authority requests without performing additional studies. Anecdotal reports indicate that in 2000, across the pharmaceutical industry, 50% of failures in clinical development have occurred in phase III. Application of M&S methodologies as a rescue operation has a high impact value, especially if successful, but does little to entrench the methodology in the mainstream of drug development.

### 12.3.4 The Learn-Confirm Paradigm

Drug development is increasingly being viewed according to the "learn-confirm" paradigm of Sheiner (47), i.e., an information-gathering process that can be thought of as two successive learn-confirm cycles. The first cycle (traditional phase 1 and phase 2a) addresses the question of whether benefit over existing therapies (in terms of efficacy/safety) can reasonably be expected. It involves learning (phase 1) what is the largest short-term dose that can be administered to humans without causing harm, and then testing (phase 2a) whether that dose induces some measurable benefit in the target patient population. An affirmative answer at this first stage provides the justification for the second cycle. This next cycle (traditional phase 2b and phase 3) attempts to learn (phase 2b) what is a good (if not optimal) drug regimen to achieve useful clinical value (acceptable benefit/risk) and ends with one or more formal confirmatory clinical (phase 3) trials of that regimen versus a comparator (47).

The learn-and-confirm approach represents a completely new paradigm in drug development and profound change in the pharmaceutical industry would be required in order to fully implement this approach. Inherent in the learn-confirm paradigm is the understanding that the design, planning, and conduct of the actual clinical trials for learning is paramount. Modelers should significantly influence those early learning studies. In this way, they can obtain high-quality data to build a reliable model of "the system" (a "drug model") and simulate the later "confirmatory" studies before they are actually conducted.

In the second of their reports on the prospects for the pharmaceutical industry in 2005 (Pharma 2005 Silicon Rally: The Race to e-R&D), Pricewaterhouse-Coopers suggested that M&S technology would be a key element in the future success of the industry as a whole (24).

### 12.3.5 Perceptions of M&S by the Nonmodelers in the Industry

"Clinical trial simulation" is becoming a new buzz-phrase in the pharmaceutical industry. It is perceived by many to be one key to the success of drug development in the future. It has been made possible by the advent of "mechanism-based" population PK/PD modeling, advances in computer power and speed, and the development of purpose built software. Clinical trial (model-based computer) simulation represents a radically different approach compared with the current empiricism used in the industry, particularly in phase II and beyond.

The model-based computer simulation of clinical trials is actually an extension of the methods and applications referred to in the preceding sections of this chapter. However, few scientists/managers within the drug industry—including the majority of people assigned to drug development project teams, their line management, as well as most senior managers in the pharma-industry today—recognize the power of the methodology.

Luckily, the core competence in PK/PD modeling and the population approach, which was built over the past decade, means that there are individuals in all large companies, as well as in the regulatory agencies, who not only recognize its potential but are in a position to take advantage of the improvements in hardware and software and to apply the methodology now. A number of examples of the application of clinical trial simulation and PK/PD modeling to drug development have been published by scientists from the drug industry (48–50), and by the regulatory authorities in support of drug registration and labeling decisions (2). Although these have been broadly reviewed in the literature (3, 5, 51), in many instances key decision makers within the industry still search for (more) evidence of the utility.

The pharmaceutical industry has been very slow in accepting computer simulation technology. The randomized controlled clinical trial is the last great innovation drug development has adopted and provides the necessary rigor to allow investigation of the question, "Does the drug work?" The clinical experts, in collaboration with the biometricians, provide the scientific evidence of safety and efficacy and if all goes well the drug industry prospers as a result. Often, however, the answer to the question is "The drug does not work 'well enough,' " and in these instances the industry's investment is lost. Modeling and simulation is not an alternative to the randomized controlled trial, but it can help ensure that drug development scientists have harvested the knowledge of the drug in such a way as to provide a tentative answer to the question, "How *well* does the drug work?" before performing the clinical trial.

The medical community by training and experience is uncomfortable with M&S. It is very hard to convince the doctor that his or her patient can be "modeled." But interactions of drug and patient beginning with a "dose" in a given regimen and resulting in a "response" pattern with quantifiable variability, is a "system" like any other and it can be modeled (at a given degree of complexity depending on the question at hand). Statisticians are the group in the pharmaceutical industry that is strongest in the application of quantitative methods. They have performed stochastic computer simulation based on statistical models (for instance, for testing the performance of statistical planning or analysis methodologies on fictive data) longer than the pharmacologists have been using PK/PD models. However, the interaction between physicians, pharmacologists, and statisticians in the pharmaceutical industry has not yet evolved to include the application of systems science to the "dose regimen–response pattern" relationship (Figure 3). This is, however, necessary for full clinical trial simulations to be done, to a degree of sophistication that goes beyond the sample size determination for a given clinical trial based on one given primary outcome.

It has been acknowledged that there is a shortage of scientists who are

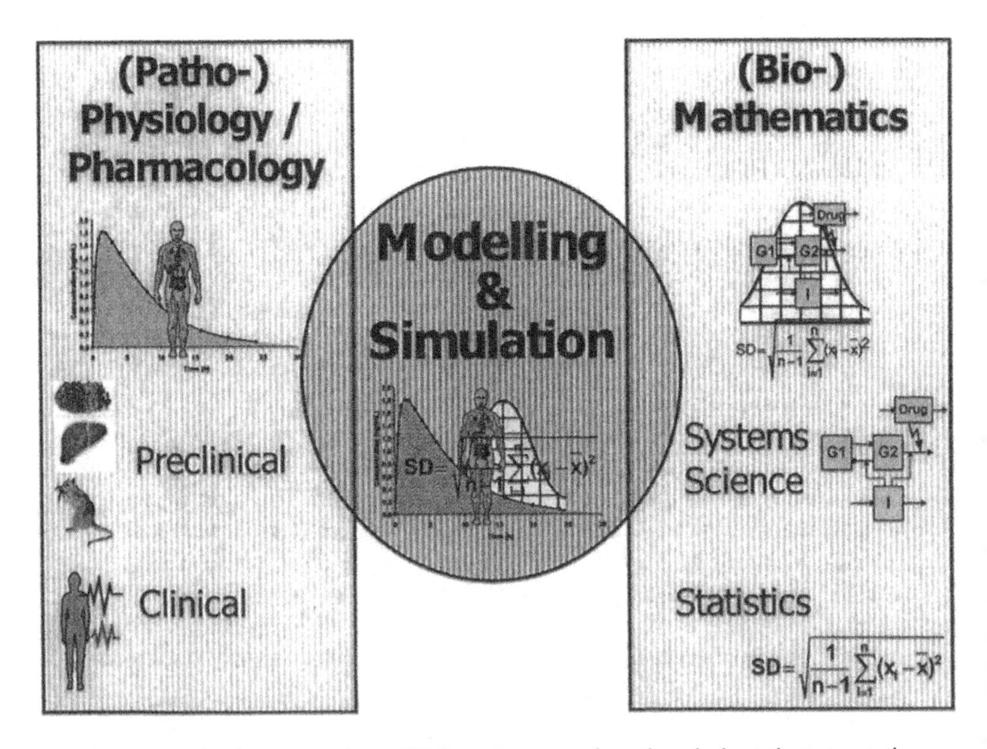


**FIGURE 3** A schema to show M&S as the cross-functional pivot that strengthens and expands the axis between the clinical development sciences of patho (physiology) and pharmacology on the one hand and the biomathematical disciplines on the other.

educated and skilled in the use of mathematical, statistical, and numerical tools required for M&S (25). While academic institutions take up the challenge of providing the training, managers within the industry grapple with how to utilize the limited available expertise. While some choose targeted M&S on key projects of a high priority to the company's portfolio, others require that all project teams identify an M&S strategy. In both schemes an increased knowledge of the methodology will be required by the decision makers in order to ensure that the expectations are manageable.

## 12.4 CLINICAL TRIAL SIMULATION

### 12.4.1 A Team Game

The ability to simulate any given clinical trial requires at its core a mathematical/statistical model, based on an underlying PK/PD hypothesis related to the (patho-) physiological process concerned and the pharmacological/toxicological effects of a drug. To speak about clinical trial simulation without first addressing pharmacokinetics, pharmacodynamics, physiology and/or pathophysiology is akin to eating a gourmet meal before the team of chefs has prepared it. Furthermore, clinical trial simulation requires the development and internal validation of a realistic pharmacostatistical model describing the set of interactions among most, or all, of these factors, with proper consideration of random elements, such as between-subject variability.

In the aerospace and automobile industry M&S has substantially reduced the costs and risks associated with expensive live tests. In the pharmaceutical industry it has the potential (given the necessary influence) to reduce the number of failed studies and late development disappointments. However, unlike the aerospace industry where variation is introduced as heavy perturbations on a series of very similar copies of a given aircraft type (where each one will react pretty much "the same"), stochastic effects in the context of pharmaceutical development are inherent in the biological system itself, since all patients are different. In addition, most engineers and scientists in the aerospace industry have, individually, a strong background in mathematics whereas this, in general, is the case only in the biometrics department of a typical pharmaceutical company.

Process optimization has been addressed in the pharmaceutical industry and the "development machinery" for progressing from one study to the next and from one phase to the next is well oiled. However, implementation of some forms of scientific innovation has lagged behind. Modeling and simulation can strengthen and expand the axis between the clinical development sciences and the biomathematical disciplines wherein it comfortably blends, and in so doing provides a key component of the scientific innovation (Figure 3).

Advances in software with which one can perform a clinical trial simulation have occurred in recent years. ACSL Biomed from MGA was developed in 1996 specifically to perform clinical trial simulation using a graphical user interface (GUI). Meanwhile, Pharsight's Trial Simulator is a sophisticated tool for performing clinical trial simulations in the industry. However, if one is willing to commit the necessary additional programming resources, any general-purpose software (e.g., SAS, MATLAB, S-PLUS) or special nonlinear mixed effects modeling software (e.g., NONMEM) can also be used. M&S comprises technical aspects (e.g., programming) for which standard operating procedures (SOPs) can be delivered, but also imagination which can hardly be fostered by SOPs.

Efficient information technology support is of course essential to take advantage of the emerging hardware and software methods. At present there is no single software package that fulfils *all* the needs of an M&S team (Figure 4) and incorporates estimation tools such as those in NONMEM, as well as sophisticated tools that can simulate complicated protocol and study design features. An often underestimated technical aspect of M&S is the resource need for data access and assembly. It is our experience that few companies have a sufficiently "flexible" data management system that can easily cope with the new ad hoc data requirements implied by the data-driven nature of M&S activities. Therefore, the proper

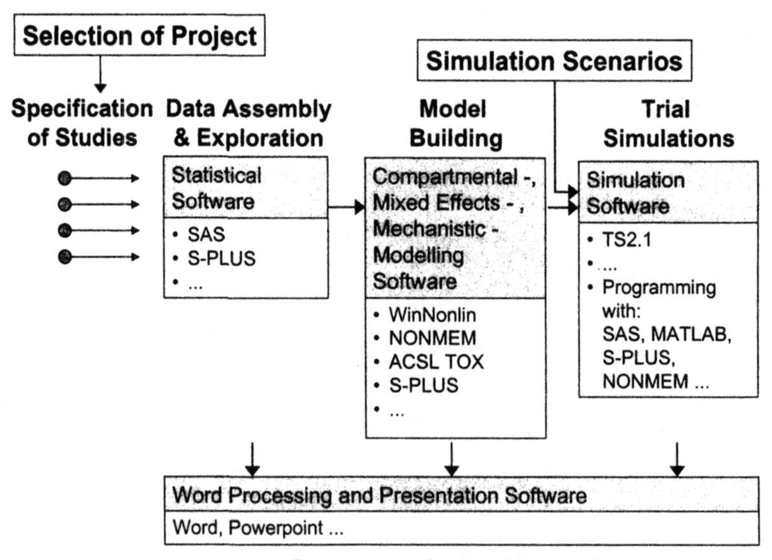


**FIGURE 4** An example of a typical organizational, communication, and information technology infrastructure for M&S.

integration of clinical trial simulation requires an integrated approach to drug development with contributions from many disciplines.

### 12.4.2 An "In Numero" Study

A key element to facilitate M&S achieving its maximum potential utility is that it should be recognized as a visible, separate work unit that is included in the planning (timelines and resources) of a drug development program. It has been proposed that this work unit be called an "in numero" study (3, 25), i.e., a protocol-driven exercise designed to extract additional information and/or answer specified drug development questions, through an integrated model-based analysis of existent raw data, often pooled across studies. The Good Practices Guidelines from the Center for Drug Development Science makes practical recommendations for the content of protocols for those in numero studies that are targeted for simulation of clinical trials (52). The advantages of formal inclusion of M&S as a trial (or drug development) planning element is that it forces the team to be explicit about the assumptions before conducting the trial. Planning and documentation are essential and help to communicate premises and results. After completion of the trial, the existence of a M&S report developed prior to the start of the study will strengthen beliefs and helps to increase the level of confidence in the findings.

### 12.4.3 Clinical Trial Simulation—Some Words of Caution

As modelers in the industry, we feel obliged to caution our colleagues and senior management that M&S is not necessarily the panacea for all things. We would therefore explicitly declare that CTS *cannot*:

- Replace clinical trial results
- Be always right. It can provide reassurance but not certainty; however, even when "wrong," M&S results help to interpret deviant results.
- Provide answers to questions that cannot be explicitly formulated via "the model"
- Be reliable in extrapolation mode
- Be simple. It is the modeler's challenge to reformulate the M&S answers "into the parameter space of the customer." Answers must be simple and accessible to other team members, even though the M&S methodology is complex.

## 12.5 MODELING AND SIMULATION: PROSPECTS

As stated earlier, there is increased regulatory demand for population PK/PD information to be included in NDA packages. This has caused industry scientists to address the challenging issue of PD endpoints, including biomarkers and surrogate endpoints (53). The empirical PD models of the past are being replaced, where possible, with more physiologically based mechanistic models. Issues such as the modeling of categorical endpoints and time dependencies introduced by underlying disease progression, toxicological manifestation, and long-term efficacy responses are being addressed (54). Mechanistic PK/PD modeling will benefit from and exploit the plethora of new information on drug transport systems and drug metabolizing enzymes. In addition, increased use of compliance, dropout, and disease state models can be expected in the near future.

An optimistic feature is the efforts by some long-standing M&S "champions" (55, 56) to encourage use of M&S as supportive evidence to enhance regulatory submissions comprising only one formal phase 3 clinical trial. Provision for this scenario already exists in the regulatory guidelines (57) but requires elucidation and understanding of the requirements.

It is envisaged that industry scientists (within the bounds of confidentiality) and academics will continue to bring evidence in the scientific literature of the utility of M&S in drug development. This will facilitate the understanding of what M&S can deliver. As the models that have been developed become more widely used and improved, the long-term cost effectiveness is likely to improve, thereby leading to an increased use.

## 12.6 CONCLUSION

Modeling and Simulation (in particular of clinical trials) is a rapidly evolving technology. The pharmaceutical industry, academia, the regulatory authorities, and most importantly, all those who use pharmaceutical products will play a role in its future. In 10 years M&S could be viewed, in retrospect, as one more promising methodology which failed to deliver. On the other hand, it could be the pivot of the drug development process. In the optimistic view of the authors the latter outcome is more likely.

## REFERENCES

1. B Reigner, P Williams, I Patel, J-L Steimer, C Peck, P van Brummelen. An evaluation of the integration of pharmacokinetic and pharmacodynamic principles in clinical drug development: experience within Hoffmann La Roche. Clin Pharmacokinet 33:142–152, 1997.
2. SC Machado, R Miller, C Hu. A regulatory perspective on pharmacokinetic/pharmacodynamic modelling. Stat Meth Med Res 8:217–245, 1999.
3. L Sheiner, J-L Steimer. Pharmacokinetic/pharmacodynamic modeling in drug development. Annu Rev Pharmacol Toxicol 40:67–95, 2000.

4. R Gieschke, J-L Steimer. Pharmacometrics: modelling and simulation tools to improve decision making in clinical drug development. Eur J Drug Metab Pharmacokin 25(1):49–58, 2000.
5. N Holford, H Kimko, J Monteleone, C Peck. Simulation of Clinical Trials. Annu Rev Pharmacol Toxicol 40:209–234, 2000.
6. WJ Kaufmann, LL Smarr. Supercomputing and the Transformation of Science. New York: Scientific American Library, 1993.
7. S Johnson. The role of simulation in the management of research: What can the pharmaceutical industry learn from the aerospace industry? Drug Information Journal 32:961–969, 1998.
8. J-L Steimer, R Kawai, S Charnick, M Lemarechal, P Francheteau, et al. Physiological pharmacokinetic/pharmacodynamic (PB-PK/PD) models in drug development? In: M Danhof, C Peck, eds. Measurement and Kinetics of In Vivo Drug Effects: Advances in Simultaneous Pharmacokinetic/Pharmacodynamic modelling. Leiden/Amsterdam Center for Drug Research, 1994, pp 42–47.
9. R Kawai, M Lemaire, J-L Steimer, A Bruelisauer, W Niederberger, M Rowland. Physiologically based pharmacokinetic study on a cyclosporin derivative, SDZ IMM 125, J Pharmacokinet Biopharm 22:327–365, 1994.
10. M Pelekis, P Poulin, K Krishnan. An approach for incorporating tissue composition data into physiologically based pharmacokinetic models, Toxicology and Industrial Health 11:511–522, 1995.
11. SB Charnick, R Kawai, JR Nedelman, M Lemaire, W Niederberger, H Sato. Perspectives in pharmacokinetics. Physiologically based pharmacokinetic modelling as a tool for drug development, J Pharmacokinet Biopharm 23:217–229, 1995.
12. K Hoang. Physiologically based pharmacokinetic models: mathematical fundamentals and simulation implementations, Toxicology Letters 79:99–106, 1995.
13. C Tanaka, R Kawai, M Rowland. Physiologically based pharmacokinetics of cyclosporine A: reevaluation of dose-nonlinear kinetics in rats. J Pharmacokinet Biopharm 27:597–623, 1999.
14. P Poulin, FP Theil. A priori prediction of tissue:plasma partition coefficients of drugs to facilitate the use of physiologically-based pharmacokinetic models in drug discovery. J Pharm Sci 89:16–35, 2000.
15. P Poulin, FP Theil. Prediction of adipose tissue:plasma partition coefficients for structurally unrelated compounds. J Pharm Sci 90:436–447, 2001.
16. N Holford, L Sheiner. Understanding the dose-effect relationship: clinical applications of pharmacokinetic/pharmacodynamic models. Clin Pharmacokinet 6:429–453, 1981.
17. M Berman. The formulation and testing of models. Ann NY Acad Sci 108:182–194, 1963.
18. L Balant, E Garrett. Computer use in pharmacokinetics. In: E Garrett, J Hirty, eds. Drug Fate and Metabolism. New York, Marcel Dekker, 1983, pp 1–150.
19. CC Peck, BB Barrett. Non-linear least squares regression programs for microcomputers. J Pharmacokinet Biopharm 7:537–545, 1979.
20. CM Metzler. Factors affecting pharmacokinetic modeling. In: KS Albert, Ed. Drug Absorption and Disposition: Statistical Considerations. Washington DC: Academy of Pharmaceutical Science, American Pharmaceutical Association, 1980, pp 15–30.

21. N Holford. The target concentration approach to clinical drug development. Clin Pharmacokinet 29(5):287–291, 1995.
22. C Minto, T Schnider. Expanding clinical applications of population pharmacodynamic modeling. Br J Clin Pharmacol 46:321–333, 1998.
23. CC Peck, WH Barr, LZ Benet, J Collins, RE Desjardins, DE Furst, JG Harter, G Levy, T Ludden, JH Rodman, L Sanathanan, JJ Schentag, VP Shah, LB Sheiner, JP Skelly, DR Stanski, RJ Temple, CT Viswanathan, J Weissinger, A Yacobi. Opportunities for integration of pharmacokinetics, pharmacodynamics, and toxicokinetics in rational drug development. Clin Pharm Ther 51(4):465–473, 1992.
24. PricewaterhouseCoopers, Pharma 2005—Silicon Rally: The Race to e-R&D, 1999, pp 1–21. http://www.pwcglobal.com
25. L Aarons, M Karlsson, F Mentre, F Rombout, J-L Steimer, A Van Peer and invited COST B15 experts. Role of modelling and simulation in Phase I drug development. Eur J Pharm Sci 13:115–122, 2001.
26. S Beal, L Sheiner. The NONMEM system. Am Stat 34:118–119, 1980.
27. L Aarons, LP Balant, M Danhof, M Gex-Fabry, UA. Gundert-Remy, MO Karlsson, F Mentre, PL Morselli, F Rombout, M Rowland, J-L Steimer, S Vozeh, eds. The Population Approach: Measuring and Managing Variability in Response, Concentration and Dose. Brussels: Luxembourg: Office for Official Publication of the European Communities, European Cooperation in the field of Scientific and Technical Research, 1997.
28. S Beal, L Sheiner. Methodology of population pharmacokinetics. Drug Fate and Metabolism. E Garrett and J Hirtz. New York: Marcel Dekker. 5:135–183, 1985.
29. T Ludden. Population pharmacokinetics. J Clin Pharmacol 28(12):1059–1063, 1988.
30. A Racine-Poon, J Wakefield. Statistical methods for population pharmacokinetic modeling. Statistical Methods in Medical Research 7(1):63–84, 1998.
31. M Rowland, L Aarons, eds. New Strategies in Drug Development and Clinical Evaluation: the Population Approach. Luxembourg, Commission of the European Communities, 1992.
32. L Sheiner. The population approach to pharmacokinetic data analysis: rationale and standard data analysis methods. Drug Metab Rev 15(1–2):153–171, 1984.
33. J-L Steimer, S Vozeh, A Racine-Poon, N Holford, R O'Neill. The population approach: Rationale, methods, and applications in clinical pharmacology and drug development. In: P Welling, L Balant, eds. Pharmacokinetics of Drugs. Heidelberg: Springer-Verlag, 1994, pp 405–451.
34. S Vozeh, J-L Steimer, M Rowland, P Morselli, F Mentre, LP Balant, L Aarons. The use of population pharmacokinetics in drug development. Clin Pharmacokinet 30: 81–93, 1996.
35. B Whiting, A Kelman, J Grevel. Population pharmacokinetics: theory and clinical application. Clin Pharmacokinet 11:387–40137, 1986.
36. S Vozeh, K Muir, L Sheiner, F Follath. Predicting individual phenytoin dosage. J Pharmacokinet Biopharm 9:131–146, 1981.
37. Population Approach Group in Europe. http://userpage.fu-berlin.de/~page/
38. Midwest Users Forum for Population Approaches in Data Analysis. http://www.kdpbaptist.louisville.edu/mufpada.html
39. EI Ette, R Miller, WR Gillespie, SM Huang, L Lesko, R Williams. The population

approach: FDA experience. In: LP Balant, L Aarons, eds. The Population Approach: Measuring and Managing Variability in Response, Concentration and Dose. Brussels: Commission of the European Communities, European Cooperation in the Field of Scientific and Technical Research, 1997.

40. H Sun, E Fadiran, C Jones, L Lesko, S Huang, K Higgins, C Hu, S Machado, S Maldonado, R Williams, M Hossain, EI Ette. Population pharmacokinetics: a regulatory perspective. Clin Pharmacokin 37:41–58, 1999.
41. Food and Drug Administration. Guidance for Industry: Population Pharmacokinetics. Washington, DC: Center for Drug Evaluation and Research, Food and Drug Administration DHHS, 1999. http://fda.gov/cder/guidance/index.htm
42. S Tett, N Holford, A McLachlan, Population pharmacokinetics and pharmacodynamics: An underutilized resource. Drug Information Journal 32:693–710, 1998.
43. Guidance for Clinical Pharmacokinetic Studies for Drugs, IYAKUSHINHATSU, No. 796, Evaluation and Licensing Division, Pharmaceutical and Medical Safety Bureau, Ministry of Health, Labor and Welfare, Japan, Jun 1, 2001.
44. Guidance for Drug-Drug Interaction Studies. IYAKUSHINHATSU, No. 813, Evaluation and Licensing Division, Pharmaceutical and Medical Safety Bureau, Ministry of Health, Labor and Welfare, Japan, Jun 4, 2001.
45. ICH Topic E4. Note for Guidance on Dose Response Information to Support Drug Registration. The European Agency for the Evaluation of Medicinal Products. Human Medicines Evaluation Unit. CPMP/ICH/378/95.
46. ICH Topic E9. Note for Guidance on Statistical Principles for Clinical Trials. The European Agency for the Evaluation of Medicinal Products. Human Medicines Evaluation Unit. CPMP/ICH/363/96.
47. L Sheiner, Learning versus confirming in clinical drug development. Clin Pharmacol Ther, 61(3):275–291, 1997.
48. R Gieschke, B Reigner, J-L Steimer. Exploring clinical study design by computer simulation based on pharmacokinetic/pharmacodynamic modelling. Int J Clin Pharmacol Therap, 35(10):469–474, 1997.
49. C Veyrat-Follett, R Bruno, R Olivares, GR Rhodes, P Chaikin. Clinical trial simulation of docetaxel in patients with cancer as a tool for dosage optimization. Clin Pharmacol Ther 68:677–687, 2000.
50. M Hale, A Nicholls, R Bullingham, R Hene, A Hoitsma, et al. The pharmacokinetic-pharmacodynamic relationship for mycophenolate mofetil in renal transplantation. Clin Pharm Therap 64:672–683, 1998.
51. M Hale, WR Gillespie, S Gupta, B Tuk, NHGHolford. Clinical trial simulation streamlining your drug development process. Applied clinical trials 35–40, August 1996.
52. Center for Drug Development Science, ed. Modeling and Simulation of Clinical Trials: Best Practices Workshop. Georgetown University Medical Center. Arlington, VA, Feb 3–5, 1999.
53. Biomarkers Definitions Working Group. Biomarkers and surrogate endpoints: Preferred definitions and conceptual framework. Clin Pharmacol Ther 69:89–95, 2001.
54. M Danhof, J-L Steimer. Measurement and kinetics of in vivo drug effects. Advances in simultaneous pharmacokinetic/pharmacodynamic modelling. Leiden: Leiden/Amsterdam Center for Drug Research, 1998.

55. LB Sheiner. Rationale for Drug Approval Based on a Single Clinical Trial + Confirmatory Evidence. Pharsight Workshop in Association with PAGE 2001. Basel, Switzerland, 5 June 2001.
56. C Peck. Confirmatory Evidence (CE) to Support a Single Clinical Trial (SCT) As a Basis for Drug Approval: An Exploratory Workshop. Washington, DC: Georgetown University Conference Center, January 15–16, 2002. http://cdds.georgetown.edu
57. Food Drug Admin. Food Drug Mod. Act 1997. Subsect. 115: Food Drug Cosmet. Act, Sect. 505D. Washington, DC: US Government, 1997.

# 13

# History-Informed Perspectives on the Modeling and Simulation of Therapeutic Drug Actions

**John Urquhart**
AARDEX Ltd/APREX Corp, Zug, Switzerland, and Union City, California, U.S.A.; Maastricht University, Maastricht, the Netherlands; and University of California at San Francisco, San Francisco, California, U.S.A.

## 13.1 INTRODUCTION

This chapter has four objectives. The first is to provide some perspectives on the origins of modeling and simulation (M&S) in physiology and endocrinology, the historical roots of which antedate commercially available computers. The second is to examine the types of problems that motivated M&S in physiology and endocrinology. The third is to emphasize characteristics of a prevalent hard nonlinearity in physiology that is an integral part of the pharmacodynamics of nifedipine in humans, corticosterone in rats, and midazolam in rats, but which is rarely looked for in conventional pharmacodynamic studies. Identification and study of this nonlinearity, called unidirectional rate sensitivity (UDRS), requires a higher than usual degree of experimental control over the rate at which drug enters the bloodstream. With such control, responses have a high degree of determinism;

without such control, responses appear chaotic. The fourth is to examine the long-dysfunctional relation between physiology and pharmacology, its implications for current research in pharmacodynamics, and the current state of M&S of pharmacodynamics, which, as later discussion suggests, has some noteworthy shortcomings.

## 13.2 ORIGINS OF MODELING AND SIMULATION IN PHYSIOLOGY

Modeling and simulation, though a very recent development in the pharmaceutical arena, has roots in physiology and endocrinology that antedate computers. A leading motivation for M&S in physiology was early recognition of the peculiarly difficult problems encountered in analyzing and describing the homeostatic, self-regulatory systems that are the essence of physiology. In these systems, causality flows circularly. Thus, the input and output are always the same, because of the circular connection that links them together, yet the control system acts to hold both within narrow limits in the face of externally imposed disturbances. The difficulty lies in the fact that circularly flowing causality defies satisfactory descriptions by language, which is inherently sequential. Semantic tricks can obscure this problem, e.g., substituting the term "venous return" for "cardiac output" in describing the regulation of the circulation, but the resulting pseudo-explanations are unsatisfactory, and, of course, not quantitative.

Indeed, a central problem in physiology is the regulation of the circulation, beginning with William Harvey's discovery in 1628 that the blood circulates, rather than ebbing and flowing, as conceived by Galen 1500 years earlier, and taught until Harvey's epochal discovery. The ubiquity of circular causality in physiology became increasingly evident as experimental physiology began to flower in the nineteenth century. Circular causality is imbedded in the term "reflex," which connotes sensory information conveyed afferently to the spinal cord, processed, and then reflected efferently as information that triggers actions which avoid or minimize the exciting event. Sherringtonian neurophysiology, which dominated the discipline during the first half of the twentieth century, is rooted in the spinal reflex.

The American physiologist, Walter B. Cannon, coined the term "homeostasis" in the 1920s, together with the concept of "homeostatic mechanisms" (1). These mechanisms collectively maintain what the nineteenth-century French physiologist, Claude Bernard, had described as "the fixity of the internal environment" in the face of external vicissitudes of temperature, nutrition, hydration, ambient oxygen tension, and the like. Bernard noted that this fixity of the *milieu interieur* was the precondition for a free and independent life (2)—a central tenet of physiology.

### 13.2.1 Pioneering Work on the Control of Blood Pressure

An important step in quantitatively analyzing reflex-based regulation was made by the German physiologist, Eberhard Koch, in his 1931 monograph, *Die Reflektorische Selbsteuerung des Kreislaufes*—the reflex self-regulation of the circulation (3). Koch reported experiments with anesthetized dogs in which he "opened the loop" (to use modern parlance) by surgically isolating the carotid sinuses from the rest of the arterial tree. That maneuver permitted him to control pressure within the carotid sinuses independently of systemic arterial pressure, and thus to observe how variations in carotid sinus pressure influence systemic arterial pressure. This key step of "loop-opening" allowed analysis of the carotid sinus reflex as a sequential process, wherein changes in the blood pressure within the carotid sinus, now no longer locked into equality with central arterial blood pressure, can influence central arterial blood pressure.

Koch's work combined advanced intellectual insight and experimental technique, not to mention extreme virtuosity in the now-obsolete art of smoked-drum kymography, the mechanical precursor to the multichannel electronic recorder. He was able to capture simultaneously, on one kymograph, experimentally induced, small stepwise increases or decreases in carotid sinus pressure (CSP) and the concomitantly prevailing arterial blood pressure, across the full physiological and pathophysiological range of both pressures. The resulting plots show how arterial pressure falls as CSP increases, or rises as CSP falls, with the maximum negative slope occurring within the usual, physiological range of arterial pressures.

Koch grasped the essential point that it was the intersection of (a) the empirical relation between carotid sinus pressure and arterial pressure, with (b) the normally prevailing equality between arterial pressure and carotid sinus pressure, that dictate the normal operating point of the intact reflex. From Koch's data, one could calculate a key parameter of the baroreceptor reflex, later known as "open loop gain." For the carotid sinus reflex, the open loop gain is 1.5–2.0, the meaning of which is that systemic arterial pressure falls or rises by 1.5–2 mmHg for every 1 mmHg rise or fall, respectively, in CSP.

Koch was among the first, if not literally the first, to use graphical techniques, plotting the inevitable pair of functional relationships that characterize a control system, to predict the closed-loop system's operating point and its dependency on the factors that modulate the open-loop characteristics of the system. These graphical techniques are widely used in many fields—including economics—to describe key features of circularly connected systems.

Thirty-five years later, Lamberti, Siewers, and I (4) studied the closed-loop dynamics of the carotid sinus reflex in intact, conscious dogs by the simple expedient of moving the dog's head through its full range of vertical motion, amounting, in long legged, large dogs, to a hydrostatic pressure bias of 25–35 mmHg, depending on the size of the dog and the vertical range of its head motion. We found that the closed-loop gain of the reflex was consistent with Koch's findings, but we also observed a very "hard" nonlinearity in the operation of the closed-loop system, of which more will be discussed in Section 13.4.

## 13.3 DEVELOPMENT OF CONTROL THEORY

In the history of ideas, it is noteworthy that Koch's work more or less coincided with that of Minorsky, Nyquist, and Terman and his students at Stanford, in analyzing self-regulating systems, or, as they came later to be called, negative feedback control systems, or simply control systems. Minorsky worked for the U.S. Navy in the early 1920s on the design of automatic steering mechanisms for large ships, to maintain direction against the vicissitudes of wind and tide (5), and later worked at Stanford. In 1932, Harry Nyquist, working at Bell Labs, published one of the basic papers in the field of automatic control (6). Though published the year following Koch's monograph, it is doubtful that Koch's work was known to Nyquist. Frederick Terman and his famous students, William Hewlett and David Packard, working at Stanford, sought ways to design, and of course analyze, control systems for AM radio circuits that could provide stable amplification of signals and automatic volume control.

Examples of purposely designed control systems go back as far as the control element on Dutch windmills that maintains their optimal positioning relative to a shifting wind, and the governor that James Watt designed to regulate steam pressure in the first steam engines, in the latter part of the eighteenth century. James Clerk Maxwell's analysis of Watt's governor, a century later, is recognized as the first formal analysis of a feedback control system (7).

World War II triggered an explosion of systematic work on design and analysis of automatic control systems. In the post-war era, work expanded, with the impetus provided by the need to control the flight of planes and rockets of unprecedented acceleration and velocity. Concomitantly, there was a dramatic expansion of capability for both measurement and control provided by advances in electronics, microcircuitry, and computers.

The fundamental idea motivating all this work in the technological domain is essentially the same as Bernard's principle of maintaining "fixity" of a certain milieu, or environment, in the face of external vicissitudes. Fixity in the literal sense is too narrow an objective; fixity in the sense of maintaining close correspondence to a defined trajectory in time and space is a better description, which includes automatic fire control in World War II, and, later, missile control, which have been main drivers of progress in the field of analyzing automatic control systems.

### 13.3.1 Feedback Control in Endocrinology

Endocrinology is the physiological field that corresponds mostly closely to pharmacology, in that blood-borne chemical substances trigger actions at a distance from the site of their secretion into the blood stream. The dynamics of hormonal actions inform pharmacodynamics, as we shall see.

During the 1930s the American endocrinologist, Philip E. Smith, adduced experimental evidence for a negative feedback regulation of the secretion of the various pituitary "tropins"—thyrotropin, adrenocorticotropin, the gonadotropins—by the secretory products of the endocrine glands whose secretions they stimulate—thyroxin and tri-iodothyronine, cortisol in humans (corticosterone in the rat), and estradiol and progesterone, respectively. Like most endocrinologists, Smith was not mathematically inclined, and put his findings into words, untaxed by experimental efforts at quantitative analysis of these endocrine control mechanisms. Such work began in the late 1950s in studies on the feedback regulation of pituitary-adrenocortical secretion (8, 9), starting a cycle of modeling-directed experimental work (10, 11), and further analysis (12).

The dynamics of insulin secretion (13) and of overall glucose regulation (14) have informed the endocrine regulation of metabolism. The Bergman minimal model (14) has improved the utility of the standard glucose tolerance test for differentiating the two types of diabetes mellitus.

### 13.3.2 Analysis of Feedback Control in Other Areas of Physiology

World War II provided a strong impetus to quantitative analyses of physiological regulation, particularly of ventilation and body temperature. One center of such work was Randolph Field, Texas, where John Gray and Fred Grodins, later professors of physiology at Northwestern University, defined the quantitative drivers of human ventilation, as specifications for inhalational gas mixtures during high altitude flight in unpressurized aircraft. James Hardy (later professor of physiology at Yale) and Harwood Belding (later professor of physiology in the School of Public Health at University of Pittsburgh) defined the limits of human tolerance to high temperatures, for the purpose of designing tanks and other vehicles for desert warfare. When World War II ended, Gray and Grodins convinced the U.S. Navy to donate to Northwestern's Department of Physiology a large analog computer, the wartime uses of which had ended. In the ensuing decade, that computer played a key role in the emergence of that department as a leading center for the analysis of physiological control systems. Fred Grodins's later book on physiological control systems (15) was a landmark contribution to the field.

Perhaps the single greatest impetus to advancing concepts of physiological control and regulation, especially focused on the cardiovascular system, has been the lifelong work of Arthur C. Guyton. Guyton's long career, from the early 1950s to the present, has been dominated by modeling-guided cardiovascular experimentation. These five decades, of course, spanned the rise and fall of analog computation and the rise and then miniaturization of the digital computer. During this time, he wrote many editions of his widely used textbook of physiology, which have conveyed the basic ideas of physiological regulation to successive generations of students of physiology and medicine.

Guyton's model of the mammalian cardiovascular system, scaled to human values, is a landmark in physiology. Consisting of over 350 functional blocks, it maps a huge body of experimental physiological literature on the many mechanisms involved in the regulation of the circulation (16, 17). The model puts each of these mechanisms into quantitative perspective, in relation to their respective strengths and the different time scales on which they operate—milliseconds, seconds, minutes, hours, days, weeks, and months. Indeed, one of the most important concepts in systems physiology is that of *time scales* on which various self-regulatory systems operate. For example, the structural changes of vascular remodeling operate on a time scale of weeks to months, but play no role in the minute to minute regulation of the circulation. In contrast, the baroreceptor reflexes play a key role in the minute to minute regulation of the circulation, but play little or no role on a time scale of weeks to months. Guyton has summarized the salient time scales of blood pressure regulation in Figure 1 (18).

In modeling various aspects of cardiovascular systems functioning, one would logically focus on a particular range of time scale, e.g., days to weeks. Doing so avoids having to represent mechanisms that operate on time scales far removed from the subject of interest.

### 13.3.3 Another Illustration of the Time-Scales Concept in Physiological Regulation

A visual example of the role of time scales in physiological regulation is provided by the adaptations of the eye to changing light intensity: the initial adaptation to a change in ambient light intensity is pupillary constriction or dilation, for increases or decreases in light intensity, respectively, thus minimizing the extent of change in the intensity of light reaching the retina. Pupillary size changes, in response to, e.g., a step change in light intensity in either direction, are completed within a few hundred milliseconds, but meanwhile there also begins a relatively much slower process of retinal bleaching or its inverse, for increases or decreases in light intensity, respectively. The retinal adaptation to sudden darkening takes 5–15 min to complete, and, as it occurs, the pupil gradually returns to its original position, poised once again to provide short-term regulation of light incident upon the retina. When the intensity of incident light increases, the retinal adaptation ("bleaching") is completed within a few seconds—much faster than dark adapta-

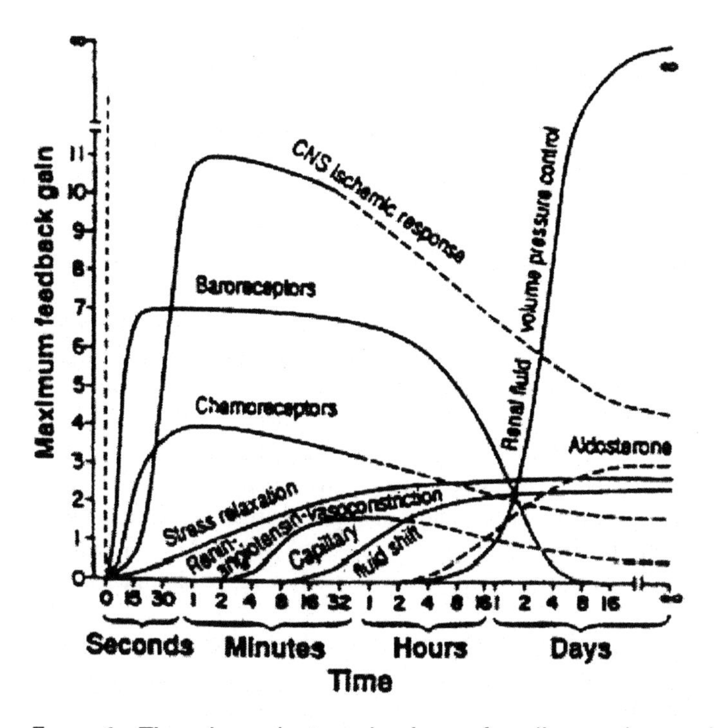


**FIGURE 1** Time-dependent mechanisms of cardiovascular regulation. The details, which are beyond our scope here, are discussed in Ref. 18. Suffice it to say that the strength (or "gain," to use the engineering term) of each of these various control actions varies over time since its activation has commenced. Baroreceptor reflexes, for example, cease to oppose changes in blood pressure that are maintained for more than about a week. In contrast, the diuretic response triggered by elevated blood pressure does not come into play for several hours, but then goes on to become a dominant regulatory mechanism. Since this figure was published one further, very important but very slowly developing regulatory mechanism is vascular and cardiac remodeling, i.e., morphological changes. These take many weeks or months to occur and include such changes as the growth of new vessels and changes in the size of the cardiac chambers and wall thickness. (Reproduced with permission from Ref. 18).

tion, but still an order of magnitude slower than the pupillary changes. Thus can several physiological control systems serving the same end cooperate to regulate—in this instance, the intensity of light incident upon the retina. The marked dynamic asymmetry between retinal adaptation to increased vs to decreased light is typical of many physiological systems. Such dynamic asymmetries naturally provide powerful constraints on modeling, as the dynamic asymmetry must be satisfactorily simulated by the model, if the model is to provide accurate simulation of the biological system's responses to various perturbations within the time scale in question.

### 13.3.4 Linear vs. Nonlinear Modeling

The earliest studies of the dynamics of physiological control systems were strongly influenced by linear control theory, imported more or less intact from systems engineering. During the first 15 years or so after the end of World War II, systems engineering was dominated by problems arising mainly within electrical engineering, many of which were amenable to analysis as linear systems. One of its first reflections in biology and medicine is the early work of Stark and his colleagues on the pupillary reflex to light and eye-tracking at the Research Electronics Laboratory at MIT (19). While recognizing the inherent nonlinearity of the pupillary reflex, Stark and his colleagues sought to use small perturbations around particular operating points, which they assumed would allow the use of so-called "piecewise" linear approximations across a larger range of system operation.

It soon became apparent, however, that this maneuver could not be used, because of a "hard" nonlinearity that defied linearization by the use of small perturbations. The nonlinearity in question, which seems to be ubiquitous in physiology, is usually referred to as "unidirectional" rate sensitivity (UDRS) (20, 21). One of its salient properties is that its nonlinear aspects become proportionately *larger* as perturbations become smaller.

## 13.4 THE UDRS NONLINEARITY

It is useful to review several examples of the UDRS nonlinearity in both physiology and pharmacology, for several reasons, not the least of which is that failure to recognize the presence of this nonlinearity can result in models that fail to simulate key dynamic attributes of the system in question. Moreover, the rate-sensitive aspect can be unwittingly triggered if experimental technique is substandard, e.g., a syringe-drive infusion pump that is afflicted with "stiction," such that the syringe(s) advance in a jerky motion. The resulting erratic triggering of the rate sensitive component of the system response can have a chaotic appearance that leads one needlessly into complex stochastic or chaos-theoretic models, when in fact the system is highly deterministic when analyzed with properly narrow tolerances on the kinetics of drug administration and entry into the bloodstream.

The seeming ubiquity of the UDRS in physiology indicates the need to test experimentally for its presence in studies on the pharmacodynamics of drugs whose actions impinge on physiological regulatory processes, which comprise most drugs, with the conspicuous exception of the anti-infective agents, the actions of which tend to be dominated by the physiological properties of the microorganisms which they attack.

### 13.4.1 UDRS in Corticosteroid Feedback Regulation of ACTH Secretion

Dallman and Yates have provided very clear evidence that the strength of the inhibitory action of corticosteroids on ACTH secretion is related not only to the absolute concentration of corticosteroid in plasma, but also its rate of increase (11). This finding in an endocrine control system (in the rat) provided a valuable precedent in the design of studies 15 years later that revealed that UDRS is a prominent feature of the pharmacodynamics of nifedipine in humans (see Section 13.4.6), which in turn led to the discovery of the UDRS nonlinearity in the pharmacodynamics of midazolam, also in the rat (see Section 13.4.7).

### 13.4.2 UDRS in the Pupillary Reflex to Light

This example has historical importance, because the pupillary reflex to light was one of the first physiological systems to be analyzed with the help of engineering control theory (19). Stark and colleagues opened the loop by restricting the diameter of an incident light beam to an area smaller than the smallest pupillary aperture (19). Thus, reflexly mediated changes in pupil diameter could have no influence on the amount of light reaching the retina.

As is well known, the pupil constricts when incident light intensity increases, and dilates when incident light intensity decreases. A briefly maintained "flash" of light causes the pupil to constrict momentarily and then return within a few hundred milliseconds to its previous size. A brief reduction in light intensity (a "darkness flash") might be expected to cause the pupil to dilate momentarily, and then return to its previous size; instead, however, the pupil *constricts* momentarily. This surprising dynamic property, first demonstrated by Clynes (20, 21), forces the abandonment of any kind of linear model, because no linear model is capable of simulating the same response to inputs of opposite direction.

The underlying dynamics are illuminated by using multiple temporal patterns of change of light intensity, including different rates of increase and decrease of light intensity, and of changes between different absolute levels of light intensity. Doing so reveals that the pupillary reflex to light can be modeled by two elements that respectively receive, (a) light and (b) its positive first time derivative, as input—a "proportional" element that triggers a degree of constriction proportional to incident light intensity, and a UDRS element that triggers a degree of constriction proportional to the magnitude of the rate of increase in incident light intensity, but gives no response to varying rates of decrease of light intensity. Thus, the rate-responsive element is strictly unidirectional, whence the term behind the abbreviation, UDRS. The paradoxical constriction triggered by the "darkness flash" reflects the pupil's response to the high rate of increase in incident light intensity as the darkness flash *ends*, whereas the experimental application of a light flash begins with simultaneous application of a high rate of increase in light intensity and an absolute increase in light intensity. The term "proportional" must be taken with a grain of salt, for strict proportionality is probably more the exception than the rule, with a static nonlinear relation between light intensity and pupil area being the more aptly descriptive. But note that the static nonlinearity is a relatively minor consideration in relation to the dynamic nonlinearity, because the latter is capable of driving the system to respond in a manner that is the complete opposite of its usual operation. Such behavior is aptly regarded as a "hard" nonlinearity, while nonproportionality between input and output is aptly regarded as a "soft" nonlinearity.

One can experimentally vary the duration of the darkness flash as well as the rates of change of the downslope and upslope of light intensity, and see that a long-maintained period of darkness would allow the pupil time to dilate, before the upstroke in light intensity triggers, in rate-sensitive fashion, constriction.

### 13.4.3 UDRS in the Carotid Sinus Reflex

Elegant studies have been made of the dynamics of individual baroreceptors in the isolated carotid sinus (22, 23). Purposefully patterned changes in intrasinus pressure, with the carotid sinuses isolated from the rest of the arterial tree, clearly show that firing rate in afferent nerve fibers, coming from the carotid sinus, is related not only to the absolute pressure but to the rate of increase of intrasinus pressure as well. In these studies, impulse traffic in individual afferent nerve fibers was measured as pressure was varied experimentally within the sinus, either as physiological pulsations in pressure, or with experimentally controlled pressures. These studies clearly show the UDRS property, operating on a time scale of tens of milliseconds.

### 13.4.4 UDRS at the Level of the Integrated Baroreceptor Reflex

In contrast to these events occurring at the level of the individual baroreceptors and their transduction of pressure into patterns of afferent nerve impulse traffic, the UDRS nonlinearity is also seen at the level of the intact, closed-loop reflex operating on a time scale of seconds, instead of milliseconds. It is revealed, as described earlier, by the experimental maneuver of moving an intact, conscious dog's head up or down, while observing central arterial pressure. As the head rises, carotid sinus pressure falls by an amount equal to the vertical change in height, expressed in hydrostatic terms as a column of blood of correspondingly varying height. In a long-legged dog, this change corresponds to a change in carotid sinus pressure of 25–35 mmHg. This hydrostatic pressure change sets in motion a compensatory rise in central arterial pressure, such that 40–50% of the initial fall in pressure at the carotid sinus is offset by an ensuing rise in central pressure, i.e., a closed-loop gain of 0.4–0.5. When the head is moved in the opposite direction, from head-up to head-down, carotid sinus pressure increases, triggering a reflex fall in central arterial pressure that offsets the initial hydrostatic change by 40–50%, but reaching its new steady state in about one-third less time than is required to reach steady state when carotid sinus pressure is initially lowered by raising the head.

The UDRS nonlinearity is suggested by this dynamic asymmetry between responses to head-down and head-up maneuvers, and clearly revealed by the responses to a transient up-down-up maneuver and a transient down-up-down maneuver. Each maneuver triggers the same response, namely a transient fall in central arterial pressure (4). That paradoxical response to inputs of opposite direction is the signature of UDRS.

A further manifestation of UDRS is seen when the head is moved in a sinusoidal pattern: frequency multiplication occurs in central arterial pressure when the head is moving at a particular frequency, which happens, in the dog, to be ca. 0.16 cycles per second. These properties and their modeling are described and illustrated in detail in Ref. 4.

An interesting sidelight on this work is that larger animals, e.g., the horse and the giraffe, lack a carotid sinus reflex. The horse's full-range head movement corresponds to a hydrostatic pressure change of about 100 mmHg (the approximate mean arterial pressure in humans, other primates, dogs, and smaller animals), and the giraffe's corresponds to over 200 mmHg. Mean central arterial pressures in the horse and giraffe are, respectively, 150 and 300 mmHg, 1.5–3 times higher than in humans. Were these larger beasts to have the same carotid sinus reflex mechanisms as the dog, there would be a very confused pair of kidneys in either species were they episodically confronted by fluctuations in central arterial pressure of 50–100 mmHg, as reflex responses from the trivial maneuver of vertical head motion. There are many other aspects of the operation of the closed-loop reflex and experimental details, as described in Ref. 4, but these are beyond our scope here.

The key point for this discussion, however, is that the UDRS nonlinearity appears on two different time scales, an order of magnitude apart: at the level of individual baroreceptors and at the level of the integrated reflex in the intact animal.

### 13.4.5 UDRS in Other Receptors

Pacinian corpuscles, which mediate touch sensation, show UDRS (24) as do various other types of mechanoreceptors (25).

### 13.4.6 UDRS in the Pharmacodynamics of Nifedipine

We now turn to a human pharmacological example of UDRS, the understanding of which led to a major cardiovascular pharmaceutical product, indeed the largest in the history of the American market until 1995. The UDRS features of nifedipine's pharmacodynamics are characterized by the evident fact that the rate of increase in the concentration of nifedipine in plasma is the trigger for the drug's principal side effect of reflex tachycardia. This pharmacodynamic property was first revealed in a small study in human volunteers, which also indicated that constant-rate administration of nifedipine could avoid rates of increase in the drug's concentration in plasma that would be high enough to trigger reflex tachycardia (26). This finding provided crucial supporting evidence for the development of a constant-rate, oral delivery system form of nifedipine. Clinical trials of the constant-rate product, involving large numbers of subjects, demonstrated the virtual elimination of the reflex tachycardia side effect of conventionally formulated nifedipine, and the product's utility as an antihypertensive. The constant-rate product went on to be the biggest yet of delivery systems-based products (27, 28). Avoidance of reflex tachycardia also allowed the constant-rate form of nifedipine to be used to treat hypertension, which had not previously been possible with rapid-release nifedipine because of its erratic cardiovascular actions (28). Comparison of the labeling of the constant-rate OROS form of nifedipine (29) and of the rapid-release capsule form of nifedipine (30) illustrate what was achieved by using a rate-controlled drug delivery system to avoid the UDRS nonlinearity in the pharmacodynamics of nifedipine.

One might have thought that this striking success would have triggered the incorporation of testing for the UDRS nonlinearity into every pharmacodynamic protocol, where feasible. For whatever reasons, that has not been the case. While this story seems to have gone unappreciated within the community of clinical pharmacologists the importance of the original study was grasped immediately by Pfizer's then-director of licensing, Dr. Robert Feeney, and the then-head of US marketing for Pfizer, William Steere. Steere made the crucial decisions necessary to develop an oral, rate-controlled form of nifedipine, using ALZA's push-pull osmotic pump technology. When that product was developed and subsequently tested in large numbers of patients during its premarket trials, the incidence of reflex tachycardia with the constant-rate form of nifedipine was essentially zero, in contrast to the 35–40% incidence of reflex tachycardia among recipients of nifedipine in conventional, rapid-release dosage forms (29, 30). The resulting commercial success of the product appears to have facilitated Steere's subsequent ascent to the chairmanship of Pfizer, from which he has only recently retired. Thus, the UDRS nonlinearity is not just an academic curiosity.

Aside from the unpleasantness of the reflex tachycardia side-effect of rapid-release nifedipine, it appears to be potentially hazardous, as it entails excessive activation of the sympathetic nervous system. Thus, it probably has a hazard profile not unlike that of the rebound effects that follow sudden discontinuation of beta adrenergic receptor antagonists that lack intrinsic sympathomimetic activity (31–33).

In respect to the prevalence of the UDRS nonlinearity in clinical pharmacology, the question naturally arises whether nifedipine is a one-off curiosity. Or has nifedipine merely been the first drug to be subjected to a sufficiently searching set of different temporal patterns of drug exposure to reveal the UDRS nonlinearity? Those who belong to the "we've never done that before, so why should we start doing it now?" school of pharmaceutical R&D can choose the former. Some of the antineoplastic drugs that manifest a high degree of "schedule-dependency" might usefully be studied to see if some of these are not reflections of the UDRS nonlinearity.

### 13.4.7 UDRS in the Pharmacodynamics of Midazolam

Recently, Danhof, who was one of the coauthors of the study that demonstrated UDRS in the PD of nifedipine, and his colleagues reported that the EC50 of midazolam's effects on the EEG of rats varied in a biphasic manner as a range of different rates of rise of midazolam concentration in plasma were tested. The highest EC50 for the drug was observed at an intermediate rate of rise, with lower EC50's at both lower and higher rates of rise. The title of their recently published paper (34) states their conclusion: "Rate of change of blood concentrations is a major determinant of the pharmacodynamics of midazolam in rats."

### 13.4.8 Summary of the Key Dynamic Features of the UDRS Nonlinearity

UDRS is a nonlinearity that cannot be linearized by limiting one's studies to very small perturbations. The reason is that reducing the range of absolute variations in, e.g., light intensity, leaves the rate-sensitive component unchanged, so that it comes to play a progressively larger role in the system's responses as the proportional part of the input signal is scaled down. This property stymies the often-used maneuver of linearizing certain types of nonlinear systems by studying their responses to small perturbations around a series of different operating points, and then linking together, across a wide range of operating points, these piecewise linear approximations. Another feature of the UDRS nonlinearity is its tendency to create frequency multiplication when a sinusoidally varying perturbation is applied. It arises when the proportional channel is driving the output in one direction, and the UDRS channel is driving the output in the opposite direction. Another important feature has to do with experimental design: one needs strict control over the rate of drug entry into the bloodstream in order to maintain control over the rate of increase of drug concentration in plasma. Without such kinetic control, i.e., when conventional, rate-uncontrolled dosage forms are used, this potentially strong pharmacodynamic variable will, if present, be actuated in a widely variable manner, thus creating the impression that deterministic modeling is inappropriate, and resort should be had to stochastic or chaos-theoretic models. There may be many circumstances in which such models are appropriate, but only after one has carried out aptly designed, kinetically controlled experiments to rule in or out the UDRS nonlinearity.

## 13.5 SOME GENERAL LESSONS FOR PHARMACODYNAMIC RESEARCH AND ITS PLACE IN DRUG DEVELOPMENT

A basic point to be taken from the foregoing is the value of exploring pharmacodynamic responses to multiple temporal patterns of drug presentation, including, of course, the rate of increase of drug concentration in plasma, and variably long lapses in dosing. We now know that actual dosing patterns are quite diverse in ambulatory pharmacotherapy, with the "holiday" pattern of multiple-day lapses in dosing recurring more or less frequently in about a third of medically unselected patients, usually beginning and ending abruptly, thus creating rapid rates of change in drug exposure. These findings come from a decade of studies with electronic monitors to compile dosing histories of ambulatory patients in virtually every field of ambulatory pharmacotherapy (35–37). It should go without saying that PD models cannot be expected to give satisfactory simulations of drug trials if they cannot reliably simulate the pharmacodynamic consequences of drug holidays and other commonly recurring patterns of partially compliant dosing. The key word is "commonly," because deviations from prescribed drug regimens are a ubiquitous fact of life in ambulatory pharmacotherapy and the drug trials that support it. Models, in any field of endeavor, whether technical or biomedical, that cannot satisfactorily simulate a system's normal modes of operation are self-evidently unsatisfactory and can only mislead.

## 13.6 OBJECTIVES OF EXPERIMENTALLY DETERMINING RESPONSES TO MULTIPLE TEMPORAL PATTERNS OF DRUG INPUT

There are multiple objectives of experimentally determining the responses to multiple temporal patterns of drug input.

### 13.6.1 First Objective: Identifying Particular Temporal Patterns of Drug Input That Can Trigger Hazardous Effects

This first objective is only indirectly concerned with dynamic modeling and simulation. It is primarily pragmatic: to identify whether, in response to interrupted or rapidly changing levels of dosing, the drug has the potential to trigger hazardous effects. These include:

- Rebound effects upon sudden discontinuation of drug exposure
- Recurrent first-dose effects when temporarily interrupted dosing resumes,
- "Rubber crutch" effects when, after a long period of drug exposure, a suddenly withdrawn drug leaves the patient with markedly down-regulated receptors that normally would mediate crucial reflex regulatory actions, thus paralyzing crucial homeostatic mechanisms
- Inappropriate activation of sympathetically mediated reflex actions, e.g., the reflex tachycardia of nifedipine

The importance of such hazardous effects is that, even though they occur in a minority of patients, they may impact on the endpoints of efficacy studies. For example, a trial of a peripheral alpha-blocker, e.g., doxazosin, would logically be designed to show the benefits of using this agent to prevent long-term complications of hypertension, i.e., stroke and the development of coronary heart disease and its various manifestations of onset of angina pectoris, acute myocardial infarction, or sudden death. Yet doxazosin is an agent with a clearly defined first-dose effect (38), which naturally indicates careful titration at the outset of its use. What is the impact, however, of a multiday holiday with this agent that may occur, e.g., after several months of more or less correct dosing? We now know that holidays occur monthly in about one-sixth of medically unselected patients, and approximately quarterly in a further one-sixth (37). That evidence raises the question: How long can dosing be interrupted before re-titration is mandatory to avoid the acute hypotension of suddenly resumed, full-strength dosing? The product's labeling says "a few days" (38). If "a few" means "2–5 days," then such episodes will occur fairly frequently in a large ambulatory trial of doxazosin. If "a few" means "10–20 days," then, because such long holidays are very infrequent, the incidence of adverse events triggered by suddenly resumed full-strength dosing will likely be too low to have material bearing on the interpretation of the trial.

The reader should bear in mind that efficacy trials are analyzed by the intent-to-treat policy, which censors information on actual drug exposure, and simply analyzes all patients according to the randomized assignment of treatment, whether or not the assigned treatment was actually received, or its regimen of administration executed well or poorly. One of the values of computer-assisted trial design, or M&S of clinical trials, is the ability to explore the sensitivity of the trial outcome to adverse effects of drug holidays and other spontaneously occurring dosing patterns that deviate from the recommended regimen. In respect to doxazosin, which is selected here merely as an example of a widely used drug with a prominent first-dose effect, its labeling cautions about orthostatic hypotension in association with the first-dose, a sudden increase in dosing, or when dosing resumes after "a few days" of interrupted therapy (38). "A few days" ought to be defined, if not from purposefully designed, controlled studies, then by observational studies.

A meta-analysis of trial outcomes with drugs that have first-dose effects would be useful to see if these agents have an unusually high likelihood of unexpectedly negative outcomes. That, of course, would be an indication for retrospective M&S—but only with pharmacodynamic models adequately constrained by their ability to simulate experimental data on "off" responses.

### 13.6.2 Second Objective: Providing Crucial Constraints on PD Modeling

Having experimental data on responses to multiple temporal patterns of drug input provides powerful constraints on dynamic modeling. Up to now, the study and understanding of pharmacodynamics have been seriously hampered by excessive focus on studies of the onset of drug action, with little experimental attention to the dynamics of the offset of drug action, after drug exposure has been maintained for various periods of time. A first strong clue that a fundamentally nonlinear dynamic model is needed is the finding of a dynamic asymmetry between onset and offset of drug action. "Asymmetry" means that the "off" response, when graphically inverted, is not congruent with the "on" response.

Two examples of markedly asymmetric on and off responses are provided by two of the presently 10 best-selling pharmaceuticals. The onset of action of omeprazole is completely developed by about 2 h after the first dose is given, but the offset of action takes 3–5 days (39). The onset of action of the antidepressant action of paroxetine takes several weeks (40), as is typical of all antidepressant drugs, but the offset of action occurs within 48 h (41). In these two examples, we have major dynamic asymmetries in opposite directions, indicating that dynamic asymmetries are (a) drug specific, and (b) unpredictable in magnitude or direction, given information only on the "on" response. No simulation experiment, with models based only on "on" responses, would indicate the true nature of these grossly different "off" responses. There is no substitute for experimental data on this key point.

Furthermore, as discussed below, the character of the drug's "off" response can be expected to change, as the duration of treatment increases, because the physiological counter-regulatory responses to the drug's actions can be expected to change as the duration of treatment lengthens. Thus, information on exposure-dependent "off" responses is not only a powerful constraint on modeling, but also may provide important pragmatic information such as changing dose requirements as duration of treatment lengthens, the potential economic importance of which is not small.

### 13.6.3 Third Objective: Understanding How PD May Change as the Duration of Exposure to the Drug Lengthens

Under the rubric of "multiple temporal patterns of drug input," one must include varying lengths of drug exposure. For drugs whose indications involve only a few days or weeks of dosing, the upper limit of duration of exposure is clear, but for the increasing number of agents indicated for essentially life-long use, the upper bound of drug exposure on which we ought to have information is measured in decades. A number of large trials are carried out for 5–7 years, but they unfortunately come to an end without taking advantage of an important opportunity to enrich pharmacodynamic understanding by making careful observations on the time course of the offset of drug actions, and to compare the results with "off" responses that occur after much shorter periods of drug exposure.

The importance of extending pharmacodynamic studies in time can be inferred from Figure 1. What Guyton shows us in this schematic figure is the way in which recognized cardiovascular regulatory mechanisms enter and leave physiological center stage on different time scales. Missing from Figure 1, of course, is information on a topic that only began coming to the fore during the 1990s, which is vascular and myocardial remodeling. These morphologically based regulatory mechanisms operate on a time scale of months to years. The reader is invited to sketch, in his or her mind's eye, one more rising curve at the right edge of Figure 1, labeled "tissue remodeling," above an abscissal label of "months-years."

Pharmacologists are invited, then, to think of pharmacodynamics as the drug's primary actions triggering and then being modulated by a succession of counter-regulatory actions, along the lines sketched in Figure 1, starting with the leftmost, which operates on the shortest time scale, and proceeding, as time passes and drug exposure continues, to engage, successively, one after another of the counter-regulatory mechanisms that Guyton has described, plus, of course, the missing one at the right edge of tissue remodeling. Struijker-Boudier provided a wide-ranging development of this argument 21 years ago (42), and again a decade ago (43), but these seminal papers have been ignored by pharmacologists and pharmaceutical scientists. They should be revisited.

M&S puts the physiological basis of pharmacodynamics in a new, rather pragmatic perspective, for the presently available PD models cannot reliably simulate the consequences of the biggest single source of variability in drug response, which is variable patient compliance with prescribed drug regimens (35–37, 44,

45). The main reason for this failure is that they are almost invariably unconstrained by experimental data on "off" responses. A point of fundamental importance for the future of pharmacodynamic research is that the cessation of drug exposure, which can occur at any time after its onset, will not only fill the aforementioned pragmatic gap, but also reveal dynamics that will sometimes teach more than we presently know of the counter-regulatory actions with which the primary actions of the drug were engaged up to the moment of discontinuation.

## 13.7 DENOUEMENT PHARMACODYNAMICS

It is useful to provide a specific term for the systematic study of drug "off" responses: "dénouement pharmacodynamics,"* because, as primary drug actions and physiological counter-regulatory actions disengage, the then presently dominant counter-regulatory actions can be expected to leave their footprints as they wind down. Furthermore, one may expect changes in the magnitudes of successively waxing and waning of counter-regulatory actions, which might be reflected in changes in the dosing regimen needed for effective, sustained drug action. Dose changes have potentially large economic importance, rising as the prices of prescription drugs rise and as economically motivated formulary committees look ever harder for ways to cut costs by cutting recommended doses.

Figure 1, as discussed above, describes the salient physiological mechanisms that operate to control blood pressure. Comparable figures are waiting to be drawn for other major organ systems, but these have not had the combination of commanding genius and lifelong persistence in research that Arthur Guyton has brought to the cardiovascular system.

## 13.8 HARD VS. SOFT NONLINEARITIES

A principal focus in pharmacodynamic research in recent years has been, not on dynamics, but on statics, i.e., the "soft" nonlinearities of nonproportionality between dose and response, and hysteresis in response to rising or falling doses.

The term "hard" nonlinearity is useful in the study of pharmacodynamic responses, to encompass paradoxical or overtly harmful drug responses that are triggered by particular temporal patterns of drug exposure. The UDRS nonlinearity is illustrative, because of its paradoxically similar responses to up- and down-going pulse inputs, and, in the case of nifedipine, to its probably hazardous triggering of intense sympathetic nerve activity. The same term is applicable to rebound effects, recurrent first-dose effects, and "rubber crutch" effects, because

* "Dénouement" is usually a literary term for the end of a drama, when the plot lines resolve.

they appear suddenly in response to a particular temporal pattern of drug input, and differ qualitatively from the therapeutic actions of the drug.

## 13.9 HOW MANY DIFFERENT TEMPORAL PATTERNS ARE ENOUGH?

Each "surprising" finding adds a valuable constraint on PD modeling if they are to simulate satisfactorily. Unfortunately, there are no a priori bounds on how diverse one must be in selecting from the potentially infinite number of different temporal patterns of drug exposure. In studying the dynamics of ACTH action on adrenocortical secretion in the 1960s (10, 46), I used the following temporal patterns, all of which were delivered by close intra-arterial administration:

- Impulse, small and large (rapid injection)
- Step increase, small and large (constant infusion)
- Step decrease, small and large (cessation of constant infusion)
- Short and long gaps between a step decrease and following step increase
- Staircase of small steps
- Sinusoidally varying rate of administration, ±40% around a low level mean, at one cycle per 50 min, per 20 min, per 10 min, per 5 min.
- Pseudo-randomly varying rate of administration, ±40% around a low level mean
- Strict constancy, at four different operating levels, from low to highest

This series of different temporal patterns of input was, of course, something of an academic tour de force. Most of the useful constraints on the modeling came from the first five patterns. The sinusoidal testing, which was very time consuming and costly, revealed only one striking result, namely that the adrenal's secretory response to ACTH, with the concentration of ACTH in adrenal arterial blood varying at one cycle per 10 min, was completely inverted, with essentially no attenuation in magnitude. This result was paradoxical in the sense that, if a naïve observer were shown just that set of experimental results and asked: "Does ACTH stimulate or inhibit the secretion of cortisol by the adrenal gland?" the answer would necessarily have been "it inhibits," which of course is not so: it is merely an instance of extreme phase lag at one particular frequency. One might also wonder if this property would not have the potential to destabilize the overall hypothalamic-pituitary-adrenocortical regulatory system, for this inversion of the usual concentration-effect relation adds a second sign inversion to the negative feedback action of cortisol on hypothalamically mediated-pituitary secretion of ACTH, creating net positive feedback. The factor preventing such instability is the relatively slow turnover of cortisol in plasma, which has a time constant of several hours or more, and completely damps a 0.1 cycle per minute signal. The pseudo-randomly varying input of ACTH added little to constrain the modeling, as described in Ref. 46. The purpose of the studies with strict constancy was to estimate the inherent variability in the gland's secretory response, revealing that the 95% confidence intervals on secretion rate under conditions of constancy span about one-seventh of the gland's operating range.

While there was a substantial element of "overkill" in the scope of this work, it has provided a useful object lesson in both what to do and what not to do. The overarching question, besides "how many patterns are enough?" is "Which temporal patterns are most likely to reveal dynamic properties that provide the strongest constraints on the modeling?" These are not trivial or academic questions, as, e.g., one now sees researchers in, e.g., the antiretroviral field, presently groping to define patterns of intermittent dosing that can provide viral suppression with least exposure to drugs (and their dose-dependent toxicities and economic costs), without triggering the emergence of drug resistant virus (47). One might reasonably expect that a well-constrained model of antiretroviral drug actions would be useful in exploring alternative choices of intermittency, and a guide to picking designs for clinical trials.

## 13.10 MINIMAL CASETTE OF TEMPORAL PATTERNS OF DRUG INPUT FOR ROBUST PD MODELING

By "minimal casette," I mean the barest minimum of different temporal patterns of drug entry into the bloodstream needed to provide a reasonably robust characterization of a systemically acting drug's salient PD. The patterns are listed in Table 1. The rationale for these patterns is to be found in the foregoing discussion.

**TABLE 1** The Minimal Cassette of Temporal Patterns of Drug Entry into the Bloodstream for Characterizing the PD of Systemically Acting Drugs

A. In the drug-naïve state, or after a few days of exposure:
   1. Dose response
   2. Sudden on
   3. Sudden off
   4. Gradual on
   5. Gradual off
   6. High vs. low rates of increase of drug concentration in plasma
B. After 90–150 days of prior exposure:
   7. Dose response repeated and contrasted with 1. If surprising, repeat 2–6.
   8. If a first-dose effect is evident, determine, after 90–150 days of exposure, how long exposure can be interrupted without the need to retitrate.

That this is only a minimal cassette, not a guarantee that all of an agent's salient nonlinear pharmacodynamics, will be found through its systematic application.

## 13.11 STARTLING PD OF GONADOTROPIN RELEASING HORMONE AND ITS ROLE IN THE CONTROL OF OVULATION

This last point is exemplified by the unusual dynamics of action of the hypothalamic releasing hormone, gonadotropin releasing hormone (GnRH), and its synthetic analogs, which have attained wide usage and substantial sales (49). The exceptional dynamics of action of GnRH in the physiological regulation of ovulation were established by Ernst Knobil and his colleagues, based on careful observation combined with astute experimentation (50). Their results showed that the physiological activation of gonadotropin secretion by GnRH is effected by a pulsatile pattern of release, with a pulse width of approximately 5 min, repeated hourly. Patterns that deviate only a little from that pattern begin a shift from activation to inhibition of gonadotropin secretion, and continuous exposure to GnRH, e.g., from a constant infusion of GnRH, acts as a profound inhibitor of gonadotropin secretion (50). The activation of gonadotropin secretion has found a therapeutic role in treatment of infertility due to primary hypothalamic failure, with GnRH given by a specially designed infusion pump. The inhibition of gonadotropin secretion has found a much more extensive clinical use in maintaining profound inhibition of pituitary gonadotropin secretion in the management of prostate, breast, or other cancers that are androgen or estrogen dependent (49).

The pulsatile pattern of GnRH secretion, operating with pulse-width and interval parameters of unprecedentedly short intervals, was only established by careful observation of normal and experimentally modified physiological function. It scarcely needs to be mentioned how completely inobvious this pattern was, and how unlikely it would have been that such an unusual pattern of secretion could have been deduced by a priori choices of test patterns of GnRH, picked without the guidance provided by observation of the physiological patterns.

It appears that the crucial underlying mechanism that determines these unusual dynamics is the up- and down-regulation of receptors for GnRH. That, of course, suggests an experimental strategy of screening to understand the time course of changes in receptor populations during continuous exposure to an agonist, complemented by observations on the time course of receptor restitution, once the agonist is withdrawn. So here we see again another potential application of dénouement pharmacodynamics, as full understanding depends upon having data from both "on" and "off" responses.

Until recently these considerations would have been only academic curiosities, but we fortunately have seen remarkable expansion in the capabilities of rate-controlled drug delivery systems, starting in the 1970 s and continuing since

(28, 49). Thus, pharmacologists no longer have their traditional excuse for limiting the methods of screening and testing to conventional dosage forms, with their primitive, uncontrolled kinetics of drug release. In effect, the technical advances in drug delivery systems technologies have outrun conventional pharmacologic testing methods (28, 49). Perhaps innovative pharmacodynamic research can lead the way, by opening new vistas in therapeutics.

## 13.12 IS THERE AN ALTERNATIVE TO EXPERIMENTAL DETERMINATION OF RESPONSES TO MULTIPLE TEMPORAL PATTERNS OF DRUG ADMINISTRATION?

An alternative is to espouse the "chaos-theoretic" point of view (51), and abandon the principle that nonlinear pharmacodynamic and physiological mechanisms are highly deterministic. The key point of differentiation lies in the hands of the experimentalist, and the art with which he/she designs and executes physiological and pharmacological experiments. When drugs or hormones are given in primitive dosage forms that provide no control over the kinetics of drug entry into the bloodstream, then one sees the drug and hormonal actions as highly variable peaks and troughs. Such experiments, although they constitute warp and woof of pharmacology, are poor choices as the starting point for modeling, partly because the actual time course of input is ill-defined or completely obscure, and partly because of the inherent dynamic complexity of the usual kinetics of drug entry into the bloodstream. It is only after one has a reasonably robust model in hand, based on experimental data from some or all of the minimal casette of input functions described above, that one can then interpret the PD responses to drug or hormone administration by the primitive dosage forms that many persevserate in using. But anyone who doubts the basically deterministic nature of drug and hormone response should (re)read the following references (4, 10–14, 20, 22–26, 29, 50).

## 13.13 CONCLUSION

Modeling and simulation have a long history in physiology, with a considerable amount of modeling-inspired experimentation. Modeling and simulation have a short history in pharmacology, pharmaceutical science, and clinical research. At present, most pharmaceutical M&S is physiologically uninformed, which is not surprising, given the intellectual and organizational separation of pharmacology from physiology (48). The advent of modeling and simulation, with its inherently integrative character, may help re-integrate these two disciplines.

Further help will hopefully come from the concepts of "dénouement pharmacodynamics" and hard nonlinearities, and the need, if simulations are to be reliable, for experimental data on the respective effects of drugs presented in a variety of temporal patterns of administration, with strict control over the rate of drug entry into the bloodstream. Together these may broaden today's perspectives on the modeling and simulation of therapeutic drug actions. There should, however, be no misunderstanding that experimental data must lead the way, for it is categorically impossible to adduce evidence for hard nonlinearities by playing simulation games with linear or underconstrained nonlinear models.

## ACKNOWLEGMENTS

I am indebted to F. Eugene Yates, Arthur S. Iberall, the late Ernst Knobil, the late Fred Grodins, Arthur C. Guyton, Richard S. Bergman, Laurence Young, Lawrence Stark, the late Manfred Clynes, John Milsum, Harry Struijker-Boudier, and Douwe Breimer for insightful discussions on technological and physiological control systems and their dynamics. I am also indebted to the editors of this book for helping me to write this chapter in terms most apt for readers from a primarily pharmacological or pharmaceutical background.

## REFERENCES

1. WB Cannon. The Wisdom of the Body. New York: W. W. Norton, 1939.
2. C Bernard. Introduction à l'Étude de la Médecine Expérimentale. Paris: Ballière, 1865.
3. E Koch. Die Reflektorische Selbststeuerung des Kreislaufes. Dresden: Steinkopff, 1931.
4. JJ Lamberti, Jr, J Urquhart, RD Siewers. Observations on the regulation of arterial blood pressure in unanesthetized dogs. Circulation Research 23:415–428, 1968.
5. N Minorsky. Directional stability of automatically steered bodies. J Amer Soc Naval Engineers 42:280–309, 1922.
6. H Nyquist. Regeneration theory. Bell Syst Tech J 11:126–147, 1932.
7. JC Maxwell. On governors. Proc Roy Soc London 16:270–283, 1868.
8. J Urquhart, FE Yates, AL Herbst. Hepatic regulation of adrenal cortical function. Endocrinology 64:816–830, 1959.
9. FE Yates, J Urquhart. Control of plasma concentrations of adrenocortical hormones. Physiol Rev 42:359–443, 1962.
10. J Urquhart. Blood-borne signals—the measuring and modelling of humoral communication and control: 14th Bowditch Lecture of the American Physiological Society. The Physiologist 13(1):7–41, 1970.
11. MF Dallman, FE Yates. Dynamic asymmetries in the coricosteroid feedback path and distribution-metabolism-binding elements of the adrenocortical system. Ann NY Acad Sci 156(2):696–721, 1969.
12. FE Yates, RD Brennan, J Urquhart. Application of control systems theory to physiology. Adrenal glucocorticoid control system. Fed Proc 28:71–83, 1969.

13. RN Bergman, J Urquhart. Pilot gland approach to the study of insulin secretory dynamics. Rec Progr Hormone Res 27:583–605, 1971.
14. RN Bergman. Toward physiological understanding of glucose tolerance: minimal-model approach. Diabetes 38:1512–1527, 1989.
15. FS Grodins. Control Theory and Biological Systems. New York: Columbia University Press, 1963.
16. AC Guyton, TG Coleman, HJ Granger. Circulation: overall regulation. Ann Revs Physiology Palo Alto: Annual Reviews, 1972, pp 13–44.
17. http://www.biosim.com
18. AC Guyton. Blood pressure control—special role of the kidneys and body fluids. Science 252:1813–1816, 1991.
19. L Stark. Stability, oscillation, and noise in the human pupil servomechanism. Prof Inst Radio Engr 47:1925–1939, 1959.
20. M Clynes. Unidirectional rate sensitivity: a biocybernetic law of reflex and humoral systems as physiologic channels of control and communication. Ann NY Acad Sci 92:946–969, 1962.
21. M Clynes. The non-linear biological dynamics of unidirectional rate sensitivity illustrated by analog computer analysis, pupillary reflex to light and sound, and heart rate behavior. Ann NY Acad Sci 98:806–845, 1962.
22. DW Bronk, G Stella. The response to steady pressures of single end organs in the isolated carotid sinus. Am J Physiol 110:708–714, 1935.
23. S Landgren. On the excitation mechanism of the carotid baroreceptors. Acta Physiol Scand 26:1–34, 1952.
24. WR Loewenstein. Rate sensitivity of a biological transducer. Ann NY Acad Sci 156(2):892–900, 1969.
25. JC Houk. Rate sensitivity of mechanoreceptors. Ann NY Acad Sci 156(2):901–916, 1969.
26. CH Kleinbloesem, P van Brummelen, M Danhof, H Faber, J Urquhart, DD Breimer. Rate of increase in the plasma concentration of nifedipine as a major determinant of its hemodynamic effects in humans. Clin Pharmacol Ther 41:26–30, 1987.
27. DD Breimer, J Urquhart. Nifedipine GITS. Lancet 341:306, 1993.
28. J Urquhart. Controlled drug delivery: pharmacologic and therapeutic aspects. J Internal Med 248:357–376, 2000.
29. PROCARDIA-XL (nifedipine) Extended Release Tablets for Oral Use. Physicians' Desk Reference, 55th ed. Montvale (NJ): 2001, pp 2512–2514.
30. PROCARDIA nifedipine CAPSULES for Oral Use. Physicians' Desk Reference, 55th ed. Montvale (NJ): 2001, pp 2510–2512.
31. MC Houston, R Hodge. Beta-adrenergic blocker withdrawal syndromes in hypertension and other cardiovascular diseases. Am Heart J 116:515–523, 1988.
32. DM Gilligan, WL Chan, R Stewart, CM Oakley. Adrenergic hypersensitivity after beta-blocker withdrawal in hypertrophic cardiomyopathy. Am J Cardiol 68:766–772, 1991.
33. Anon. Long-term use of beta blockers: the need for sustained compliance. WHO Drug Information 4(2):52–53, 1990.
34. A Cleton, D Mazee, RA Voskuyl, M Danhof. Rate of change of blood concentrations

is a major determinant of the pharmacodynamics of midazolam in rats. Br J Pharmacol 127:227–235, 1999.
35. JA Cramer. Microelectronic systems for monitoring and enhancing patient compliance with medication regimens. Drugs 49:321–327, 1995.
36. H Kastrissios, TF Blaschke. Medication compliance as a feature in drug development. Ann Rev Pharmacol Toxicol 37:451–475, 1997.
37. J Urquhart. The electronic medication event monitor—lessons for pharmacotherapy. Clin Pharmacokinet 32:345–356, 1997.
38. CARDURA (doxazosin mesylate) tablets. Physicians' Desk Reference, 55th ed. Montvale (NJ): 2001, pp 2473–2477.
39. PRILOSEC (omeprazole) DELAYED-RELEASE CAPSULES, Physicians' Desk Reference, 55th ed. Montvale (NJ): 2001, pp 587–591.
40. PAXIL brand of paroxetine hydrochloride tablets and oral suspension, Physicians' Desk Reference, 55th ed. Montvale (NJ): 2001, pp 3114–3120.
41. JF Rosenbaum, M Fava, SL Hoog, RC Ascroft, WB Krebs. Selective serotonin reuptake inhibitor discontinuation syndrome: a randomized clinical trial. Biol Psychiatry 44:77–87, 1998.
42. HAJ Struyker-Boudier. Dynamic systems analysis as a basis for drug design: application to antihypertensive drug action. In: EJ Ariens, ed. Drug Design, vol X. New York: Academic Press, 1980, pp 145–191.
43. HAJ Struyker-Boudier. Kinetics and dynamics of cardiovascular drug action: introduction. In CJ Boxtel, NHG Holford, M Danhof, eds. The In Vivo Study of Drug Action. Amsterdam: Elsevier, 1992, pp 219–231.
44. JG Harter, CC Peck. Chronobiology: suggestions for integrating it into drug development. Ann NY Acad Sci 618:563–571, 1991.
45. J Urquhart. Pharmacodynamics of variable patient compliance: implications for pharmaceutical value. Adv Drug Delivery Revs 33:207–19, 1998.
46. J Urquhart. Physiological actions of adrenocorticotropic hormone. In: E Knobil, WH Sawyer, eds. The Pituitary Gland and Its Neuroendocrine Control, part 2. Washington DC: American Physiology Society, 1974, chap 27, pp 133–157 (Handbook of Physiology, Endocrinology, sec 7, vol IV).
47. A Fauci. Research sheds light on treatment interruption. AIDS Alert 16:114–146, 2001.
48. B Uvnas. From physiologist to pharmacologist—promotion or degradation? Fifty years in retrospect. Ann Rev Pharmacol Toxicol 24:1–18, 1984.
49. J Urquhart. Can drug delivery systems deliver value in the new pharmaceutical marketplace? Brit J Clin Pharmacol 44:413–419, 1997.
50. E Knobil. The discovery of the hypothalamic gonadotropin-releasing hormone pulse generator and of its physiological significance. Endocrinology 131:1005, 1992.
51. A Dokoumetzidis, A Iliadis, P Macheras. Nonlinear dynamics and chaos theory: concepts and applications relevant to pharmacodynamics. Pharm Res 18:415–426, 2001.

# 14

# Evaluation of Random Sparse Sampling Designs for a Population Pharmacokinetic Study

## Assessment of Power and Bias Using Simulation

**Matthew M. Hutmacher**
Pharmacia Corporation, Skokie, Illinois, U.S.A.

**Kenneth G. Kowalski**
Pfizer, Inc., Ann Arbor, Michigan, U.S.A.

## 14.1 INTRODUCTION

This chapter presents a case study in which simulation was used to evaluate candidate population pharmacokinetic (PPK) substudy designs for a phase III clinical trial in black/Caucasian hypertensive patients. The primary endpoint of the clinical trial was to compare the antihypertensive effect of orally administered eplerenone, an aldosterone antagonist, with placebo and an active control. A sparse sampling strategy was desirable in this outpatient setting to facilitate patient recruitment by eliminating lengthy clinic visits that burden the patient and study sites.

Aarons et al. (1) have acknowledged that sparse sampling designs can fail to support models derived in the data-rich phase I and II environments. Therefore, it is imperative that the analyst assesses sparse PPK designs for their ability to yield meaningful results. For this substudy, it was desirable to ensure valid inference by screening candidate designs for bias in key pharmacokinetic (PK) parameters. Since eplerenone is prescribed chronically, inference on apparent clearance (CL/F) was the primary focus. Furthermore, although the primary endpoint of the efficacy analysis is focused on black/Caucasian differences, it was of clinical interest to assess other patient factors (covariates) that could influence the eplerenone PK profile. To this end, the PPK substudy was powered to detect a clinically relevant change in CL/F in an arbitrary subpopulation of a certain size (i.e., covariate subpopulations).

The following case study is an extension of work published by Kowalski and Hutmacher (2) on an evaluation of a sparse sampling design via simulation. To assess the candidate designs via simulation, a PPK model and a sample time distribution are necessary. In this chapter, the formulation of the PK model from healthy volunteer (phase I) data is discussed in Section 14.2. The development and assessment of the sample time distribution is presented in Section 14.3. The structure of the proposed phase III clinical trial and how the substudy is integrated therewith is described in Section 14.4. Section 14.5 details how the simulations were conducted to assess the designs and their results. The chapter concludes with a brief discussion in Section 14.6.

## 14.2 POPULATION PK MODEL

### 14.2.1 Data

The population PK model used to design this PPK substudy was developed from three phase I healthy volunteer clinical trials. These trials were rich in within-subject data and consisted of a single-dose tolerability, a multiple-dose tolerability, and a food-effect study. A subset of the data, comprised of 556 observations from 44 patients receiving doses of 10, 50, and 100 mg (single dose) and 100 mg daily (steady-state), was modeled. These dosing regimens were selected because they maintain dose proportionality and bracket the set of possible doses/regimens proposed for the PK substudy (50 and 100 mg daily). It should be noted that eplerenone does not exhibit a food effect.

### 14.2.2 Modeling Methods

Nonlinear mixed-effect methodology as implemented in the NONMEM software (3) was used to fit the PK model. Specifically, the approximate maximum likelihood technique known as the first-order conditional estimation (FOCE) method was used to estimate the model parameters. The model was parameterized in terms of apparent clearances and volumes. The intersubject variability of these parameters was modeled using the submodel equation

$$\theta_i = \theta_0 \exp(\eta_i) \tag{1}$$

where $\theta_0$ denotes the parameter vector of the population's typical values and $\eta_i$ represents patient $i$'s vector of random effects. The $\eta_i$ are assumed to have a zero mean and variance-covariance matrix $\Omega$. The square roots of the diagonal elements of $\Omega$ are approximate coefficients of variation (CV) for the interpatient variability of the elements of $\theta_i$.

The log transform was assumed to stabilize the residual variability and the transform-both-sides approach was used to maintain the integrity of the structural model. The intrasubject stochastic model was postulated as

$$\ln(y_{ij}) = \ln(c_{ij}) + \varepsilon_{ij} \tag{2}$$

where $i$ indexes the subjects and $j$ indexes time, $y_{ij}$ represents the observed eplerenone plasma concentration, $c_{ij}$ represents the expected value as postulated by the model, and $\varepsilon_{ij}$ represents an independent and identically distributed random variable with assumed zero mean and variance $\sigma^2$. The parameter $\sigma$ can be interpreted as an approximate CV.

### 14.2.3 Results

#### 14.2.3.1 Model Fit

A two-compartment open model with first-order absorption (NONMEM subroutine ADVAN4) demonstrated a better fit to the data than a one-compartment open model fit by a large reduction in the extended least squares objective function ($\Delta$ELS = 165.642). The disposition parameters were modeled using apparent clearance (CL/F), apparent central volume (Vc/F), apparent intercompartment clearance (Q/F), and apparent peripheral volume (Vp/F) (NONMEM subroutine TRANS4). An absorption rate constant (ka) and a lag-time parameter (tlag) were used to characterize the absorption process.

Throughout the course of model development, models differing in the structure of $\Omega$ were compared. A full $\Omega$ structure was attempted, but difficulties in convergence lead to the conclusion of model overparameterization. A subset of possible models with $\Omega$'s of lower dimensionality was evaluated using the ELS objective function. Results suggested that the absorption variance components (tlag, ka) were correlated and the disposition variance components (CL/F, Vc/F, Q/F, Vp/F) were correlated. Results further suggested that the variance components for Q/F and Vp/F were 100% correlated. To avoid this singularity, Q/F and Vp/F were forced to share the same variance component. This variance component was scaled to allow the variance of Vp/F to be different from Q/F; i.e., these components were linearly related by the submodel

$$\eta_i^{V_p/F} = \phi \eta_i^{Q/F} \tag{3}$$

**TABLE 1** Estimates of the Healthy Volunteer Population Pharmacokinetic Parameters (Fixed Effects and Variance Components)

| Parameter | Estimate ± SE | %CV ($\omega$) | Interindividual correlations | | | | | |
|---|---|---|---|---|---|---|---|---|
| | | | tlag | ka | CL/F | Vc/F | Q/F | Vp/F |
| tlag (h) | 0.192 ± 0.045 | 53.7 | 1 | | | | | |
| ka (liters/h) | 1.94 ± 0.21 | 40.6 | 0.434 | 1 | | | | |
| CL/F (liters/h) | 10.8 ± 0.6 | 27.3 | 0.0 | 0.0 | 1 | | | |
| Vc/F (liters) | 44.2 ± 1.3 | 12.8 | 0.0 | 0.0 | 0.593 | 1 | | |
| Q/F (liters/h) | 0.856 ± 0.343 | 80.0 | 0.0 | 0.0 | 0.414 | 0.813 | 1 | |
| Vp/F (liters) | 42.1 ± 38.2 | 145 | 0.0 | 0.0 | 0.414 | 0.813 | 1 | 1 |

$\sigma$(~%CV) = 25.0 ± 4.5.

where $\phi$ represents the scale parameter. This parameterization restricts the random effects of Q/F and Vp/F to be 100% correlated, and forces the correlations between the random effects for the other PK parameters with Q/F or Vp/F to be identical. The estimates of the fixed effects and variance components are displayed in Table 1.

Figure 1a displays the observed and predicted eplerenone concentration-time profiles by regimen. For compactness, the single-dose plasma concentrations have been normalized to 100 mg. Median plasma concentrations (also scaled) are plotted to provide a benchmark for the model fit; i.e., they provide a measure of central tendency for the distribution of concentrations at each time point. The model predictions show good concordance with the medians throughout the time curve. Figure 1 also indicates that the terminal elimination phase occurs near the lower limit of quantification. The apparent biphasic elimination is not an artifact of assay sensitivity. Doses greater than 100 mg, which do not have assay sensitivity issues, exhibit this terminal elimination phase also. The data from doses greater than 100 mg were not included in the model fit because these doses are not dose proportional due to saturable dissolution. Plots of individual and population weighted residuals (NONMEM's IWRES and WRES) were also examined (not shown). These plots were assessed in conjunction with plots of the empirical Bayes predictions of the $\eta$'s (not shown), and no lack of fit of the model was indicated. Semilog plots of the absolute values of the individual residuals vs. the individual predictions (not shown) were used to confirm that the log transformation had stabilized the intrasubject variability (residual variance).

### 14.2.2 Simulation Model Validation

To assess whether the proposed PK model had captured the stochastic features of the data, the PK profiles of 44 patients with doses and sample times identical

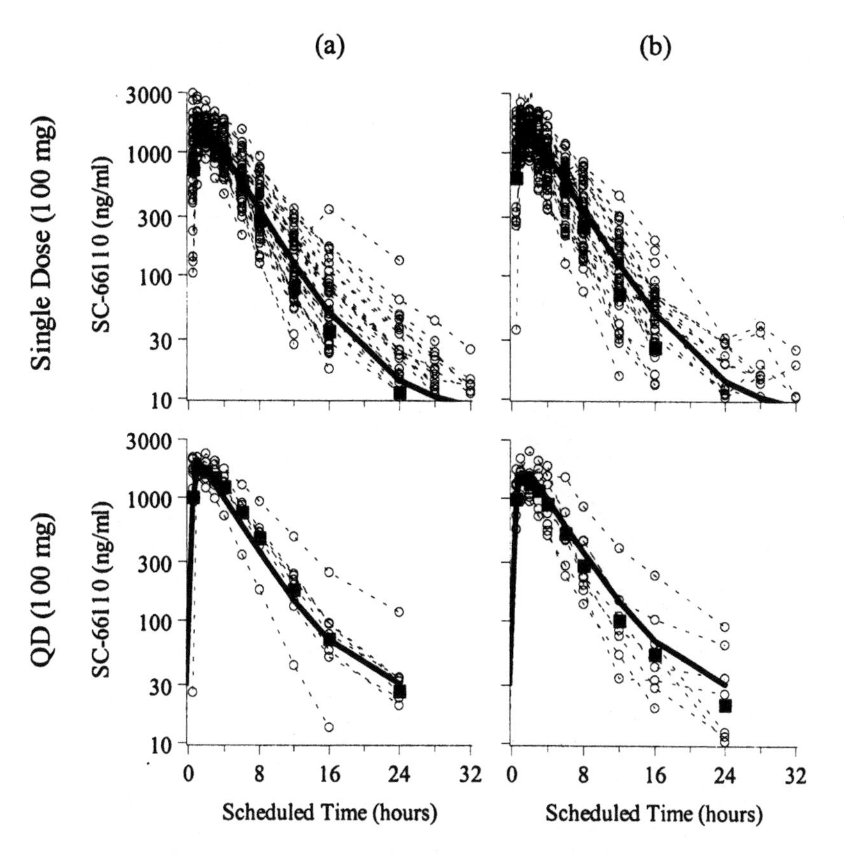


**FIGURE 1** (a) Individual drug plasma concentration-time curves (-○-), medians (■), and population mean two-compartment model fit (▬) in healthy volunteers. (b) Simulated individual drug plasma concentration time curves, medians, and population mean two-compartment model fit. Plasma concentrations are normalized to a 100 mg dose. Concentrations determined below the lower limit of quantification (10 ng/ml) were set to 0 for purposes of calculating the median.

to the healthy volunteer PK data were simulated using the parameter estimates, variances, and covariances reported in Table 1; the $\eta$'s were simulated from a multivariate normal distribution with mean 0 and covariance matrix $\Omega$. An example of a simulated data set is displayed as Figure 1b. The single-dose concentrations have been scaled to a 100 mg dose as previously described. Qualitatively, the observed and simulated data appear similar. More formally, in the spirit of the posterior predictive check (4), 10 independent data sets were simulated. The

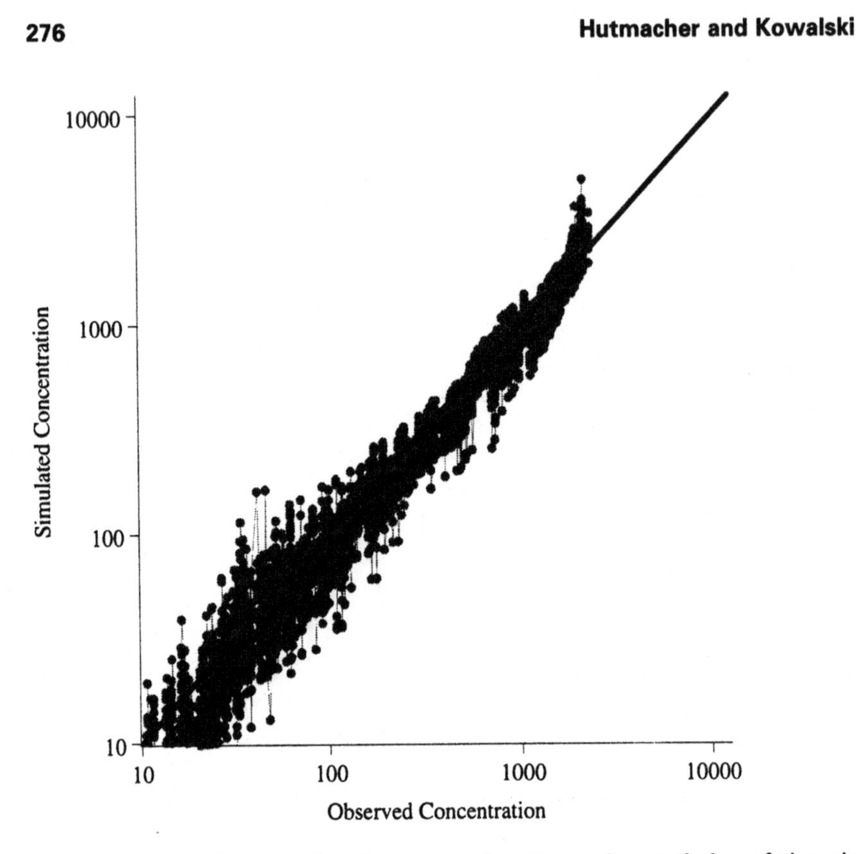


**Figure 2** Quantile-quantile plot comparing the order statistics of the observed and simulated ($N = 10$) plasma concentrations in healthy volunteers. The solid line denotes the line of concordance.

10 data sets were then merged (by data set) with the original data by their order statistics. A quantile vs. quantile (Q-Q) plot was then constructed to assess the concordance of the simulated and actual data distributions. Figure 2 displays the outcome, and suggests that the model adequately captures the stochastic features of the data. It should be noted that all the data in the Q-Q plot has been collapsed for clarity. Initially, it is good practice to bifurcate the Q-Q plots by dose/regimen and between the absorption and elimination phases to determine where the lack of concordance, if any, resides.

## 14.3 SAMPLE TIME MODEL

### 14.3.1 Data

A sample time distribution from a phase III eplerenone clinical trial was not available. Therefore, data from two Celebrex population PK studies consisting of 195 patients were used to model a sampling time distribution. The Celebrex studies comprise a patient population (arthritis) different from the hypertensive population considered in this trial. The proposed eplerenone PK substudy visits are similar in nature to those observed in the Celebrex studies, however. The designs are both based on the patient taking the scheduled medication in the morning and visiting the clinic at a pseudorandom (essentially not restricted) time during the day. Therefore, it is anticipated that the distribution of sample times relative to dose would be comparable between the two populations.

### 14.3.2 Modeling Methods

The first sample times relative to dose for the Celebrex studies were used to model the random sampling distribution. Let $T$ denote the random variable that represents the first sample time relative to dose (or the sampling distribution). To model the sampling density for $T$, the observed sampling distribution was partitioned using key features of the data. Complier status was one feature of the data. Let $X$ denote a Bernoulli random variable that represents complier status, where $P(X = c) = \pi_c$ and $P(X = nc) = \pi_{nc} = 1 - \pi_c$. Patients, who took their medication on the day of the visit, were considered "compliers," i.e., $X = c$. A second feature of the data was that the sampling distribution $T$, conditioned on $X = c$, was bimodal. The early and latter modes of this distribution were interpreted as morning and afternoon sampling, respectively, since Celebrex maintains a morning dosing regimen. Let $Z$ denote the random variable that represents morning/afternoon sampling status, where $P(Z = m) = \pi_m$ and $P(Z = a) = \pi_a = 1 - \pi_m$ ($m$ denotes morning and $a$ denotes afternoon). The conditional distribution of the sampling distribution given the patient was a complier was postulated using a mixture distribution as

$$\begin{aligned} P(T|X = c) &= P(Z = m)P(T|X = c, Z = m) \\ &\quad + P(Z = a)P(T|X = c, Z = a) \end{aligned}$$

where $P(T|X = c, Z = m)$ denotes the conditional distribution of $T$ given $X = c$ and $Z = m$ and $P(T|X = c, Z = a)$ denotes the conditional distribution of $T$ given $X = c$ and $Z = a$. The distribution, $P(T|X = c)$, was assumed to be a mixture of gamma distributions since this family of distributions is flexible and positive in value. More specifically, it was assumed that $P(T|X = c, Z = m) \sim \Gamma(\mu_m, \sigma_m^2)$ and $P(T|X = c, Z = a) \sim \Gamma(\mu_a, \sigma_a^2)$, where $\Gamma$ denotes the gamma distributions for morning and afternoon sampling, respectively (with means $\mu$ and variances $\sigma^2$). The mixing parameter was $P(Z = m) = \pi_m$. An extended least squares procedure was used to estimate the parameters $\pi_m$, $\mu_m$, $\mu_a$, $\sigma_m^2$, and $\sigma_a^2$ of the mixture distribution.

A very small percentage of patients were noncompliers ($X = nc$), so interpretation of the sampling distribution conditioned on $X = nc$ was difficult. Therefore, a simplifying assumption was made that the sampling distribution, conditioned on $X = nc$, was $P(T|X = nc) = P(T + 24|X = c)$; i.e. the sampling distribution for noncompliers was the same as the compliers plus 24 h (1 day latter). The probability $\pi_{nc}$ was estimated using a frequency-based estimator, i.e., the number of patients that took their dose on the day before divided by the total number of patients. The overall sampling distribution was thus modeled as

$$P(T) = P(X = c)P(T|X = c) + P(X = nc)P(T|X = nc)$$

### 14.3.3 Results

The mixing probability of the two gamma distributions was estimated as $\pi_m = 0.786$, with the means of the two distributions estimated as $\mu_m = 1.78$ and $\mu_a = 5.61$ h (i.e., an estimated 78.6% of the data had an estimated mean of 1.78 h and an estimated 21.6% of the data had an estimated mean of 5.61 h). A pooled estimate of variability was necessary, since the data were unable to support separate variability estimates for each distribution. The pooled variability was estimated as $\sigma_m^2 = \sigma_a^2 = 1.08$. Figure 3 displays a histogram of the median of 10

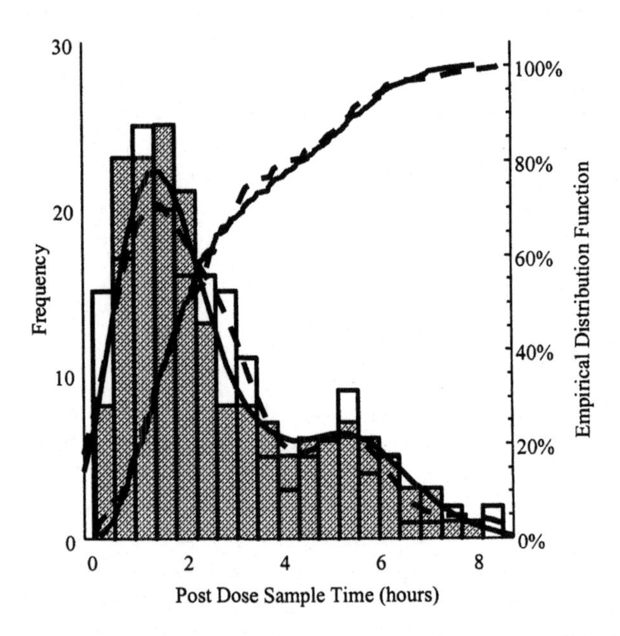


**FIGURE 3** Histogram comparing the observed (black bars) and simulated (gray filled bars) distribution of the postdose sample times in complying patients. The bimodal features are captured by the density smooth [observed (▬) and median (- -) of $N = 10$ simulations]. The empirical (cumulative) distribution functions of the two densities are also shown [observed (▬) and median (- -) of $N = 10$ simulations].

simulations of time of first sample relative to dose and the actual times for the complying patients. The mixture model captures the bimodality, which is interpreted as morning (a.m.) and afternoon (p.m.) sampling. The graph also reflects the greater frequency of sampling times in the earlier sample time distribution (1.78 h); this feature is captured by the estimate of the mixing probability (78.6%). The probability of not complying was estimated to be $\pi_{nc} = {}^{15}/_{195}$ (i.e., 15 out of 195 patients or 7.7% did not take their dose on the morning of their visit).

## 14.4 POPULATION PK SUBSTUDY DESIGN

The PPK substudy comprises a subset of the patients enrolled in this phase III, double-blinded, parallel group, randomized trial. To ensure suitable power for the efficacy analysis, a minimum of 540 black and Caucasian patients with mild to moderate hypertension were considered necessary (only 140 patients in each of the three treatment arms were expected to complete the study). These patients are scheduled to enter into the study in an approximately a 2:1 ratio, balanced, and stratified within each study center.

Table 2 lists the sequence of events and how the PPK substudy will be integrated into the overall trial. The trial consists of three periods: a screening period, a placebo run-in period, and a double-blind treatment period. Eligible patients begin the double-blind treatment period of the trial randomized to eplerenone 50 mg, active comparator, or matching placebo. If a patient's blood pressure is uncontrolled at certain, prespecified study visits, his dose of study medication will be doubled (see Table 2). Thus, patients failing to respond to eplerenone 50 mg are titrated to eplerenone 100 mg. The placebo and active comparator arms are treated in a similar manner. The protocol does allow for an additional dose escalation, but this can occur only after the last scheduled visit of the PPK substudy.

To ensure a distribution of samples times able to support a population model, special PK visits were necessary. The efficacy and safety measurement

**TABLE 2** PK Substudy Schedule of Visits and Observations

| | Screen | Single-blind treatment (Placebo Run In) | | | Double-blind treatment | | | | | |
|---|---|---|---|---|---|---|---|---|---|---|
| Week ± 3 days | −6 to −3[a] | −4 to −2[a] | −2[a] | 0[a,b] | 1[c] | 2[a] | 4[a,d] | 5[c] | 6[a] . . . | 17[a] |

[a] Typical study visit.
[b] PK substudy visit: A 1 h post dose blood sample on day 0.
[c] PK substudy visit: two pseudorandom blood samples 1 h apart.
[d] Escalation to double the dose is possible.

**TABLE 3** Summary of Design Descriptions for the Population PK Substudy for the Black/Caucasian Antihypertension Trial

| Design | Day 0 | Week 1[a] | Week 5[b] |
|---|---|---|---|
| Standard | — | $T$, $T + 1$ | $T$, $T + 1$ |
| Standard + 1 h | +1 | $T$, $T + 1$ | $T$, $T + 1$ |
| Stratified | — | $T$, $T + 1$ (a.m., p.m.) | $T$, $T + 1$ (p.m., a.m.) |
| Stratified + 1 h | +1 | $T$, $T + 1$ (a.m., p.m.) | $T$, $T + 1$ (p.m., a.m.) |

$T$ = random sampling time
+1 = 1 h sample post initial dose
Stratified = stratification to one morning and one afternoon visit
[a] 50 mg daily (steady state)
[b] 0.4 probability of 50 mg daily or 0.6 probability of 100 mg daily (steady state)

visits require the patient to withhold his medication until arrival at the clinic. If the PK sampling were integrated into these visits, lengthy clinic stays, or returns to the clinic on that day, would be required to maintain the necessary breadth of sampling times. Most likely, the trial would require a fixed-time sampling design, which is considered cumbersome and inconvenient for the patient. The special PK visits allow the patient to dose according to his regular schedule and visit the clinic at a convenient time during the day. This pseudorandom sampling plan uses the stochastic nature of patients visiting the clinic and the resulting distribution of times to support the PPK model.

The PPK substudy visits are scheduled for weeks 1 and 5 of the double-blind treatment period with the potential for a sample after the first dose of double-blind study medication on day 0. Four designs with this framework are considered for the PPK substudy. The designs consist of a "standard," "standard + 1 h," "stratified," and "stratified + 1 h." Table 3 lists a description of the possible designs. The standard design consists of a pseudorandom sample and a sample 1 h later at each of the two study visits (weeks 1 and 5). The standard + 1 h design is identical to the standard design with the inclusion of a 1 hour sample following the initial dose of double-blind study medication (day 0). The stratified and stratified + 1 h designs are similar to the standard and standard + 1 h designs, except the sampling times are stratified by visit (weeks 1 and 5) to either morning or afternoon sampling. More details regarding the stratification are presented in the next section.

## 14.5 PPK SUBSTUDY SIMULATIONS

### 14.5.1 Methods

The sample time model previously described was used to simulate the random sample times for the hypothetical trial designs. The random sample times for the standard design were generated using the mixture of gamma distributions and the binomial distribution for complier status. More specifically, based on the mixing probability, 78.6% ($\pi_{nc} = 0.786$) of the data were expected to be simulated using the gamma distribution with a mean of 1.78 h and the remaining 21.4% were expected from the gamma distribution with a mean of 5.61 h. The gamma distributions were both simulated using a variance of 1.08 (from the sample time model fit). A Bernoulli random variable with $\pi_{nc} = 0.077$ was used to determine if the patient was a noncomplier (an expected 7.7% of the data). Twenty-four hours were added to the noncomplier sample times to emulate a missed dose on the day of the visit (see Section 14.3.2).

The standard + 1 h design was simulated identically to the standard design except for the inclusion of an additional sample 1 h after the initial dose of study medication (day 0). The stratified and a stratified + 1 h were simulated similarly to the standard and standard + 1 h designs, respectively, except for one key feature. The stratified designs were simulated by randomly allocating each visit (weeks 1 and 5) to a morning and afternoon sampling time for each patient; i.e., 50% of the data was simulated from the gamma distribution with mean 1.78 h (morning) and 50% was simulated from the gamma distribution with mean 5.61 h (afternoon). A 1.08 variance was used for these distributions as well. The stratified + 1 h design has a 1 h postdose sample on day 0, similar to the Standard + 1 h design.

For each design and simulation, PK profiles were simulated based on the PPK model previously discussed in Section 14.2 using the simulated random sample times. Patients at day 0, i.e. for designs with a "+1 h" sample, were simulated using a single dose of 50 mg eplerenone. Steady state was assumed for all the designs at weeks 1 and 5. For week 1, all patients were simulated at 50 mg daily. To reflect dose escalation, a Bernoulli random variable was used to simulate whether a patient was dose escalated to 100 mg daily by week 5. Clinical judgment expected the probability for dose escalation to be 0.6; i.e., 60% of the simulated patients were expected to dose escalate to 100 mg daily (see Table 3).

Two hundred data sets were simulated for each design/sample size with a 40% reduction in clearance for an arbitrary 5 or 10% subpopulation; a greater than 40% reduction was considered to be clinically relevant to eplerenone's dosing regimen. Thus, the subpopulation was simulated with a CL/F of 6.48 liters/h compared to the reference population's CL/F of 10.8 liters/h (see Table 1). A base model and a covariate model were fit to each simulated data set using the first-order (FO) estimation method of the NONMEM software. The base model restricts the reference population's and the subpopulation's estimates of CL/F to be identical, while the covariate model allows for separate estimates. Likelihood ratio tests (LRTs) were then performed on each data set. The proportion of significant results ($\chi^2$ test with $p$ value $< 0.05$) for each design/sample size was used as an estimate of power.

It was anticipated that a one-compartment model approximation to the two-compartment model would alter the asymptotic distribution of the LRTs. To assess the potential aberrance, simulations were conducted with no reduction in CL/F between the two subpopulations, i.e., under the null hypothesis. An empirical distribution of the test statistics (LRTs) from these simulations was used to assess the actual significance level (type I error) of the test.

The parameter estimates from the covariate model for each simulation were used to calculate the bias for the design/sample size. Percent bias was estimated using the formula

$$\%_{\text{bias}}(\hat{\xi}_i) = \frac{\hat{\xi}_i - \xi}{\xi} \times 100\%$$

where $\xi \in \{\theta_0, \Omega, \sigma\}$ is the true parameter (fixed effects and variance components) corresponding to the two-compartment model and $\hat{\xi}_i$ represents the corresponding estimate of $\xi$ from the one-compartment model for simulation $i$, $i \in \{1, \ldots, 200\}$.

### 14.5.2 Results

An example of a simulated data set of the stratified + 1 h design of sample size 100 with 40% change in clearance in a 10% subpopulation is displayed as Figure 4. The concentrations and predictions in this plot are scaled to a 50 mg dose for better acuity. In addition, the population predictions with and without the reduction in CL/F are displayed. The decrease in the predicted terminal slope from the covariate model relative to the noncovariate model demonstrates the difference in CL/F between the two subpopulations. Preliminary modeling confirmed that there was insufficient information to fit a two-compartment model. A one-compartment model was therefore used to approximate the two-compartment model. The approximation was assessed for each design by the median bias of key model parameters (i.e., CL/F).

The LRTs based on one-compartment model approximation showed an inflated type I error using a $\chi^2$ with 1 degree of freedom. The empirical distribution's 95th percentile of the LRT under the null hypothesis was 5.726. This cutoff value is much larger than the $\chi^2(1)$ cutoff value of 3.84 for $\alpha = 0.05$. Figure 5 displays the empirical distribution, its cumulative distribution function, and the cumulative distribution function of a $\chi^2(1)$ random variable. The empirical distribution's cutoff value was similar to that of a $\chi^2$ with 2 degrees of freedom (95th percentile of 5.991), however. Therefore, a $\chi^2(2)$ test was used to calculate significance and maintain the type I error ($\alpha = 0.05$).

The estimates of power for each design are displayed in Table 4. A general trend across the designs is apparent. The estimates increase with increased sample

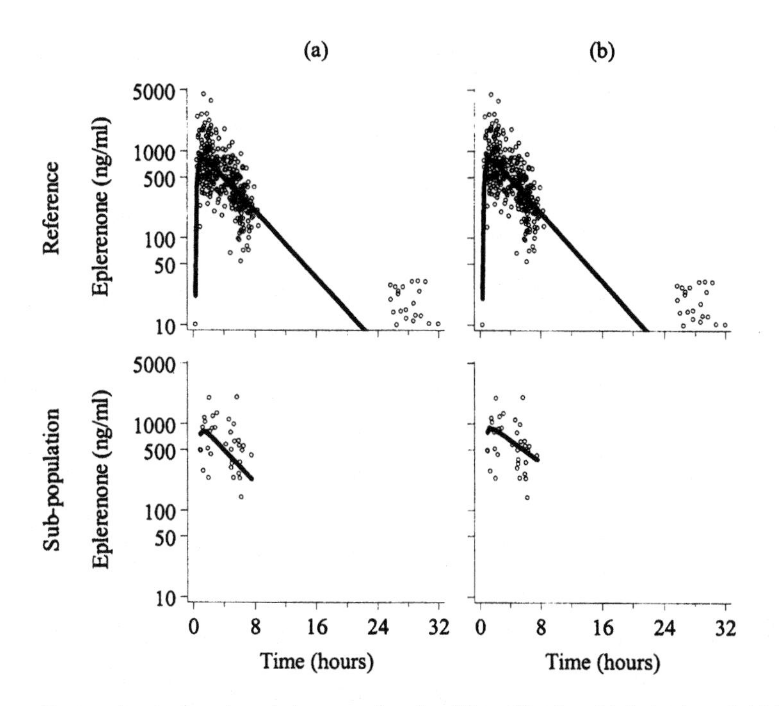


**FIGURE 4** A simulated data set for the "Stratified + 1 h" design. (a) No covariate was added to the model to account for the subpopulation's difference in CL/F. (b) A covariate was added to the model to account for the subpopulations's difference in CL/F. The symbol ○ represents the simulated data and ▬ represents population mean prediction. Only the steady-state data are presented. Concentrations simulated below the lower limit of quantification (10 ng/ml) are not shown.

size, increased proportion of the subpopulation, with the addition of the "+1 h" time point, and with the stratification to one morning and one afternoon visit. The stratified + 1 h design is recommended, since greater than 90% power is achieved with a 10% subpopulation. The sample size necessary to detect a 40% reduction in apparent clearance with at least 80% power in a 5% subpopulation is prohibitive since it exceeds the expected sample size for the primary safety and efficacy analyses of 140 per treatment arm.

For the proposed design (stratified + 1 h), Table 5 displays median percent bias, power, and significance levels for the one-compartment model's approximation of the true two-compartment model. The results are displayed for both the FO and FOCE estimation methods. The symbol Δ represents the parameter that

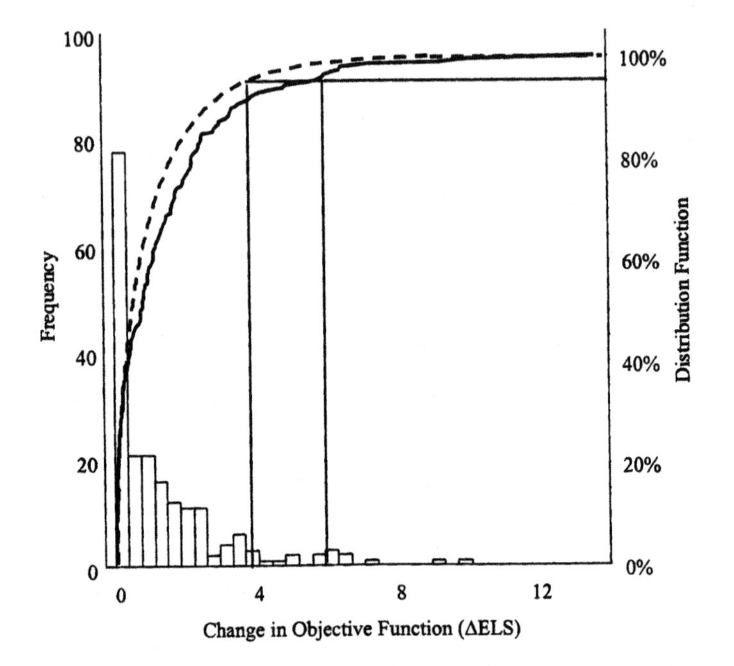


**FIGURE 5** Histogram of the empirical distribution ($N = 200$) of the LRTs for a one-compartment model fit to data simulated from a two-compartment model under the null hypothesis. The empirical distribution function of the LRTs (—) and the distribution function of a $\chi^2(1)$ random variable (- -) are plotted also. Vertical lines identify the 95th percentiles for a $\chi^2(1)$ (3.84) and the simulated test statistic (5.726).

**TABLE 4** Estimates of Percent Power by Design and Sample Size (FO Method)

| Sample size | $N = 100$ | | $N = 140$ | |
|---|---|---|---|---|
| Subpopulation | 5% | 10% | 5% | 10% |
| Standard | 42.5 | 68.0 | 55.5 | 81.5 |
| Stratified | 60.0 | 88.5 | 74.5 | 94.5 |
| Standard + 1 h | 48.5 | 71.0 | 63.0 | 87.5 |
| Stratified + 1 h | 66.0 | 91.0 | 77.0 | 98.0 |

**TABLE 5** Percent Bias and Power of the Recommended Design

| | Bias (%) | |
|---|---|---|
| Parameter | First-Order (FO) | First-Order Conditional (FOCE) |
| tlag (h) | 21.2 | 36.3 |
| ka ($h^{-1}$) | 35.5 | 49.5 |
| CL/F (liters/h) | 6.51 | −0.40 |
| Vc/F (liters) | 14.1 | 17.0 |
| Vss/F (liters) | −41.6 | −40.1 |
| Δ | −5.39 | −1.45 |
| σ(∼% CV) | 1.98 | 9.06 |
| Var(ka) | 266 | 98.2 |
| Var(CL/F) | 170 | 158 |
| Covar(CL/F-Vc/F) | 730 | 735 |
| Var(Vc/F) | 1810 | 1870 |
| Power (alpha level) | 91.0 (4.5) | 90.0 (3.0) |

models the percent reduction in apparent clearance for the subpopulation relative to the reference population. The small median percent bias observed in estimating apparent clearance and Δ indicates that using a one-compartment model with this design is adequate for making inference on CL/F. The parameter, ka, demonstrated the greatest median percent bias of the fixed effects parameters.

Due to the likelihood of observing a highly biased estimate of ka, it was of interest to assess how data sets, which lead to highly biased estimates of ka, would affect the estimate of CL/F. Figure 6 shows scatter plots of the 200 FO estimates from the recommended design with estimates of ka greater than 3.5 bolded. Inspection of the plots reveals a uniform distribution of CL/F and Δ estimates for estimates of ka > 3.5; i.e., the estimates of CL/F and Δ are not highly correlated with the estimate of ka. Therefore, it can be concluded that the estimation of ka has little impact on the estimation of CL/F or Δ. It is interesting to note that the estimate of Vc/F does show an increased degree of bias with extremely biased estimates of ka (not shown).

It is also interesting to note that the FOCE method shows greater bias in the estimation of the absorption parameters and of σ, but a slightly decreased bias in the estimation of CL/F and Δ. Both methods show greatly biased estimates of the variance components with FOCE providing the best estimate for the variance component of ka. Both methods demonstrate similar power and type I error rates. It also should be noted that the beta phase for eplerenone occurs late in

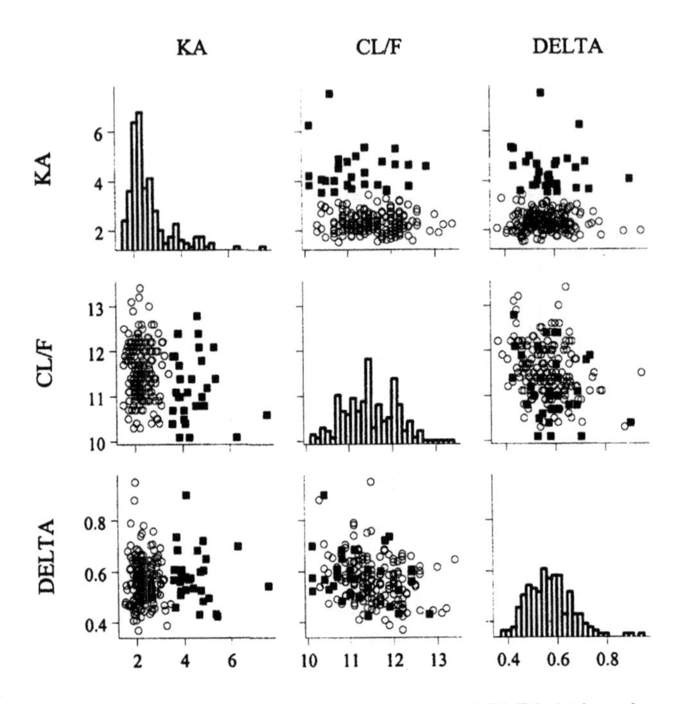


**FIGURE 6** Scatter plot of ka, CL/F, and DELTA (Δ) estimates from the one-compartment model fit to data simulated from the proposed design ($N$ = 200). The symbol ○ represents estimates when ka was estimated to be ≤3.5, and ■ represents estimates when ka was estimated to be >3.5. The histograms reflect the distribution of the estimates.

the concentration-time profile and close to the sensitivity limit of 10 ng/ml for the doses considered. Therefore, the estimate of volume for the one-compartment model is closer to the apparent central compartment volume of the two-compartment model (Vc/F) as opposed to the apparent steady-state volume of distribution (Vss/F = Vc/F + Vp/F).

The FOCE method was not used to screen the designs for power and bias. Since it was known a priori that a one-compartment model would be necessary to approximate the true two-compartment model, it was unclear that the FOCE method could provide improved, if not worse, results than the FO method. In addition, the FOCE method requires an increase in computation time and can suffer from difficulty in convergence. Table 5 demonstrates the two methods are similar, however, with respect to power and type I error for the recommended design.

## 14.6 DISCUSSION

This PPK substudy was designed to make inference on patient factors affecting eplerenone exposure (CL/F) in hypertensive patients. The inferential properties of the chosen design were established by demonstrating little bias in estimating CL/F and requiring a sufficient sample size for adequate statistical power in detecting a clinically relevant change in CL/F for an arbitrary, certain sized subpopulation. Phase III PPK studies are typically both confirmatory and exploratory in nature. Covariates that influence CL/F in phase I healthy volunteer studies need to be confirmed in the population for which the drug will be prescribed. However, many covariates observed in patients are not observed in healthy volunteers. Other factors such as various disease states and concomitant medication usage could influence the patients' exposure to the drug. Therefore, it is necessary to have a sample size suitable to detect a relevant difference in a meaningfully sized, arbitrary subpopulation. To help facilitate the recruitment of this desired sample size, this PPK substudy was designed with a convenient sampling plan to the patient and clinic.

Self-contained studies using sparse sampling strategies and less complicated models are not the only available designs/analyses for PPK studies. Aarons et al. (1), at a meeting in Brussels in March 1995, discussed mixing data-rich phase I and phase II studies to support structurally more complex models. No consensus was reached regarding this issue. In our opinion, the pooling of phase I and II studies and the resulting data heterogeneity are problematic. Unexplained study-to-study differences, increased interindividual variability in patient studies even after accounting for patient factors, and missing covariates due to the failure to collect this information in all the studies are a few of the issues. In our experience, a single self-contained study results in cleaner data analysis and better inference on the patient population in which the study was conducted and in which the drug was intended to be administered.

Simulation is a powerful and flexible tool in which to assess the type of designs previously discussed. Indeed, many examples, where simulations were used to design PPK studies, can be found in the literature (5–8). Other, notable, theoretically appealing methodologies exist for designing PPK studies. Methods that maximize the Fisher information (D-optimal) have been proposed for designing PPK trials (9–12). These designs require fixed sample times or sampling windows, which may burden the patient and the clinic in an outpatient setting. These inconveniences could hinder recruitment, diminishing the pool of patient factors on which to draw inference. Furthermore, CL/F or exposure was considered the primary parameter that could affect the dosing regimen for this drug. Methods that maximize the information matrix with respect to all parameters do not specifically address inferences on key parameters. The simulation strategy discussed above, allows one to power the study in a clinically meaningful way (with respect to covariate subpopulations and hypothesis tests) and assess bias in key parameters. Furthermore, simulation facilitates testing the design's sensitivity to assumptions and their ramifications on bias and power (13, 14).

## REFERENCES


1. L Aarons, LP Balant, F Mentre, PL Morselli, M Rowland, J-L Steimer, S Vozeh. Practical experience and issues in designing and performing population pharmacokinetic/pharmacodynamic studies. Eur J Clin Pharmaco 49:251–254, 1996.
2. KG Kowalski, MH Hutmacher. Design evaluation of a population pharmacokinetic study using clinical trial simulations: a case study. Stat Med 20:75–92, 2001.
3. SL Beal, LB Sheiner, AJ Boeckmann. NONMEM User's Guide, parts I–VIII: NONMEM Project Group, University of California, San Francisco, CA, 1998.
4. A Gelman, X-L Meng, H Stern. Posterior predictive assessment of model fitness via realized discrepancies. Statistica Sinica 6:733–807, 1996.
5. MK Al-Banna, AW Kelman, B Whiting. Experimental design and efficient parameter estimation in population pharmacokinetics. J Pharmacokin Biopharm 18:347–360, 1990.
6. LB Sheiner, Y Hashimoto, SL Beal. A simulation study comparing designs for dose ranging. Stat Med 10:303–321, 1991.
7. Y Hashimoto, LB Sheiner. Designs for population pharmacodynamics: value of pharmacokinetic data and population analysis. J Pharmacokin Biopharm 19:333–353, 1991.
8. M Hale, WR Gillespie, S Gupta, B Tuk, NHG Holford. Clinical trial simulation as a tool for increased drug development efficiency. Applied Clin Trials 5:35–40, 1996.
9. GEP Box, HL Lucas. Design of experiments in non-linear situations. Biometrika 46:77–90, 1959.
10. F Mentré, A Mallet, D Baccar. Optimal design in random-effects regression methods. Biometrika 84:429–442, 1997.
11. M Tod, F Mentré, Y Merlé, A Mallet. Robust optimal design for the estimation of hyperparameters in population pharmacokinetics. J Pharmacokin Biopharm 26:689–716, 1998.
12. S Retout, S Duffull, F Mentré. Development and implementation of the population Fisher information matrix for the evaluation of population pharmacokinetic designs. Comput Methods Programs Biomed 65:141–151, 2001.
13. X Jia, and JR Nedelman. Errors in time in pharmacokinetic studies. J Biopharm Stat 6:303–318, 1996.
14. P Girard, LB Sheiner, Kastrissios[HK1], and TF Blaschke. Do we need full compliance for population pharmacokinetic analysis? J Pharmacokin Biopharm 24:265–282, 1996.

# 15

# Use of Modeling and Simulation to Optimize Dose-Finding Strategies

**Jaap W. Mandema and Wenping Wang**
Pharsight Corporation, Mountain View, California, U.S.A.

## 15.1 INTRODUCTION

An important step in drug development is to identify the dose or dose range of a compound that offers adequate efficacy with minimal safety risk. Phase II often fulfills a critical step in learning about the dose-response relationship in the patient population to be treated. The main objective of phase II dose-finding trials is to identify dosing procedures and target patient populations for further clinical development. The hope is to identify one or two doses with the appropriate marketable characteristics that can be confirmed in phase III of the development program, or to stop development. Therefore, the design of the phase II dose-finding trials should be optimized for the trial's ability to quantify the key features of the dose-response relationship and the trial's ability to determine the "right" dose(s), if any, for further development.

Clinical trial modeling and simulation quantifies a trial's ability to show specific features of the dose-response relationship in a given patient population. By setting a clear treatment target up front, the phase II dose-finding strategy can be optimized to have the highest likelihood to identify the optimal dose and dosing frequency to achieve that target in a certain patient population. The benefit is a conclusive dose-finding strategy, reducing the uncertainty of a successful outcome of phase III.

Modeling and simulation is a powerful new tool to optimize the phase II dose finding strategy (1, 2). Recently, modeling and simulation of trial and treatment outcomes, also known as computer-assisted trial design (CATD), has emerged as a novel methodology to improve the scientific decision process with respect to trial design and drug development strategy (3, 4). The simulations are based on a pharmacologically (e.g., placebo-response model, PK/PD relationship) and statistically (e.g., within- and between-subject variations) realistic model of drug action in the patient population.

This drug model characterizes the probability distribution of trial outcomes conditional on trial design and drug, patient, and disease characteristics. The objective is to integrate as much prior knowledge into the model as possible and to carefully document the assumptions. Literature, preclinical and clinical data of the new chemical entity (NCE) in development and related compounds in the therapeutic area can all be used to derive the model. The model is applied to quantify the uncertainty in trial outcome due to model uncertainty and patient variability as well as to quantify the sensitivity in trial outcome to controllable (e.g., treatment arms, sample size) and uncontrollable (e.g., drug potency, patient compliance) trial design factors. This provides a clear, consistent, scientific, and quantitative rationale for selecting the best treatment and trial strategies. For example, this can be used to derive a robust trial design given the uncertainty in the drug model and uncontrollable trial features. Or it can be used to compare the information yield of competing trial strategies vs. time, costs, and trial complexity. This enables the development team to balance the value vs. cost of additional information derived from the trial. The probabilistic trial outcome models can be leveraged into a decision-making framework that can be used to optimize the overall development strategy from a financial value perspective.

In this manuscript a case study is presented to illustrate the use of modeling and simulation to optimize the phase II dose-finding strategy for a new 5-$HT_{1D}$ agonist indicated for the relief of moderate to severe migraine pain. A drug model is derived for the new compound on the basis of literature (including summary basis for approval), preclinical and early clinical data of the NCE, and other 5-$HT_{1D}$ agonists such as sumatriptan, zolmitriptan, naratriptan, and rizatriptan. This model is used to explore the likely range of treatment outcomes for the NCE in comparison to the other 5-$HT_{1D}$ agonists. This provides initial insight on dose selection and marketable efficacy targets. The model is subsequently used to design a dose-ranging study in terms of number of treatment arms, spacing of active doses, and sample size with the objective to confirm efficacy and to support dose selection with the appropriate marketable profile for further development.

## 15.2 DEVELOPMENT OF THE DRUG MODEL

Migraine is a common disorder affecting 9–16% of the adult population, with women having approximately a two-fold higher incidence than men. Recently, several 5-HT$_{1D}$ agonists, such as sumatriptan, zolmitriptan, naratriptan, and rizatriptan, have become available for the treatment of migraine pain. The severity of migraine headache is often measured on a four point scale with 0 = none, 1 = mild, 2 = moderate, and 3 = severe. Efficacy endpoints include the fraction of patients with headache relief (defined as none or mild pain) at a specific time point after the start of treatment (often 2 h) and time to remedication or recurrence. The fraction of patients with headache relief was selected as the primary endpoint.

The available clinical trial data for oral treatment with these 5-HT$_{1D}$ agonists was extracted from the literature and summary basis of approval (6–11). Dose-ranging data from the larger trials, including at least 350 patients, were included in the analysis. Response data on 7835 patients from seven dose-ranging trials was available for analysis. Table 1 gives an overview of the available data. The data was analyzed to derive the relationship between 5-HT$_{1D}$ agonist dose and fraction of patients having migraine pain relief at 2 h after treatment (sumatriptan, zolmitriptan, and rizatriptan) or 4 h after treatment (naratriptan). The data was also used to quantify the trial-to-trial differences in response magnitude.

The following logistic regression model is proposed to link the dose of a compound and the fraction of patients with pain relief, Pr(Pain Relief), in a specific trial:

$$\begin{aligned} \text{Pr(Pain Relief)} &= g\{\beta + \frac{E_{\max} \cdot D_T^n}{D_T^n + ED_{50,T}^n} + \eta\} \\ g\{x\} &= \frac{e^x}{1 + e^x} \end{aligned} \tag{1}$$

where $\beta$ is the intercept, reflecting placebo response; $E_{\max}$ is the maximal drug effect; $D_T$ is the dose of 5-HT$_{1D}$ agonist $T$; $ED_{50,T}$ is the dose of 5-HT$_{1D}$ agonist $T$ to achieve 50% of $E_{\max}$; $n$ is the factor reflecting steepness of dose-response relationship; and $\eta$ is a trial specific random effect assumed to be normally distributed with mean zero and variance $\omega^2$. The likelihood of observing $N_{r,i}$ patients with migraine pain relief in trial $i$ with a total of $N_{t,i}$ patients is given by

$$L(N_{r,i}, N_{t,i}) = P^{N_{r,i}} \cdot (1 - P)^{N_{t,i} - N_{r,i}} \tag{2}$$

The Laplacian approximation to the log likelihood as implemented in the NONMEM program (NONMEM version V[12]) is used to provide maximum likelihood estimates of the model parameters. Model selection was done on the basis

**TABLE 1** Dose-Response Data of 5-$HT_{1D}$ Agonists at 2 or 4 h (Naratriptan) Posttreatment

| Drug (reference) | Dose (mg) | Number of patients | Number of patients with pain relief | Percent of patients with pain relief (%) |
|---|---|---|---|---|
| Rizatriptan (5) | 0 | 304 | 106 | 35 |
| | 5 | 458 | 284 | 62 |
| | 10 | 456 | 324 | 71 |
| Rizatriptan (6) | 0 | 85 | 15 | 18 |
| | 10 | 89 | 46 | 52 |
| | 20 | 82 | 46 | 56 |
| | 40 | 121 | 81 | 67 |
| Naratriptan (7) | 0 | 602 | 199 | 33 |
| | 0.25 | 591 | 230 | 39 |
| | 1 | 591 | 337 | 57 |
| | 2.5 | 586 | 398 | 68 |
| Sumatriptan (8) | 0 | 91 | 25 | 27 |
| | 25 | 286 | 123 | 43 |
| | 50 | 285 | 165 | 58 |
| | 100 | 277 | 161 | 58 |
| Sumatriptan (9) | 0 | 205 | 55 | 27 |
| | 100 | 305 | 204 | 67 |
| | 200 | 283 | 207 | 73 |
| | 300 | 299 | 200 | 67 |
| Zolmitriptan (10) | 0 | 121 | 41 | 34 |
| | 1 | 125 | 66 | 53 |
| | 2.5 | 260 | 169 | 65 |
| | 5 | 245 | 164 | 67 |
| | 10 | 248 | 166 | 67 |
| Zolmitriptan (11) | 0 | 88 | 18 | 20 |
| | 5 | 179 | 109 | 61 |
| | 10 | 191 | 128 | 67 |
| | 15 | 194 | 130 | 67 |
| | 20 | 188 | 139 | 74 |

of the log likelihood criterion ($p < 0.05$) and visual inspection of the fits. The difference in $-2$ times the log of the likelihood ($-2$LL) between a full and reduced model is asymptotically $\chi^2$ distributed with degrees of freedom equal to the difference in number of parameters between the two models. A decrease of more than 3.84 in $-2$LL is significant at the $p < 0.05$ level for one additional parameter. Standard errors of the parameter estimates are approximated using the asymptotic variance-covariance matrix.

Figure 1 shows the fit of the dose-response model to the data. It indicates an adequate summarization of the data by the proposed model. The symbols reflect the "data-derived estimates" of the fraction of patients with pain relief at 2 (sumatriptan, zolmitriptan, and rizatriptan) or 4 h after treatment (naratriptan) at each evaluated dose in each trial. The data-derived estimates are adjusted to a placebo response of 0.28 (mean placebo response) to account for the impact of trial-to-trial differences. This adjustment allows the evaluation of the fit of the dose response part of the model to the data. In principle, Figure 1 is a residual plot, but, instead of being positioned around zero, the residuals are shown relative to the predicted response for the specific drug and a placebo response of 0.28. The data-derived estimates are the sum of the mean of the residuals of the model fit (for each drug, trial, and dose) and the model prediction. The vertical line around each of the symbols reflects a 95% confidence interval on the data-derived estimates of the fraction of patients with pain relief. Because of the logarithmic dose axis, the placebo treatment groups are plotted at the lowest dose value shown in each graph. Table 2 shows the estimates of the model parameters.

The analysis of the dose-response data for the 5-$HT_{1D}$ agonists yields several important conclusions. First, a consistent dose-response relationship is found across the 5-$HT_{1D}$ agonists. The compounds were found to differ only in their potency ($ED_{50}$) and not in their efficacy ($E_{max}$) or shape of the dose-response relationship. Including a different maximal effect for each of the 5-$HT_{1D}$ agonists in the model does not result in a significant improvement of likelihood ($\Delta - 2LL = 0.42, p = 0.94$). Similar efficacy profile is expected for compounds with a similar mechanism of action. Given the fact that the NCE's mechanism of action is similar to the 5-$HT_{1D}$ agonists, only an estimate of the new compound's potency needs to be obtained to scale the model to the novel compound. The efficacy and shape of the dose-response relationship of the NCE is likely to be the same as observed for the analogues. The relative potency of the 5-$HT_{1D}$ agonists can be derived from the $ED_{50}$ estimates. 2.5 mg zolmitriptan is about equivalent to 10 mg rizatriptan and 75 mg sumatriptan in providing pain relief at 2 h after treatment. At the recommended doses of these compounds, about 60% of the patients have adequate pain relief at 2 h after the start of treatment.

The maximum response to any 5-$HT_{1D}$ agonist is 70% at a mean placebo response of 28%. This provides insight in the product profile that needs to be

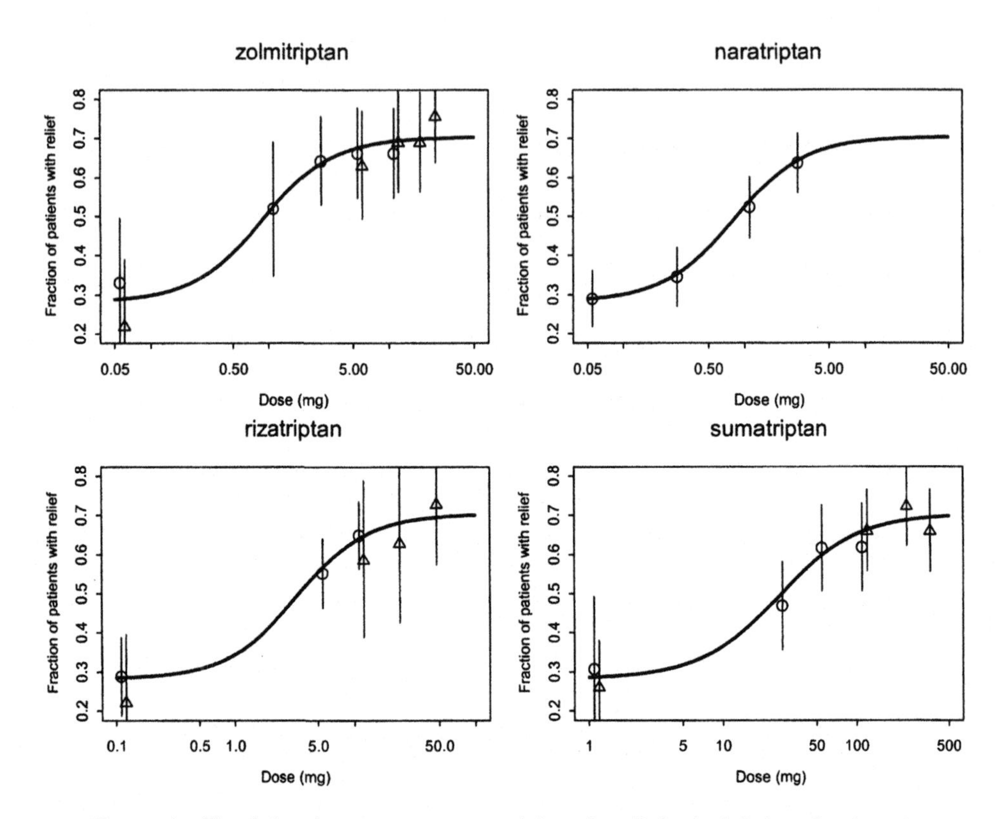


**Figure 1** Fit of the dose-response model to the clinical trial data for four 5-$HT_{1D}$ agonists. The symbols reflect the data derived estimates of the fraction of patients with pain relief at 2 (sumatriptan, zolmitriptan, and rizatriptan) or 4 h after treatment (naratriptan) at each evaluated dose in each trial. The data-derived estimates are the sum of the mean of the residuals of the model fit (for each drug, trial, and dose) and the model prediction (using a mean placebo response of 0.28). The vertical line around each of the symbols reflects a 95% confidence interval on the data-derived estimates of the fraction of patients with pain relief. Because of the logarithmic dose axis, the placebo treatment groups are plotted at the lowest dose value shown in each graph.

obtained for the NCE. The NCE needs to provide pain relief in at least 60% of the patients at the final marketed dose for the consideration of registration. The 60% threshold is the response to the marketed 5-$HT_{1D}$ agonists at their recommended doses. The maximum response of 70% shows that there is not much room for improved effectiveness by increasing the dose if the compound had a

**TABLE 2** Model Parameter Estimates and Standard Errors of the Analysis of the Relationship Between 5-HT$_{1D}$ Agonists Dose and Fraction of Patients with Adequate Pain Relief

| Parameter | Value | SE |
|---|---|---|
| $\beta$ | −0.932 | 0.124 |
| $n$ | 1.39 | 0.26 |
| $E_{max}$ | 1.8 | 0.18 |
| $ED_{50,\text{zolmitriptan}}$ | 0.984 | 0.19 |
| $ED_{50,\text{naratriptan}}$ | 0.815 | 0.12 |
| $ED_{50,\text{rizatriptan}}$ | 3.32 | 0.63 |
| $ED_{50,\text{sumatriptan}}$ | 25.9 | 3.5 |
| $\omega^2$ | 0.038 | 0.026 |

better safety profile. The only differentiation between the NCE and marketed 5-HT$_{1D}$ agonists that could be obtained is a faster speed of onset due to better pharmacokinetic properties, a lower incidence of recurrence and reduced side-effect liability at similar effectiveness.

The model also shows that the between-trial variability in response is larger than expected on the basis of a binomial distribution (pain relief or not). Removing the trial-to-trial random effect from the model results in a significant change in the likelihood ($\Delta$ $-2LL = 12.53$). The trial-to-trial variability in response is expected due to differences in the patient populations for each of the seven trials. Including a trial-to-trial random effect on $E_{max}$, in addition to the random effect in Eq. (1), does not significantly improve the fit. This is an important finding, suggesting that trial-to-trial variability in response to drug treatment is correlated with the response to placebo. What this means is that the trial-to-trial differences in placebo response are predictive for trial-to-trial variations in response to active treatment. This implies that a placebo-controlled trial would suffice to compare across the 5-HT$_{1D}$ agonists without the need to include an active control to adjust for potential trial-to-trial differences in $E_{max}$.

The scaling of the dose-response model for the 5-HT$_{1D}$ agonists to the NCE requires an estimate of the potency of the novel compound. An estimate is obtained by assuming that the relative potency between the NCE and other 5-HT$_{1D}$ agonists on the basis of free drug plasma concentrations in animals is similar in humans. On the basis of this assumption, preclinical and phase I pharmacokinetic studies of the NCE suggest a relative dose potency of 3.0 compared to sumatriptan with a standard deviation of 0.5. The standard deviation includes the uncertainty in the scaling from animal to man. The phase I studies of the NCE also showed that 75 mg is the maximum dose that would yield acceptable side effects from a marketing perspective.

## 15.3 EXPLORATION OF THE LIKELY DOSE-RESPONSE RELATIONSHIP FOR THE NCE

The first application of the model is to explore the likely dose-response relationship for the NCE given data up to phase I and dose-response data of the other 5-$HT_{1D}$ agonists. To characterize the uncertainty in the dose-response relationship of the NCE, a sample of 1000 sets of model parameters are drawn from a multivariate normal distribution with mean and variance-covariance matrix obtained from the model-building step outlined in Section 15.2. This set of model parameters reflects the uncertainty in the model for the NCE. The dose-response relationship for the NCE is calculated for each set of parameters, yielding a distribution of likely response rates as a function of dose for the novel compound.

Figure 2 shows the distribution, expressed in percentiles, of the likely response to the NCE as a function of dose. Note that this figures expresses the

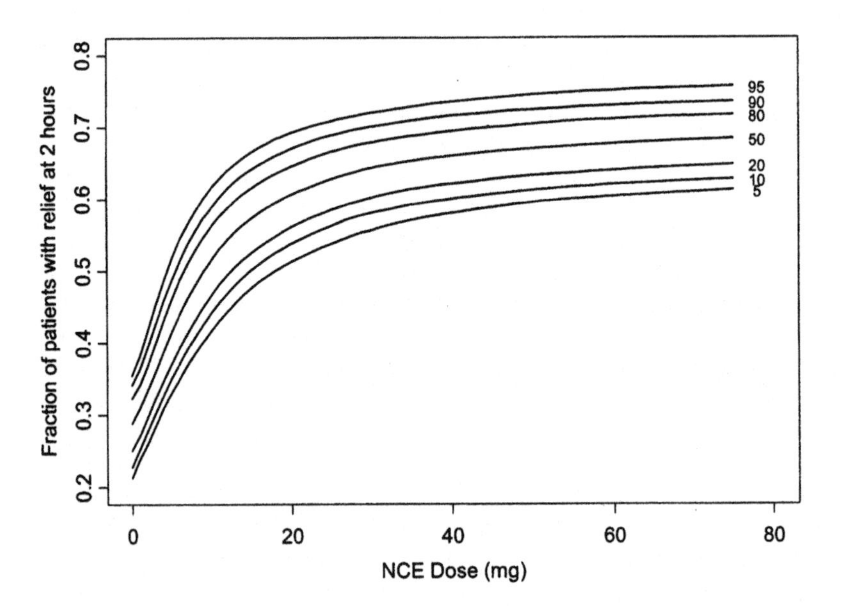


**FIGURE 2** Distribution of the likely response at 2 h posttreatment of the NCE as a function of dose. Shown are the 5th, 10th, 20th, 50th, 80th, 90th, and 95th percentiles of the distribution.

**TABLE 3** Probability That a Certain Dose of the NCE Will Exceed the Target Response[a]

| | Fraction of patients with relief | | |
|---|---|---|---|
| Dose | 0.55 | 0.6 | 0.65 |
| 10 | 0.251 | 0.071 | 0.015 |
| 20 | 0.849 | 0.535 | 0.171 |
| 30 | 0.961 | 0.793 | 0.422 |
| 40 | 0.977 | 0.882 | 0.565 |
| 50 | 0.986 | 0.927 | 0.663 |

[a] Three target response rates are shown, namely 55, 60, and 65% of the patients achieving adequate pain relief at 2 h after drug administration. The probabilities are derived from the distribution of the likely dose-response relationship for the NCE given data up to phase I and dose response data of the other 5-$HT_{1D}$ agonists (see Figure 2).

uncertainty in the dose-response relationship for the NCE and not between-subject variability. The figure provides a view of the expected dose-response relationship for the NCE before any clinical efficacy data are obtained on this compound. The figure provides an initial insight in appropriate doses to be included in the phase II dose-ranging trial in patients. The figure can be used to evaluate the probability of achieving a certain response at a selected dose. Conversely, the figure can be used to evaluate the probability of a certain dose, achieving a selected response target. For example, the expected response at 20 mg is 61% with 90% chance that the response is between 51 and 69%. On the other hand, the expected dose to achieve pain relief in 60% of the patients is 19 mg with 80% chance that the dose is between 11 and 40 mg.

A different view of the likely dose to achieve a certain response is given in Table 3. The table shows the probability of a certain dose to exceed a certain response target. From Figure 2 and Table 3 it can be concluded that 20 and 40 mg seem to be appropriate doses to be included in the dose-finding study. These doses have a 54 and 88% chance, respectively, to exceed the minimum marketing target of migraine pain relief in 60% of the patients at 2 h.

## 15.4 STUDY POWER CONSIDERATIONS

Study power at a given sample size is an important design aspect. One of the objectives of the trial is to confirm that the NCE provides a significant improvement in effect when compared to placebo. Traditionally, investigators derive a

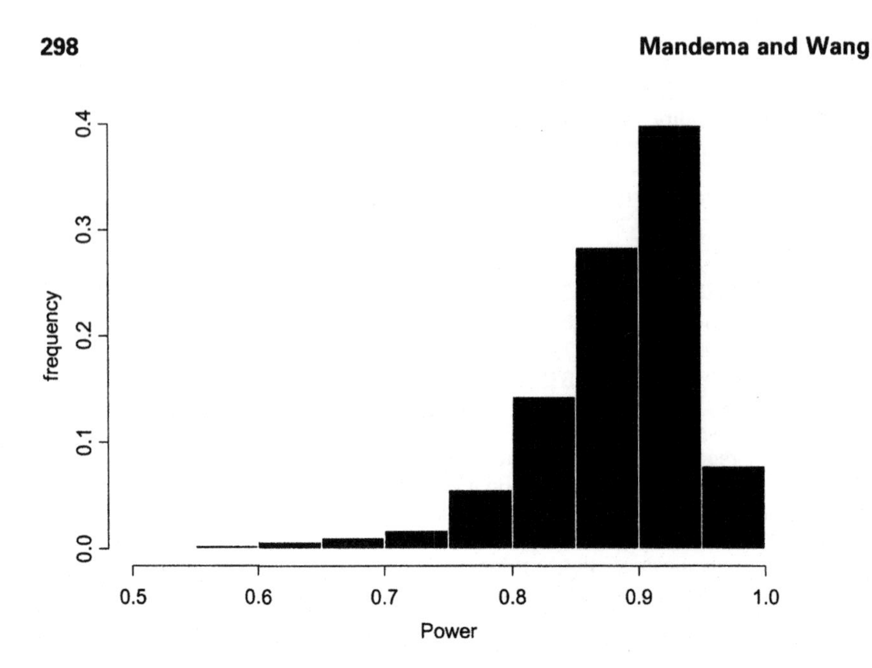


**FIGURE 3** Distribution across model uncertainty of the power of obtaining a significant difference between placebo and 20 mg at a sample size of 50 patients per treatment arm and alpha of 0.05.

point estimate of the effect size based on a clinically relevant effect size, prior information, or literature data and then compute the sample size required to provide adequate power. Figure 2 shows that the effect size at a specific dose is a distribution due to uncertainty in the model. This implies that study power is a distribution due to model uncertainty. Figure 3 shows the distribution, across model uncertainty, of the power of obtaining a significant difference between placebo and 20 mg at a sample size of 50 patients per treatment arm and alpha of 0.05. Sample size could be determined to provide a certain mean power across model uncertainty.

However, a more conservative alternative approach could be to determine the sample size to provide 80% chance that the power is at least 0.8. Figure 4 shows these two criteria. The top panel shows the average power to determine a significant difference between placebo and active treatment (at alpha of 0.05) as a function of NCE dose and sample size. The bottom panel shows the profile for an 80% chance that the power will be greater than the plotted power value ($y$ axis) as a function of NCE dose and sample size. The two graphs provide the insight that 50 patients per treatment arm would provide adequate power. At this sample size there is an 80% chance that the power to show a difference between placebo and active treatment is at least 0.84 for any dose greater than 20 mg.

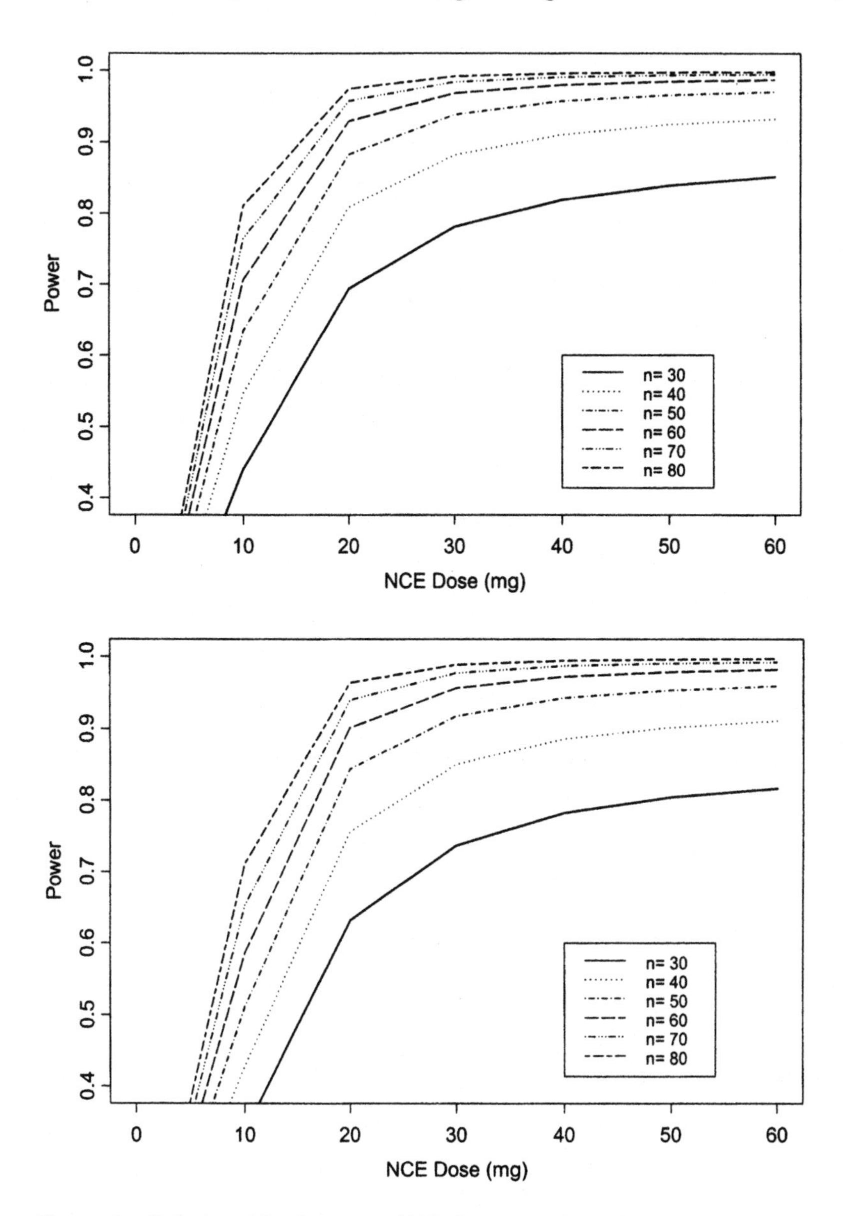


**FIGURE 4** Relationships between NCE dose, number of patients per treatment arm, and study power. The top panel shows the average power (mean of the power distribution) to determine a significant difference between placebo and active treatment (at alpha of 0.05) as a function of NCE dose and sample size. The bottom panel shows the profile for an 80% chance that the power will be greater than the plotted value as a function of NCE dose and sample size (20th percentile of the power distribution).

The mean power is at least 0.88 for any dose greater than 20 mg. Apart from providing a reasonable sample size, the graph also allows for the appropriate trade-off to be made if cost and duration of the trial are limiting. For example, a sample size of 30 still provides a mean power of 0.8 for doses greater than 40 mg. One of the advantages of modeling and simulation is that a sensitivity analysis can be performed to identify trial design factors (as is done in Figure 4), as

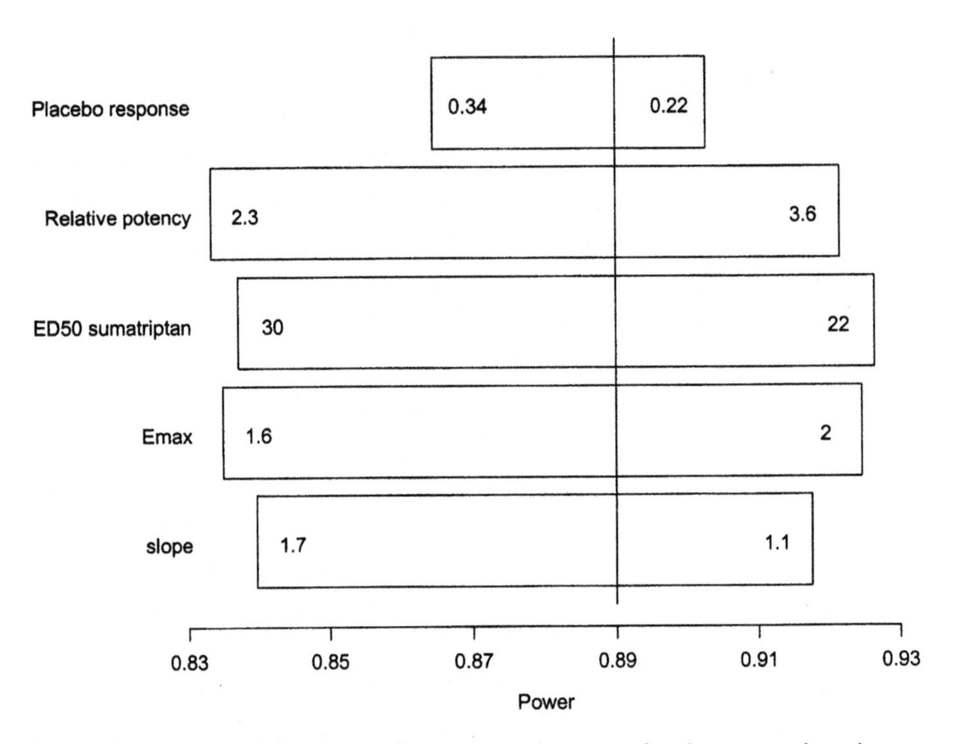


**FIGURE 5** Chart of the change in mean study power for the comparison between placebo and 20 mg at a sample size of 50 patients per treatment arm when conditioning on the 10th and 90th percentiles of the distribution of the model parameters. The bars show the impact of the uncertainty in the model parameter on the study power. The numbers plotted inside the bar reflect the 10th and 90th percentile of the parameter. The bar indicates the mean power conditional on the specific value of the model parameter (all other parameters are varied according to the conditional uncertainties). Placebo response refers to the expected placebo response in the trial's patient population, relative potency refers to the relative potency estimate between the NCE and sumatriptan, ED50 sumatriptan refers to the sumatriptan dose yielding 50% of the maximal response, $E_{max}$ refers to the maximal drug effect, and slope refers to the power parameter of the dose response relationship $n$.

well as specific model uncertainties/assumptions that have a large impact on the study power. Figure 5 shows a chart of the change in mean study power for the comparison between placebo and 20 mg at a sample size of 50 patients per treatment arm when conditioning on the 10th and 90th percentiles of the marginal distribution of the model parameters. For example, the second bar shows the impact of the uncertainty in the relative potency estimate between the NCE and sumatriptan on the study power. The numbers plotted inside the bar reflect the 10th and 90th percentile of the relative potency. The edges of the bar indicate the mean power conditional on the specific value of the relative potency. The mean power is the mean of the power distribution generated by conditioning on the specific value of relative potency, but varying all other parameters according to their conditional uncertainties. The graph could reveal where additional information should be acquired if certain parameters have a disproportional large impact on outcome of the study. For example, if the relative potency estimate between sumatriptan and the NCE had a disproportional large impact on outcome, additional preclinical or phase I surrogate studies would be warranted to reduce that uncertainty. An interesting observation in this example is that the uncertainty in expected placebo response (which could vary from 22 to 34%; note that this is the expected placebo response in the patient population represented in a specific trial and not the observed placebo response in the trial) has very little impact on the study power. The possible reason for this is that the observed trial-to-trial variability in placebo response is correlated to the response to active treatment.

## 15.5 POWER TO DETECT A DOSE-RESPONSE RELATIONSHIP

The main objective of the phase II dose-finding trial is to identify dosing procedures and target patient populations for further clinical development. The hope is to identify one or two doses with the appropriate marketable characteristics that can be confirmed in phase III of the development program, or to stop development. Therefore, the design of the phase II dose-finding trials should be evaluated for the trial's ability to quantify the key features of the dose response relationship and the trial's ability to determine the "right" dose(s), if any, for further development. The next step is to design a dose-ranging study in terms of number of treatment arms, spacing of active doses, and sample size with the objective to support dose selection with the appropriate marketable profile for further development. The questions to be addressed are: How many treatment arms should be included? Should there be a placebo arm? What is the highest dose in the study? How should I space the active arms? How many patients are needed for this study? Should I allocate patients evenly across treatment arms?

A first step in identifying the appropriate dose(s) for further development is to show that a dose-response relationship can be established on the basis of the phase II trial data. Several alternative trial strategies are evaluated for their power to show a dose-response relationship. For each sample of model parameters used to evaluate the uncertainty in the dose-response relationship of the NCE, a potential trial outcome is simulated according to a certain sample size and dose group assignment. This yields the predictive distribution of potential trial outcomes given a certain trial design and incorporates uncertainty in the outcome due to sample size as well as model uncertainty. A total of 1000 trials are simulated for each design. Each of the trials is analyzed using logistic regression, using an $E_{max}$ model or using a model that assumes that all active doses have a similar response ($ED_{50}$ approximates 0). The difference in $-2LL$ of the two models is evaluated using a $\chi^2$ test to calculate the $p$ value of the preference of the $E_{max}$ model vs. the model assuming all active doses to be similar. If the $E_{max}$ model provides a significant improvement ($p < 0.05$) it is concluded that the trial shows a dose-response relationship.

Table 4 summarizes the power of several trial strategies to detect a dose-response relationship. The previous steps have identified 20 and 40 mg as potential doses and 50 subjects as an appropriate sample size to satisfy a traditional evaluation of efficacy. Therefore, a first logical design choice would include placebo, 10, 20, and 40 mg at 50 patients per treatment arm. Despite the fact that at the proposed sample size 20 and 40 mg are highly likely to show a significant difference from placebo, the design is poorly powered to establish a dose-response relationship. This implies that it would be impossible to select a dose from the phase II trial data alone, because no trend among the active treatment arms can be observed. Increasing the top two doses by 25% as is done in design 2 provides a minimal increase in power. On the other hand, changing the lowest dose from 10 to 5 mg has a tremendous impact on the power to establish a dose-response relationship. It is anticipated that including a lower dose increases the chance to detect a dose-response relationship because 10 mg is still above the expected $ED_{50}$ of about 8 mg for the NCE (steep part of dose-response relationship). The magnitude of the power increase, however, is not anticipated. The

**TABLE 4** Power of Several Trial Strategies to Detect a Dose-Response Relationship

| Design | Dose groups (mg) | Sample size | Power |
|---|---|---|---|
| 1 | 0, 10, 20, 40 | 50, 50, 50, 50 | 0.34 |
| 2 | 0, 10, 25, 50 | 50, 50, 50, 50 | 0.37 |
| 3 | 0, 5, 25, 50 | 50, 50, 50, 50 | 0.79 |
| 4 | 0, 5, 10, 25, 50 | 40, 40, 40, 40, 40 | 0.71 |
| 5 | 0, 5, 10, 25, 50 | 50, 33, 33, 33, 50 | 0.68 |

difference in power between designs 2 and 3 shows how a relatively small change in the design can have a profound impact on the performance of the trial, which in this case can only be explored by simulation. Increasing the number of treatment arms at a similar total sample size, as is done in design 4, does not provide any benefit in the ability to detect a dose-response relationship. The performance is actually worse. The reason is that the shape of the dose-response relationship is well known and the trial is only looking for the location. This design also jeopardizes the power to show a significant difference between placebo and each of the treatment groups because of the smaller sample size per treatment groups. The last design maintains the power to show a difference between placebo and the highest dose by allocating 50 patients to these groups and subdividing the remainder of the patients equally across the other treatment groups. This design is also not preferred to establish the dose-response relationship. Design 3 seems an appropriate design to satisfy the objective to show a significant difference at the two highest doses when compared to placebo and to determine a dose-response relationship.

## 15.6 ABILITY TO SELECT DOSES FOR FUTURE DEVELOPMENT

Establishing the dose-response relationship is not the only goal of the phase II dose-finding trial. The trial should quantify the relevant part of the dose-response relationship with such precision to allow the selection of a dose with the appropriate marketable profile for further development. This is a complex assessment. Figure 6 shows nine possible (simulated) outcomes of the phase II trial under design 3. The symbols reflect the fraction of patients with pain relief at each dose group, whereas the line reflects the fit of an $E_{max}$ dose-response model to the data (using logistic regression). The figure shows the difficulty of interpreting the results with respect to dose selection. Some of the trials suggest almost linear dose-response relationships (right top and bottom panels), whereas others suggest a clear saturation of response (middle top and bottom panels). Some of the trials suggest a maximal effectiveness much lower than observed for the other 5-$HT_{1D}$ agonists (middle bottom panel), whereas others suggest a chance of higher efficacy (left middle and bottom panels).

Figure 7 shows the performance of several trial strategies to identify the dose required to achieve a certain response rate. The metric to judge the trial's performance to "correctly" identify the dose required to achieve a certain response rate is the distribution of the ratio of the dose estimated from the simulated phase II trial to achieve a certain response and the "true" dose to achieve this response. The observed dose is the dose derived from a regression analysis of the simulated phase II data, whereas the "true" dose is the dose derived from the model on which the simulation is based. Note that the "true" dose varies for each

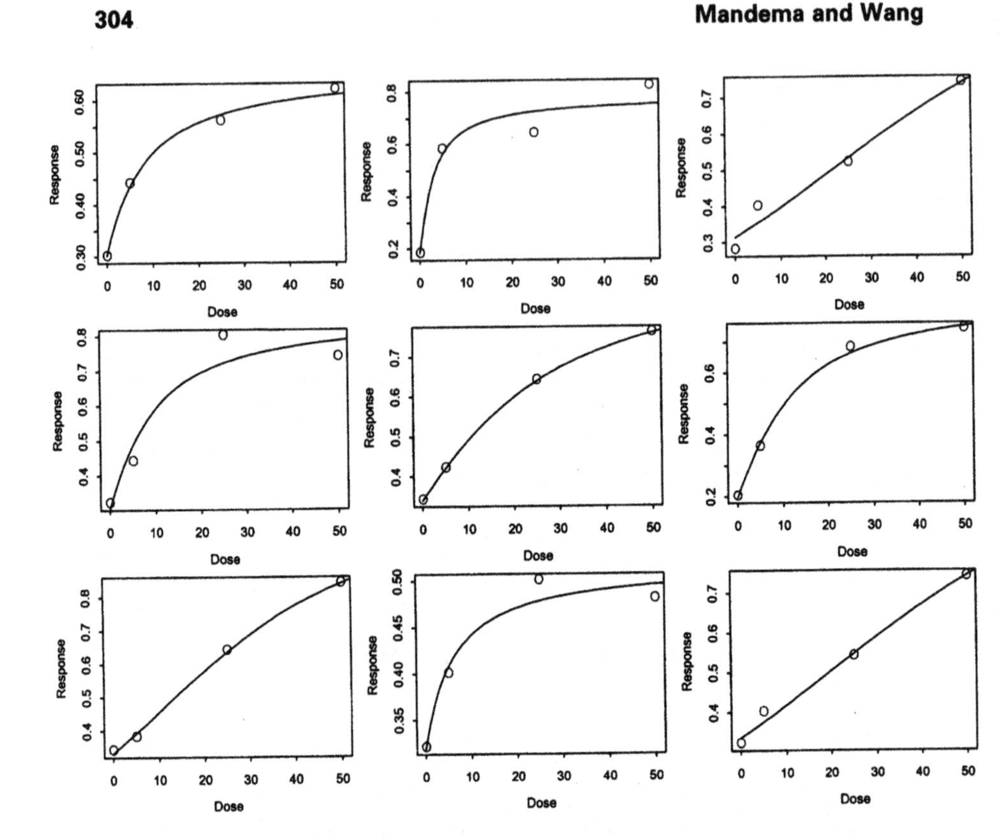


FIGURE 6 Overview of nine possible (simulated) outcomes of the phase II trial under design 3. The symbols reflect the fraction of patients with pain relief at each dose group, whereas the line reflects the fit of an $E_{max}$ dose-response model to the data (using logistic regression).

replication because of the uncertainty in the model parameters. The ratio is chosen to normalize for the magnitude of the dose to achieve the response. The figure shows the distribution (expressed as percentiles) of the ratio of the observed dose and the true dose to reach a certain response. This distribution shows the bias and precision with which the trial of a certain design and sample size can detect the (true) dose to achieve a certain response. The 50th percentile shows that there is no bias in estimating the dose to achieve a range of responses for the different designs. The outer lines of the graphs span the 5th to 95th percentile, indicating the precision of estimating the dose to achieve a certain response rate. The narrower the distance between the 5th and the 95th percentiles, the better the design can identify the dose required to achieve a certain target.

As may be expected, the trials with a larger power to determine a dose-

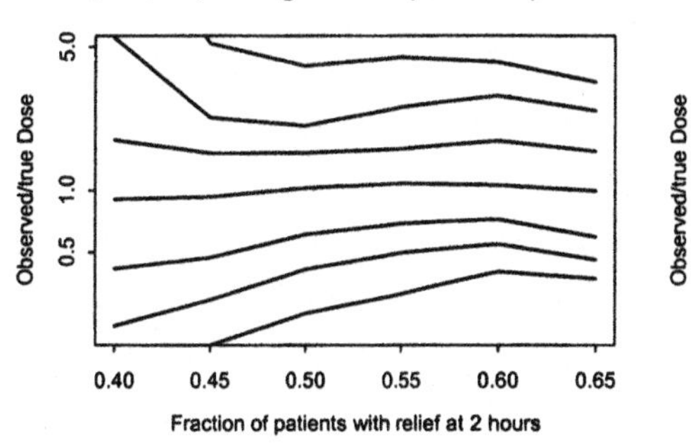


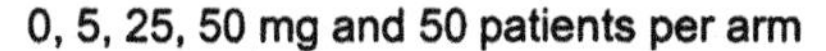


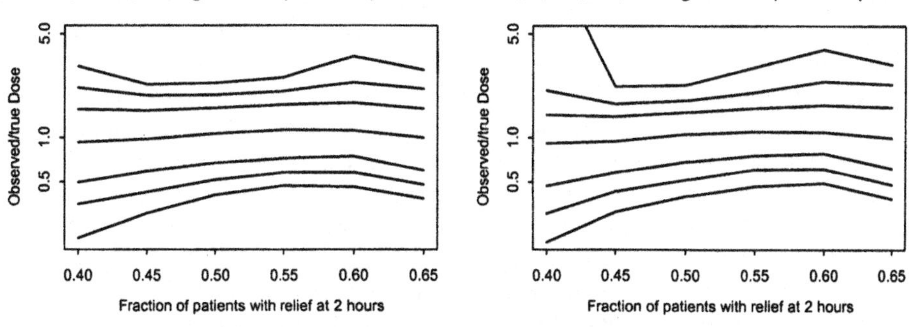


**FIGURE 7** Performance of several trial strategies to identify the dose required to achieve a certain response rate in the patient population. The figure shows the distribution (expressed as percentiles) of the ratio of the observed dose and the true dose to reach a certain response. The observed dose is the dose derived from a regression analysis of the simulated phase II data, whereas the "true" dose is the dose derived from the model on which the simulation is based. Shown are the 5th, 10th, 20th, 50th, 80th, 90th, and 95th percentiles of the distribution.

response relationship also perform better in determining the dose to achieve a certain response rate. For example, for design 3 (0, 5, 25, and 50 mg) there is an 80% chance that the dose to relieve 60% of the patients of their pain observed in the trial is between 0.58 and 2.3 times the true dose. For design 1 (0, 10, 20, and 40 mg), the corresponding values are 0.55 and 2.9. Please note that this assessment of performance is done across the model uncertainty. What this means is that the performance of the trial is the average performance across the likely range of dose-response relationships weighted by their (prior) probability to occur.

The actual result of the trial of this design may not be satisfying. Even though the design is very likely to show a significant difference at the two highest doses when compared to placebo and the design is very likely to determine a dose-response relationship, the design is not that powerful in identifying one dose with the adequate marketing characteristics. There is a 20% chance that the true dose to achieve 60% response is less than half or more than twice the observed dose. Actually, the precision with which the phase II trial for the NCE can determine the dose to achieve a certain response is almost similar to the precision with which the model, established before any phase II data is available, can determine this dose. Based on the model for the NCE, there is an 80% chance that the dose to provide pain relief in 60% of the patients is within 11 and 40 mg. This indicates that based on the NCE's phase II results alone, one can select the dose to achieve a certain response target just as well as on the basis of the model derived from clinical data of the other 5-$HT_{1D}$ agonists and preclinical and phase I data of the NCE.

Unfortunately, the doses to be taken forward into phase III development cannot be selected from a continuous range of doses. Only several dose options will be available to choose from. Selecting the best set of dose options to choose from, as well as the criteria to choose between them, is also an important aspect of the trial design. Figure 8 shows the ability of design 3 to select the smallest dose that will achieve pain relief in at least 60% of the patients from the following options: none, 1, 2.5, 5, 10, 25, 50, or 75 mg. The *x* axis shows the difference between the selected dose bin (selected on the basis of the simulated trial results) and the true dose bin (selected on the basis of the model on which the simulation was based). For example, 0 means that the correct dose is selected, 1 means that the dose selected is 1 bin higher than the true dose and the dose selected is too high. The figure shows that given design 3, the dosing options to select from, and dose selection criteria, there is only a 42.5% chance that the "right" dose is selected. This does not seem very satisfactory.

However, how often the "right" dose group is selected does not seem an appropriate criterion to evaluate the performance of the phase II trial. This criterion is dependent on the number and location of the dose groups and good performance can be guaranteed if there is a small number of dose options or bad positioning of the dose options. Furthermore, if the targeted response is on the flat part of the dose-response relationship, it may not be relevant that the dose is a bit too high or too low. Since the drug will be evaluated on the basis of effectiveness and safety in phase III, a better performance criterion seems to be the distribution of the expected response rate in the phase III patient population at the selected dose. In this way the performance of the phase II trial is tested for its ability to guarantee a successful phase III outcome.

The left panel of Figure 9 shows the distribution of the expected response rate in the patient population at the selected dose for design 3. The lowest dose

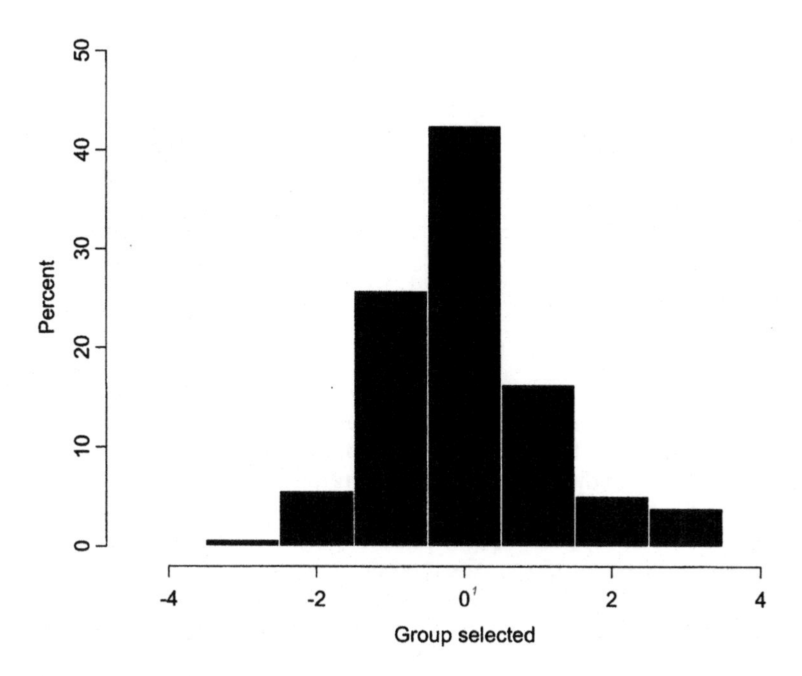


**FIGURE 8** Ability of design 3 to select the smallest dose that will achieve pain relief in at least 60% of the patients from the following options: none, 1, 2.5, 5, 10, 25, 50, or 75 mg. The graph shows the distribution of the difference between the selected dose bin (selected on the basis of the simulated trial results) and the true dose bin (selected on the basis of the model on which the simulation was based). For example, 0 means that the correct dose is selected, 1 means that the dose selected is one bin higher than the true dose and the dose selected is too high.

that achieved pain relief in at least 60% of the patients was selected from the following options: none, 1, 2.5, 5, 10, 25, 50, or 75 mg. The graph shows that there is an 80% chance that the true percentage of patients with pain relief at 2 h at the dose selected for phase III is between 52 and 68%. It is interesting to note that while design 2 performs bad among the designs to establish a dose-response relationship, it serves well the current study objective: For design 2, there is an 80% chance that the true response at the dose selected for phase III is between 56 and 68%. This observation shows that different study objectives require different designs. To select the most appropriate design requires a careful balancing of these different objectives.

Apart from the trial design, the available dose options and dose selection criterion can be modified to improve the performance of the phase II dose-finding

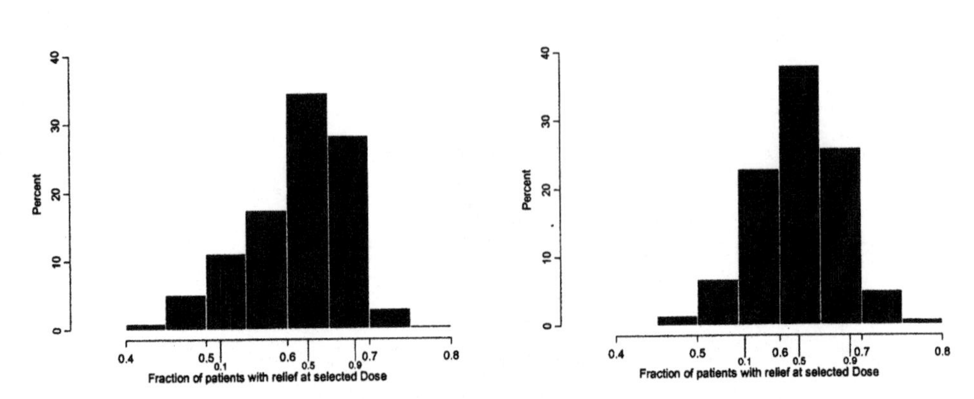


**FIGURE 9** Distribution of the expected response rate in the phase III patient population at the selected dose for two dose-finding strategies. The left panel shows the distribution of the expected response rate in the patient population at the selected dose for trial design 3. The lowest dose that achieved pain relief in at least 60% of the patients was selected from the following options: none, 1, 2.5, 5, 10, 25, 50, or 75 mg. The long tick marks on the *x* axis indicate the 10th, 50th, and 90th percentile of the distribution. The right panel shows the distribution of the expected response rate if the 25 mg option is selected on the basis of the prior model and the phase II results are ignored.

trial. Table 5 summarizes the performance of several dose selection criteria for two sets of available dose options under design 3. The table shows the median and 10th and 90th percentile of the percentage of patients with pain relief at 2 h at the dose selected for phase III. If two doses are selected, the distribution is shown of the response rate closest to the target of the two response rates. The table shows that selecting the smallest available dose option that is observed to provide pain relief in at least 60% of the patients performs better than selecting the dose option that is closest to the observed dose reaching this target. Selecting two doses to be taken forward to phase III performs a little bit better, but the improvement is minimal considering the associated additional costs for the phase III trials. Having more dosing options available to select from does not improve the performance either. This is due to the fact that the precision to determine the "correct" dose is small relative to the distance between the dose groups. The rationale for selecting two doses for phase III anyway could be to have an alternative dose available for those patients that do not respond to the primary dose.

There is also the question on what one can learn from the phase II trial results alone vs. what is already known from the prior model used for the simulations. The right panel of Figure 9 shows the distribution of the expected response rate if the 25 mg option is selected on the basis of the prior model and the phase II results are ignored. Under this strategy there is an 80% chance that the true

**TABLE 5** Performance of Several Dose Selection Criteria for Two Sets of Available Dose Options Under Trial Design 3[a]

| Dose options (mg) | Criterion | Fraction of patients with relief | | |
|---|---|---|---|---|
| | | Median | 10% | 90% |
| 0, 1, 2.5, 5, 10, 25, 50, 75 | First dose higher than observed dose | 62 | 52 | 68 |
| | Dose closest to observed dose | 54 | 42 | 65 |
| | Two doses around observed dose | 60 | 52 | 65 |
| | Two doses higher than observed dose | 63 | 57 | 68 |
| 0, 1, 2.5, 5, 10, 17.5, 25, 35, 50, 75 | First dose higher than observed dose | 61 | 52 | 67 |
| | Two doses around observed dose | 60 | 52 | 65 |

[a] The table shows the median and 10th and 90th percentile of the percentage of patients with pain relief at 2 h at the dose selected for phase III. If two doses are selected, the distribution is shown of the response rate closest to the target of the two response rates. The observed dose is the dose estimated from the phase II trial results that provides pain relief in at least 60% of the patients in the simulated NCE trial.

percentage of patients with pain relief at 2 h at the dose selected for phase III is between 57 and 69%. This strategy seems to perform at least as good, if not better, as basing the dose selection on the phase II results alone. The willingness to take this strategy depends on the validity of the relative potency estimate derived for the NCE on the basis of animal studies. Because clinical as well as preclinical information on several 5-$HT_{1D}$ agonists is available, the validity of this assumption is confirmed by the correlation between relative potency in animal studies and clinical studies of sumatriptan, zolmitriptan, naratriptan, and rizatriptan. Therefore, the best strategy for the NCE may be to minimize the extent of the phase II dose finding, select the doses for phase III now, and perform a trial at two doses with the opportunity to greatly speed up the development of this compound.

In this particular example the dose selection for phase III is solely based on the phase II results (apart from the upper limit of 75 mg), because this is the only trial evaluating the effect of the NCE in the target patient population and because this is a path commonly taken by drug companies. The analysis strategy of the phase II trial could be modified to include the prior information. This would ensure a dose selection based on the current knowledge combined with the phase II results for the NCE. The precision of dose selection could be increased or the scope of the phase II trial could be reduced if this analysis strategy was taken, especially because so much is known on the location of the maximal effect for $5\text{-HT}_{1D}$ agonists.

## 15.7 DISCUSSION

An approach is presented to quantitatively evaluate and optimize the phase II dose-finding trial strategy. The goal of phase II dose-finding trials is to identify one or several doses with the appropriate marketable characteristics for further clinical development, or to stop development. Phase II dose-ranging trials are often powered with respect to showing a significant difference between placebo and active treatment. However, the trials are not powered with respect to the trial's ability to quantify key features of the dose-response relationship or the trial's ability to select an adequate dose, if any, for further development. Modeling and simulation offers a powerful new approach to optimize the phase II dose-finding strategy with respect to the latter, as well as the former objectives.

In this chapter a case study is presented to illustrate the use of modeling and simulation to optimize the phase II dose-finding strategy for a new $5\text{-HT}_{1D}$ agonist indicated for the relief of moderate to severe migraine pain. A model was developed for the relationship between the dose of the NCE and the fraction of patients with pain relief at 2 h after dosing. The model was based on trial outcome data of several $5\text{-HT}_{1D}$ agonists, such as sumatriptan, zolmitriptan, rizatriptan, and naratriptan, and scaled to the NCE on the basis of animal studies and phase I pharmacokinetic data.

The model quantified the uncertainty in trial outcome due to model uncertainty as well as patient variability. The model was used to explore the likely range of treatment outcomes for the NCE in comparison to the other $5\text{-HT}_{1D}$ agonists to provide initial insight on dose selection and marketable efficacy targets. The model was subsequently used to design a dose-ranging study in terms of number of treatment arms, spacing of active doses, sample size, and analysis strategy with the objective to confirm efficacy and to select a dose with the appropriate marketable profile for further development. A careful evaluation of several trial strategies is crucial because minor modifications in the design were found to have a profound impact on performance, especially with respect to dose selection. Model uncertainty was accounted for in the trial simulations to ensure a trial robust to this uncertainty.

The presented approach is generally applicable in drug development. The peculiarity of the example presented is that a lot of prior information is available on compounds with a similar mechanism of action. Actually, so much information is available that the best strategy may be not to embark on an extensive dose-finding trial because the trial could provide, at best, as much information as can be inferred from the prior model. In most situations the uncertainty about the potency, safety, and efficacy of an NCE will be much larger, increasing the importance of a carefully evaluated dose-finding strategy.

Modeling and simulation is a general tool to evaluate complex systems that are hard to track analytically. The promise and applicability of modeling and simulation is certainly far wider than demonstrated in this example. This methodology can be virtually incorporated with every step in the drug development process to improve efficiency and assist decision making. For instance, it can be used to compare the information yield of competing trial strategies vs. time, costs, and trial complexity. This enables the development team to balance the value vs. cost of additional information derived from the trial. The probabilistic trial outcome models can be leveraged into a decision-making framework that can be used to optimize the overall development strategy from a financial value perspective. This link is important because the trial simulations can only determine the relationship between drug features, trial features, and power or precision to make a certain decision. It cannot, however, determine what level of confidence one would like to achieve at a certain point in development.

## REFERENCES

1. MB Schoenhoff, JW Mandema. Clinical trial simulation: retrospective analysis of dose ranging trials in detrusor instability with tolterodine. Clin Pharmacol Ther 65: 203, 1999.
2. PA Lockwood, EH Cox, JW Mandema, J Koup, W Ewy, RJ Powell. Computer assisted trial design (CATD) to support dose selection for Cl-1008 in chronic neuropathic pain trials. Pharm Res 16:188, 1999.
3. LB Sheiner, JL Steimer. Pharmacokinetic/pharmacodynamic modeling in drug development. Annu Rev Pharmacol Toxicol 40:67–95, 2000.
4. NHG Holford, HKimko C, JPR Monteleone, CC Peck. Simulation of clinical trials. Annu Rev Pharmacol Toxicol 40:209–234, 2000.
5. J Teall, M Tuchman, N Cutler, M Gross, E Willoughby, B Smith, K Jiang, S Reines, G Block. Rizatriptan (MAXALT) for the acute treatment of migraine and migraine recurrence. A placebo-controlled, outpatient study. Rizatriptan 022 Study Group. Headache 38:281–287, 1998.
6. WH Visser, GM Terwindt, SA Reines, K Jiang, CR Lines, MD Ferrari. Rizatriptan vs sumatriptan in the acute treatment of migraine. A placebo controlled, dose-ranging study. Dutch/US Rizatriptan study group. Arch Neurol 53:1132–1137, 1996.
7. NT Mathew, M Asgharnejad, M Peykamian, A Laurenza. Naratriptan is effective and well tolerated in the acute treatment of migraine. Results of a double-blind, placebo-controlled, crossover study. The Naratriptan S2WA3003 Study Group. Neurology 49:1485–1490, 1997.
8. V Pfaffenrath, G Cunin, G Sjonell, S Prendergast. Efficacy and safety of sumatriptan tablets (25 mg, 50 mg, and 100 mg) in the acute treatment of migraine: defining the optimum doses of oral sumatriptan. Headache 38:184–190, 1998.
9. C Dahlof, C Edwards, S Ludlow, PDOB Winter, M Tansey, and the oral sumatriptan

dose-defining study group. Sumatriptan- an oral dose-defining study. Eur Neurol 31: 300–305, 1991.
10. AM Rapoport, NM Ramadan, JU Adelman, NT Mathew, AH Elkind, DB Kudrow, NL Earl. Optimizing the dose of zolmitriptan (Zomig, 311C90) for the acute treatment of migraine. A multicenter, double-blind, placebo-controlled, dose range-finding study. The 017 Clinical Trial Study Group. Neurology 49:1210–1218, 1997.
11. NDA 20-768, Zolmitriptan efficacy review, Study 008, CDER, FDA, 1997, pp 49–53.
12. SL Beal, AJ Boeckman, LB Sheiner. NONMEM users guide. NONMEM project group. University of California, San Francisco, 1996.

# 16

# Model-Based Integration for Clinical Trial Simulation and Design

## A Phase II Case Study for Naratriptan

**Ivan A. Nestorov*, Gordon Graham, and Leon Aarons**
University of Manchester, Manchester, United Kingdom

**Stephen B. Duffull**
University of Queensland, Brisbane, Queensland, Australia

## 16.1 INTRODUCTION

There is recent and growing interest in using computer simulation of clinical trials to improve clinical trial design. Simulating clinical trials, applying techniques commonly used in other technology-based industries such as the aerospace industry, can allow clinical trial designers to thoroughly test their designs and analyze the simulated results before actually conducting a clinical trial. Herein we describe investigations into the feasibility and utility of model-based clinical trial simulation (CTS) applied to the clinical development of naratriptan with effect measured on a categorical scale.

Naratriptan is a novel $5HT_{1B/1D}$ agonist indicated for the acute treatment of migraine. It has been developed for oral treatment and followed the development

* *Current affiliation*: Amgen, Inc., Seattle, Washington, U.S.A.

of the "first-in-line" triptan-sumatriptan (1, 2). Naratriptan was chemically designed to have a better oral bioavailability and longer elimination half-life than sumatriptan. We have reported results from initial work on the modelling and analysis of computer simulations applied to the naratriptan project elsewhere (3). Briefly, a PK/PD model for naratriptan was developed using information gathered from previous naratriptan and sumatriptan preclinical and clinical trials. The phase IIa naratriptan data were used to check the PK/PD model in terms of its ability to describe future data. A further PK/PD model was developed using the phase IIa naratriptan data, and a phase IIb trial was designed by simulation, with constraints set by logistics and regulation. The design resulting from CTS was compared to that derived using D-optimal design. The PK/PD model showed reasonable agreement with the data observed in the phase IIa naratriptan clinical trial. CTS resulted in a design with 4 or 5 arms at 0, 2.5 and/or 5, 10, and 20 mg, PD measurements to be taken at 0, 2, and 4 and/or 6 h and at least 150 patients per arm. A sub-D-optimal design resulted in 2 dosing arms at 0 and 10 mg and PD measurements to be taken at 1 and 2 h. The research was carried out retrospectively and was not used to influence the development of naratriptan.

In the present report we describe extensions to the work described above. Specifically, preclinical and phase I data for naratriptan and sumatriptan were used to construct a model of subcutaneous naratriptan to simulate the phase IIa clinical trial for naratriptan. In addition, optimal design methodology was used to explore a number of design issues. In the next section the methodology that was used is detailed together with a description of the data and information that was used to build the models. Then results from the simulations and optimal design studies are presented. Finally, an analysis of resources, in terms of time, software, and experience needed to carry out the CTS and design exercise are presented.

## 16.2 MODELS

The core of any CTS and design exercise is a PK/PD model for the drug of interest. The PK part of the PK/PD model used for naratriptan was a classical two-compartment absorption model (Figure 1) with an effect compartment, accounting for the delay between the concentration in the central compartment and the observed effect.

The pharmacodynamic model developed was based on the clinical end point of naratriptan-migraine pain severity. Pain severity was measured on a four-point ordinal scale with 0 = no pain, 1 = mild pain, 2 = moderate pain, and 3 = severe pain. Regulatory requirements define "pain relief" as a measure of "success" of the drug to reduce pain severity at 2 h "significantly," i.e. a success of the therapy is achieved when a reduction from categories 3 or 2 to 1 or 0 is observed. This definition leads to the dichotomisation of the categorical responses into pain relief (success, value 1) and no pain relief (failure, value 0).

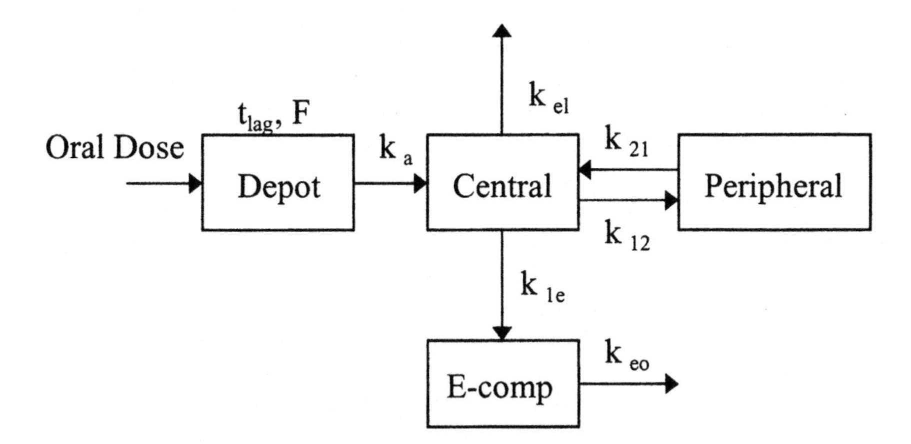

**FIGURE 1** PK model for orally administered naratriptan.

The PD part of the naratriptan PK/PD model was taken from a study (4) where phase II sumatriptan pharmacodynamic data after nasal administration was analyzed. It is defined in Eq. (1) as a binary logistic model for the dichotomization (pain relief/no pain relief) and is expressed in two parts:

$$\mathbf{logit}(\Pr(Y_{ij} = 1 \mid \text{score}_0)) = \begin{cases} \theta_1 + \theta_2 \log(\text{time}_{ij}) + \dfrac{\theta_3 C_{eij}}{\theta_4 + C_{eij}} + \eta_i & \text{score}_0 = 2 \\ \phi_1 + \phi_2 \log(\text{time}_{ij}) + \dfrac{\phi_3 C_{eij}}{\phi_4 + C_{eij}} + \xi_i & \text{score}_0 = 3 \end{cases} \tag{1}$$

The first line of Eq. (1) corresponds to a baseline pain severity score of 2 observed in 55% of migraineurs. The second line corresponds to a baseline pain severity score of 3 observed in 45% of migraineurs. The placebo model is given by log (time) and the drug effect is defined as an $E_{\max}$ type model, where $C_e$ is the concentration of the drug at a hypothetical effect site (5) as shown in Figure 1. The random effects are denoted by $\eta_i$(baseline score of 2) and $\xi_i$ (baseline score of 3) and are normally distributed with mean zero and variance given by $\omega_\eta^2$ and $\omega_\xi^2$, respectively.

## 16.3 PK MODEL PARAMETER SPECIFICATION

Information available for the modeling and simulation included preclinical and clinical data accumulated by Glaxo Wellcome up to 1995 and publications in the literature. The phase I studies included intravenous ($n$ = 8), subcutaneous ($n$ =

**TABLE 1** Parameter Values for the Naratriptan Oral Solution Absorption PK Model[a]

| Parameter | Estimated Value | | |
|---|---|---|---|
| | Value | CV (%) | pdf |
| CL (liters/h) | 22.7 | 27.9 | lognorm |
| $V_1$ (liters) | 17.2 | 20.7 | lognorm |
| $V_2$ (liters) | 98.4 | 26.3 | lognorm |
| $Q$ (liters/h) | 101.4 | 25.5 | lognorm |
| ka(solution) | 1.30 | 32.9 | lognorm |
| $F$(solution) | 0.69 | n.a.[b] | n.a. |

[a] The parameter probability distribution function (pdf) assumed for the NONMEM parameter estimation is also shown.
[b] n.a. = not available from analysis.

8), and oral solution ($n$ = 10) routes of administration from a total of 26 healthy male volunteers. This dataset was the basis of the development of the PK model. After a population pharmacokinetic analysis (software: NONMEM) of the Phase I naratriptan data a two-compartment open model provided the best fit to the data. Intravenous infusion, subcutaneous, and oral administration data were fitted simultaneously so that population estimates of rate constants of absorption and availability could be determined. The parameter values for the oral solution absorption model are given in Table 1. The last column of Table 1 shows the parameter probability distribution function (pdf) assumed for the NONMEM parameter estimation.

At this point, no experimental data related to the oral absorption of the tablet in migraine patients were available. Therefore, we needed to modify the absorption parameters (given in the last two rows of Table 1) in an appropriate way, in order to construct a model for the oral tablet form of naratriptan. It should also be noted that we attempted to estimate variability on $F$ and ka, but only variability on ka could be estimated.

### 16.3.1 Absorption Lag Time

The observed mean lag time in the phase I oral solution data was 11 min; the range was 2.5–37.5 min. When the absorption of the oral tablet is considered, this lag time should be extended to include the dissolution time. Coulie et al. (6) report a prolonged lag phase of the gastric emptying, measured for both liquids and solids in healthy volunteers treated with sumatriptan. It may be expected that the oral tablet of naratriptan has a similar effect. When a patient population is concerned, several authors (7, 8) report a significant increase in the gastric-emptying lag phase, related to migraine. All of the above considerations indicate that for the PK model in tablet-treated migraine patients, the lag time, observed in healthy volunteers after administration of an oral solution of naratriptan, should be significantly increased. Therefore, for the purpose of the simulations, we assumed that the pharmacokinetic lag time in patients receiving oral naratriptan tablets during a migraine attack is between 15 and 45 min, with the lower limit of the interval seeming too optimistic. As the true variability is hard to predict, we assumed a uniform probability distribution of the lag time.

### 16.3.2 Bioavailability

It is likely that the oral bioavailability of a tablet form could be lower than the bioavailability of the solution. Furthermore, there is abundant evidence that migraine attacks severely impair drug absorption (7–12), the reasons being attributed to gastric stasis and reduced rate of gastric emptying. A decrease in the bioavailability during migraine attacks is reported by Tokula et al. (8); Hussey et al. (11) report this decrease to be approximately 20%. At the same time, an increase in the intersubject variability is also observed. Based on the above considerations and the estimated value of bioavailability for oral solution being 0.6 (Table 1), the bioavailability of the oral naratriptan tablet was assumed to be in the interval 0.4–0.7. Again, as there was no information of the true variability, we assumed a uniform probability distribution of the bioavailability.

### 16.3.3 Absorption Rate

It is also probable that the absorption rate of a tablet should be lower than the solution. In addition it appears that migraine attacks also affect the rate of absorption of chemicals (7, 9–11). The variability is also likely to be increased significantly. Therefore, we assumed a lognormally distributed absorption rate with a mean of 0.8 $h^{-1}$ and a CV of 75%.

## 16.4 LINK AND PD MODEL PARAMETER SPECIFICATION

### 16.4.1 Rate Constant of the Effect Compartment

Naratriptan and sumatriptan share the same mechanism of action. Therefore, the time offset between the plasma concentration and the effect should be similar for both drugs. Consequently, the values of $k_{e0}$ for naratriptan was assumed to be the same as that published for the sumatriptan model (4)—$k_{e0}$ = 0.78 $h^{-1}$ for baseline pain severity score 2, and $k_{e0}$ = 2.04 $h^{-1}$ for baseline pain severity score 3. No variability was assigned to either rate constants, since no information regarding their variability was available.

### 16.4.2 Logistic Model Parameters

The first parameters of the PD model, Eq. (1), are the respective cutpoints $\theta_1$ and $\phi_1$. As the cutpoints do not relate to the influence of either the time or the dose on the effect, their values should be drug independent. Therefore, the sumatriptan cutpoints (4) were adopted for the naratriptan model, i.e., $\theta_1 = -2.06$ for baseline pain severity score 2, and $\phi_1 = -4.38$ for baseline pain severity score 3. No variability was assigned to either parameter, since no relevant information was available.

The second parameters of the PD model, $\theta_2$ and $\phi_2$, describe the placebo effect during migraine. As the placebo effect should not depend on the drug administered, the sumatriptan parameters were adopted for the naratriptan model; i.e., $\theta_2 = 1.93$ for baseline pain severity score 2 and $\phi_2 = 2.32$ for baseline pain severity score 3. No variability was assigned to either parameter, since no relevant information was available.

The drug effect is described by the $E_{max}$ type component in Eq. (1), where $\theta_3$ and $\phi_3$ correspond to the "maximum effect" parameter, while $\theta_4$ and $\phi_4$ correspond to the EC50 parameter. To carry out the scaling between sumatriptan and naratriptan, published preclinical in vitro and in vivo sumatriptan data were used (2, 13). The preclinical studies included 5HT receptor binding assays in COSM6 cells that were transfected with the $5HT_{1D}$ gene (2). The in vitro and in vivo comparative naratriptan-sumatriptan studies (13) included comparison of the effect of naratriptan and sumatriptan on isolated dog basilar and middle cerebral artery preparations (experiment 1); comparison of the effect of naratriptan and sumatriptan on isolated human coronary artery preparations from explanted hearts (experiment 2); measurement of the effect of naratriptan and sumatriptan on the carotid vascular resistance in anaesthetised dogs in vivo (experiment 3) ; and characterizing the inhibition of the trigeminal nerve activity by measuring the effect of neurogenically medicated inflammation in the dura (experiment 4). The results reported were as follows: (a) experiment 1—the relative potency of naratriptan to sumatriptan 3.25 (basilar artery) and 2.65 (middle cerebral artery), the maximum contraction being similar; (b) experiment 2—the relative potency of naratriptan to sumatriptan 4.3, the maximum contraction being similar; (c) experiment 3—the relative potency of naratriptan to sumatriptan 2, the maximum contraction being similar; and (d) experiment 4—sumatriptan and naratriptan were approximately equipotent at inhibiting neurogenically mediated dural plasma extravasation. The overall conclusion in Ref. 13 was that the pharmacological profile of naratriptan is very similar to that of sumatriptan, but naratriptan is about 2–3 times more potent, the maximum effect of the two drugs being identical.

In order to be on the conservative side, we assumed an uniformly distributed relative potency of naratriptan with respect to sumatriptan, varying between 1.5 and 3.5. Consequently, we divided the respective values of $\theta_4$ and $\phi_4$ for

**TABLE 2** Parameter Values for the Naratriptan PD Model

| | Estimated value | | |
|---|---|---|---|
| Parameter | Value | SD | pdf |
| Baseline score 2 | | | |
| $k_{e0}(h^{-1})$ | 0.78 | 0 | Fixed |
| $\theta_1$ | −2.06 | 0 | Fixed |
| $\theta_2$ | 1.93 | 0 | Fixed |
| $\theta_3$ | 3.96 | 0 | Fixed |
| $\theta_4$ | [1–2.34] | n.a. | Uniform |
| $\eta_1$ | 0 | 4.23 | Normal |
| Baseline score 3 | | | |
| $k_{e0}(h^{-1})$ | 2.04 | 0 | Fixed |
| $\phi_1$ | −4.38 | 0 | Fixed |
| $\phi_2$ | 2.32 | 0 | Fixed |
| $\phi_3$ | 9.85 | 0 | Fixed |
| $\phi_4$ | [13.1–30.6] | n.a. | Uniform |
| $\xi_i$ | 0 | 4.42 | Normal |

n.a., not available from analysis.

sumatriptan (4) by the relative potency in order to obtain the respective values in the naratriptan model. Hence, for naratriptan, $\theta_4$ and $\phi_4$ were sampled uniformly from the intervals [1–2.34] and [13.1–30.6], respectively. As the maximum effect of naratriptan is the same as with sumatriptan, the values of $\theta_3$ and $\phi_3$ (4) remained unchanged (at 3.96 and 9.85, respectively).

The random effects for naratriptan were assumed to be identical to the random effects with sumatriptan, expressed by $\eta_i$ (baseline score of 2) and $\xi_i$ (baseline score of 3) in Eq. (1). The random effects were assumed to have normal distributions with mean zero and variances $\omega_\eta^2$ and $\omega_\xi^2$ of 17.7 and 19, respectively (4).

The parameter values used in the naratriptan PD model are presented in Table 2.

## 16.5 SIMULATIONS

In order to characterize the dose-effect relationship for the naratriptan oral tablet and to determine the asymptotically "correct" values (nominal values) which served as a basis for detecting significant differences in the power studies, an intensive trial simulation was undertaken, with seven doses and placebo (0, 0.1, 0.25, 0.5, 1, 2.5, 5 and 10 mg) of naratriptan. Two hundred trials were simulated

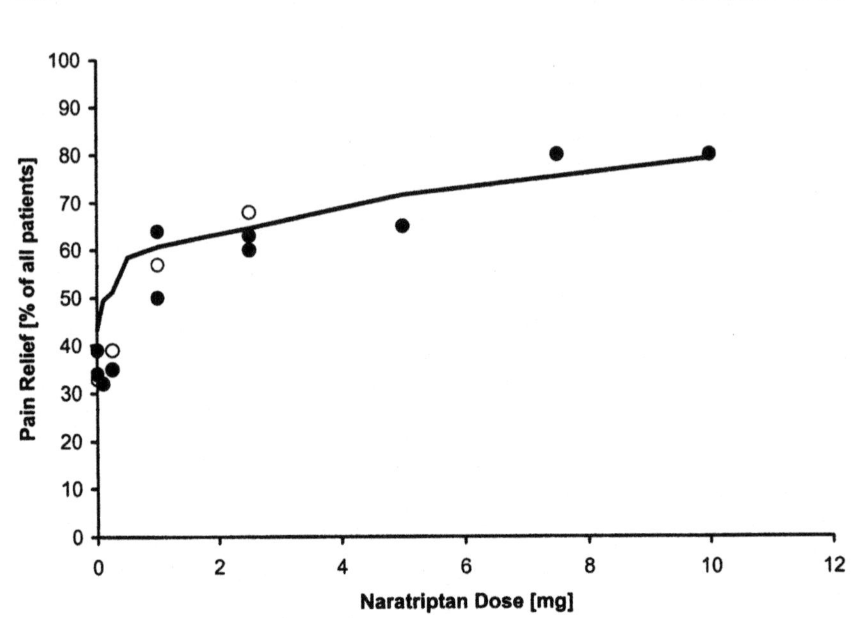


**FIGURE 2** Oral naratriptan dose-effect relationship at 4 h postadministration. The solid line represents the simulated trial results from phase I data and sumatriptan data and the symbols represent experimental data (● from Ref. 1, ○ from Ref. 14).

with the proposed PK/PD model, each with 1000 patients per dose, which can be considered to be a sufficiently large sample.

In Figure 2, the simulation results are compared to actual data taken from the literature (1, 14). Reference 1 reports results from two oral naratriptan dose-ranging studies with 643 and 613 patients respectively; Ref. 14 reports a naratriptan study with 595 migraine patients. Effect was sampled at 4 h post administration of the tablet.

The simulation shown in Figure 2 is in good agreement with the reported actual clinical trial results, with a slight overestimation (around 10%) of the effect at small doses—below 0.5 mg—which has no clinical significance.

## 16.6 ELEMENTS OF CLINICAL TRIAL DESIGN

### 16.6.1 Optimal Design

The design selected from the simulations can be compared to the design resulting from the application of D-optimality to the model specified in Eq. (1). It was not the intention of these analyses to propose a D-optimal design per se but rather the theoretical design was used for computation of relative alternative design efficiency. The PK/PD model is conditional on the baseline pain severity score (2 or 3) resulting in four parameters for each conditional part of the model. The D-optimal design was therefore determined assuming that there was one model with two sets of parameter values. Assuming the doses were allowed to take any nonnegative value and time any positive value [because of the log(time) component of the model], the D-optimal design was dose = (0.0, 0.0, 0.89, 191423) mg and time = (0.54, 0.85, 0.85, 4.5) h for a baseline score of 2 and dose = (0.0, 0.001, 3.04, 32845) mg and time = (0.1, 1.4, 2.9, 10.1) h for a baseline score of 3. The D-optimal design was determined from 16 dose-time combinations (that is, 4 doses and 4 times), the same number that was in the original study. For both baseline models, the maximum dose was much greater than that which could be considered safe. A constrained optimization routine was run where the upper limit on dose was set to 20 mg. The sub-D-optimal design is given in Tables 3 and 4 for baseline models 2 and 3, respectively, with the relative efficiencies of three alternative designs resulting from the simulations and the design used in Ref. 1. The D-optimal design results in replicate dose-time pairs because the structural model includes only fixed effect parameters, and each observation is assumed to be independent within and between each subject. Although replicates may be impossible to collect in practice, such designs are useful for comparison with alternative designs, from a modeling perspective.

The efficiency of a particular design, compared to an optimal design is defined as the amount of information in that design relative to the optimal design. For example, if a design has an efficiency of 50% compared to the optimal design, then doubling the number of patients in this design will provide the same information as the optimal design. The relative efficiencies for the designs were approxi-

**TABLE 3** Relative Efficiencies for the Designs Obtained from the Simulations and the Experimental Design with Respect to the Sub-D-optimal Design for Baseline Score of 2

| Design | Doses | Times | Replicates | Efficiency |
|---|---|---|---|---|
| Sub-D-optimal | 0, 0, 0.48, 20 | 0.61, 0.62, 1.16, 4.35 | 1 | 1.0 |
| Alternative 1 | 0, 0.5, 2.5, 10 | 2, 4 | 2 | 0.462 |
| Alternative 2 | 0, 0.75, 2.5, 10 | 2, 4 | 2 | 0.407 |
| Alternative 3 | 0, 1, 2.5, 10 | 2, 4 | 2 | 0.361 |
| Experimental design | 0, 0.1, 0.25, 1, 2.5, 5, 7.5, 10 | 2, 4 | 1 | 0.424 |

*Source*: Fuseau et al. (1).

**Table 4** Relative Efficiencies for the Designs Obtained from the Simulations and the Experimental Design with Respect to the Sub-D-optimal Design for Baseline Score of 3

| Design | Doses | Times | Replicates | Efficiency |
|---|---|---|---|---|
| Sub-D-optimal | 0, 0, 4.73, 20 | 0.6, 0.6, 3.68, 9.23 | 1 | 1.0 |
| Alternative 1 | 0, 0.5, 2.5, 10 | 2, 4 | 2 | 0.55 |
| Alternative 2 | 0, 0.75, 2.5, 10 | 2, 4 | 2 | 0.56 |
| Alternative 3 | 0, 1, 2.5, 10 | 2, 4 | 2 | 0.57 |
| Experimental design | 0, 0.1, 0.25, 1, 2.5, 5, 7.5, 10 | 2, 4 | 1 | 0.53 |

*Source*: Fuseau et al. (1).

mately 40 and 55% for baseline scores 2 and 3, respectively. Based upon these efficiencies, it seems sensible to choose the design with highest efficiency. For the model with baseline score 2, alternative design 1 (Table 3) gave the highest relative efficiency, whereas for the model with baseline score 3, all the alternative designs have approximately the same relative efficiency (~56%). Based upon these results it seems reasonable to choose alternative design 1.

### 16.6.2 Sample Size Determination

The number of patients per dose (sample size) was determined from the results of the simulations based on the requirement to observe a statistically significant difference of the effect. The statistical significance of the difference was assessed at a probability level of 0.05, and the required power of the test was set as 80% as standard. For comparative purposes, the sample size was calculated by a formula from statistical tables, derived using a normal approximation to the binomial distribution (15). The results of the sample size calculations for comparison at different effect sampling times are given in Table 5. The guiding line when selecting the sample size should be the 2 h effect sampling time, which is required by the regulator.

An analysis of the results in Table 5 shows that the formula is conservative, expressed mainly in quite high sampling sizes for the small dose comparisons (e.g., 0.5 mg vs. placebo). In principle, a larger sample is required when comparing proportions (such as probability of pain relief) which have similar values. The number of patients here is larger, compared to real needs, for two probable reasons. One is the slight overprediction of the PK/PD model for the lower doses (the initial part of the simulated dose-effect curve in Figure 2) with regard to the observed values. The other could be the normal distribution approximation used

TABLE 5 Calculated Sample Sizes for Dose Comparisons to Observe a Statistically Significant Difference of the Effect with 80% Power

| | Effect sampling times (h) | | | |
|---|---|---|---|---|
| Dose comparisons | 1 | 2 | 4 | 6 |
| 0.5 mg vs. 0 mg | 1600 | 490 | 396 | 564 |
| 2.5 mg vs. 0 mg | 164 | 73 | 71 | 90 |
| 10 mg vs. 0 mg | 46 | 26 | 27 | 32 |
| 2.5 mg vs. 0.5 mg | 351 | 194 | 220 | 248 |
| 10 mg vs. 0.5 mg | 67 | 43 | 49 | 55 |
| 10 mg vs. 2.5 mg | 204 | 152 | 171 | 192 |

(15). The same effect has been observed and reported by us before (3). From the results above, a reasonable compromise with logistics and the existing limitations would seem to give a sample size of 200–250 patients per dose, which is likely to guarantee the required power of the trial.

## 16.7 DISCUSSION

Based on the above results, the following parallel oral naratriptan phase II dose-ranging trial is recommended: 4 arms—placebo, 0.5 or 0.75 or 1, 2.5, 10 mg—with 250 patients per arm. As seen from Tables 3 to 5, the suggested trial contains about 250 patients less and is logistically much simpler than the one which was actually carried out (the combinations of dose-ranging studies 1 and 2 in Ref. 1).

It can be argued that the most important purpose of a CTS and design exercise is to study the influence of the uncontrollable trial factors (such as assumptions, models structures, parameters) on the controllable trial factors (such as dosing and sampling schedules, sites and modes, sample size). In our case, we showed an example of how a PK/PD model for simulation and design can be developed integrating information from various sources. The PK part of the model was identified from phase I data, adjusted for the expected change in the kinetics due to the dosage form and the migraine. The PD part of the model was scaled from in vitro relative potency data with a similar (first in line) drug. The PK/PD model developed adequately describes the dose-response relationship obtained after oral tablet administration of naratriptan to migraine patients.

The application of D-optimal design resulted in a theoretically optimal design where the criterion is to estimate the model parameters with minimum uncertainty. The designs resulting from the simulations were compared in terms of their ability to minimize the parameter confidence region by assessing their relative

**TABLE 6** Resources Needed to Carry Out the Naratriptan CTS and Design

| Software | Time | Experience |
|---|---|---|
| • Model identification<br>Logistic regression:<br>BMDP, SPSS, S+<br>NONMEM<br>• Simulations<br>Any programming language<br>ACSL, MATLAB, etc.<br>• Analysis<br>Excel, etc. | • Data formatting and manipulation, 40% of total duration<br>• Modeling and parameter estimation, 25% of time<br>• Simulation, 15% of time<br>• Analysis and trial design, 20% of time<br>• Total duration of the whole exercise, 3 months | • PK/PD modeling<br>• Mixed-effect modeling<br>• Statistics<br>• Software implementation and development<br>• Interdisciplinary knowledge |

efficiencies with respect to the sub-D-optimal design. Although the designs based on D-optimality are usually logistically difficult to implement, they allow the more realistic designs obtained by CTS&D to be assessed by other criteria other than statistical significance and power.

Table 6 shows our estimation of the resources necessary for the CTS and design procedure carried out. The data in the table show that, generally, CTS and design are not very demanding with respect to resources. The software needed is usually available in a commercial environment. The same is true with respect to the duration of the procedure. The most time-consuming element for us was the data formatting and manipulation. However, not working in a pharmaceutical company, we did not have access to a data management resource. On the other hand, all the modeling, simulation, and analysis can be done in parallel with the conventional modeling activities, further reducing the time requirements for trial design.

In terms of experience, however, the CTS and design processes are most demanding, since these require a combination of highly sophisticated and interdisciplinary skills in the area of population and general PK/PD modeling and statistics.

## REFERENCES

1. E Fuseau, R Kempsford, P Winter, M Asgharnejad, N Sambol and CY Liu. The integration of the population approach into drug development: a case study, naratriptan. In: L Aarons et al., eds. The Population Approach: Measuring and Managing

Variability in Response, Concentration and Dose. COST B1 Medicine. European Commission, Brussels, 1997, pp 203–214.

2. C Dahlof, L Hogenhuis, J Olesen, H Petit, J Ribbat, J Schoenen, D Boswell, E Fuseau, H Hassani, P Winter. Early clinical experience with subcutaneous naratriptan in the acute treatment of migraine: a dose-ranging study. Eur J Neurol 5:469–477, 1998.
3. I Nestorov, G Graham, S Duffull, L Aarons, E Fuseau, P Coates. Modelling and simulation for clinical trial design involving a categorical response: a Phase II case study with naratriptan. Pharm Res 18:1210–1219, 2001.
4. O Petricoul, E Fuseau. Meta-analysis of the exposure/efficacy relationship for sumatriptan nasal spray. Poster at Population Approach Group in Europe, Saintes, France, 1999.
5. N Holford, LB Sheiner. Kinetics of pharmacological response. In: M Rowland, G Tucker, eds. Pharmacokinetics: Theory and Methodology. Oxford, UK: Pergamon Press, 1986, pp 189–212.
6. B Coulie, J Tack, B Maes, B Geypens, M de Roo, J Janssens. Sumatriptan, a selective 5-HT1 receptor agonist, induces a lag phase for gastric emptying of liquids in humans. Am J Physiol 272:G902–G908, 1997.
7. RA Tokola, PJ Neuvonen. Effect of migraine attack and metoclopramide on the absorption of tolfenamic acid. Br J Clin Pharmacol 17:67–75, 1984.
8. R Royle, PO Behan, JA Sutton. A correlation between severity of migraine and delayed gastric emptying measured by an epigastric impedance method. Br J Clin Pharmacol 30:405–409, 1990.
9. RA Tokola, PJ Neuvonen. Effect of migraine attacks on paracetamol absorption. Br J Clin Pharmacol. 18:867–871, 1984.
10. J Parantainen, H Hakkarainen, H Vapaatalo, G Gothoni. Prostaglandin inhibitors and gastric factors in migraine. Lancet 1(8172):832–833, 1980.
11. GN Volans. Migraine and drug absorption. Clin Pharmacokinet 3:313–318, 1978.
12. EK Hussey, KH Donn, MA Busch, AW Fox, JR Powell. Pharmacokinetics of oral sumatriptan in migraine patients during an attack and while painfree (abstract). Clin Pharmacol Ther 49:PI–46, 1991.
13. HE Connor, W Feniuk, DT Beattie, PC North, AW Oxford, DA Saynor, PPA Humphrey. Naratriptan: biological profile in animal models relevant to migraine. Cephalalgia 17:145–152, 1997.
14. NT Mathews, M Ashgharnejad, M Peykamian, A Laurenza. Naratriptan is effective and well tolerated in the acute treatment of migraine. Neurology 49:1485–1490, 1997.
15. S Lemeshow, DW Hosmer, J Klar, SK Lwanga. Adequacy of sample size in health studies. New York: World Health Organization, 1990.

# 17

# Prediction of Hemodynamic Responses in Hypertensive and Elderly Subjects from Healthy Volunteer Data

## The Example of Intravenous Nicardipine

**Patrice Francheteau,* Henri Merdjan,† Madeleine Guerret,* Stéphane Kirkesseli,‡ Paolo Sassano,§ and Jean-Louis Steimer¶**

Sandoz Pharmaceuticals, Rueil-Malmaison, France, and Basel, Switzerland

## 17.1 INTRODUCTION

Pharmacodynamic (PD) modeling has been presented as a means of "improving the scientific knowledge base for early decision making in developing new drugs" (1). This chapter illustrates the use of integrated pharmacokinetic-pharmacodynamic (PK/PD) modeling of multiple physiological responses in order to predict

* Retired.

† *Current affiliation*: Servier Research and Development, Fulmer, Slough, United Kingdom.

‡ *Current affiliation*: Aventis Pharmaceuticals, Bridgewater, New Jersey, U.S.A.

§ *Current affiliation*: Novartis Pharmaceuticals, Rueil-Malmaison, France.

¶ *Current affiliation*: F. Hoffmann-La Roche, Basel, Switzerland.

drug response in *patients* based on a model established in *healthy volunteers*. This application of PK/PD modeling is especially relevant to early clinical development, particularly during transition from phase 1 to phase 2.

Central to the PK/PD model is a mechanistic model that mathematically describes the essential features of the physiological system (in the present case the acute cardiovascular feedback regulation), *both* in healthy volunteers and in patients. The mathematical structure of the theoretical system is the same in both populations of subjects, but the parameters which describe the system (steady-state conditions as well as control parameters) are different. The example in this chapter is the action of the calcium-channel blocker, nicardipine, a vasodilating drug used as an antihypertensive agent. This was a retrospective study intended to demonstrate the feasibility of extrapolating the expected drug response from healthy volunteers to patients, through mimicking the corresponding prospective situation.

The aims of this chapter are

1. To give a summarized description of the mathematical model
2. To illustrate the modeling approach with application to literature data with intravenous nicardipine in hypertensive patients and elderly patients with elevated blood pressure
3. To perform computer simulations mimicking the prospective application of the modeling approach in a context of drug development, namely, by predicting response in patients from healthy volunteer data
4. To discuss the results with special emphasis on the potential applicability of this modeling approach in new drug development

## 17.2 BACKGROUND—THE MATHEMATICAL MODEL

The mechanistic and parsimonious model of the cardiovascular system, as proposed by Francheteau et al. (2), was evolved using the physiological bases of cardiovascular drug action. It incorporates within a closed-loop system a restricted number of selected hemodynamic variables.

For clarity of the present description, the number of equations will be limited to a minimum, and the presentation of the model will be restricted to its concepts. Readers interested in mathematical aspects are invited to refer to Francheteau et al. [2], where a detailed derivation can be found.

The essential feature of the model is that it is divided into two submodels: (a) a conventional PK/PD model relating plasma concentration of a vasodilator drug to the primary effect, i.e., a decrease in total peripheral resistance (TPR), and (b) a physiological model linking changes in TPR to the changes in other hemodynamic variables, taking into account the existence of a feedback loop in

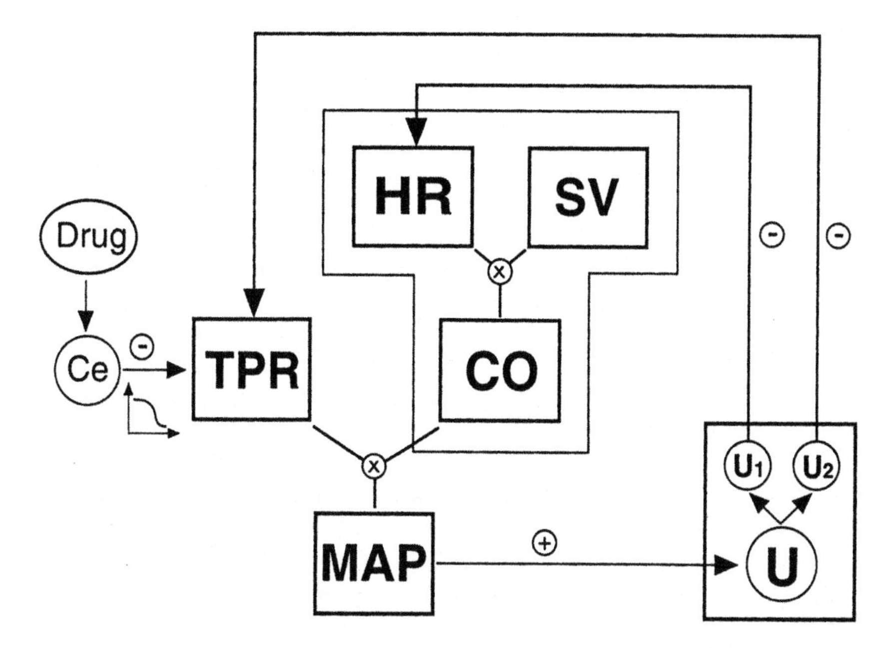


**FIGURE 1** Physiologically based PK/PD model for a vasodilator drug. HR, SV, CO, TPR stand for heart rate, stroke volume, cardiac output and total peripheral resistance, respectively. On this scheme MAP stands for the difference between mean arterial pressure and right atrial pressure. $U$, $U_1$, and $U_2$ are three auxiliary variables. $C_e$ is the drug concentration in the effect compartment. The sign + and − indicate whether the variation of an input variable makes the output variable to vary in the same or opposite way, respectively. The connection sign × indicates when an output variable is the product of the two input variables. [From Francheteau et al. [2] with permission.]

order, in the normal state, to control blood pressure. Additional assumptions are summarized below.

The first model (left-hand side of Figure 1) assumes that a vasodilator drug acts primarily by decreasing TPR. The intensity of this effect $E$ is related to drug plasma concentrations $C_p$ by means of (a) an effect compartment linked to the central compartment through a first-order rate-constant $k_{e0}$, and (b) a sigmoidal effect model between $E$ and the concentration at the effect site $C_e$:

$$E = E_{\max} \frac{C_e^\gamma}{C_e^\gamma + I_{50}^\gamma} = (\text{TPR}_{\text{eq}} - \text{TPR}_{\text{min}}) \frac{C_e^\gamma}{C_e^\gamma + I_{50}^\gamma} \tag{1}$$

Note that in Eq. (1), the maximum intensity of effect $E_{max}$ is expressed as a difference between the TPR value at equilibrium ($TPR_{eq}$) and the minimal reachable value for TPR ($TPR_{min}$). Parameter $I_{50}$ is the steady-state concentration in the central compartment that produces an effect equal to half the maximal effect, and $\gamma$ is the parameter allowing for sigmoidicity of the relationship.

The second model (right-hand side of Figure 1) is a closed-loop system, incorporating a cascade of physiological events supposed to represent, in a parsimonious but still mechanistic manner, some essential features of cardiovascular regulation. It links TPR to other hemodynamic variables, namely, mean arterial pressure (MAP), right atrial pressure (RAP), heart rate (HR), cardiac output (CO), and stroke volume (SV). An auxiliary variable $U$ is also defined. Variable $U$ can be understood as the "output signal" of a "proportional-derivative controller" (i.e., a mathematical representation of the baroreceptors, for instance) in response to a change in MAP. It is assumed that time changes of $U$ are proportional (a) to deviations of MAP from its equilibrium value $MAP_{eq}$ through a so-called "deviation-sensitive" parameter, noted $a_c$, and (b) to the rate of change of MAP with time, through a so-called "rate-sensitive" parameter noted $b_c$. Equilibrium delays between changes in MAP and changes in $U$ are accounted for by a time constant $\tau$: the larger the value of $\tau$, the longer the delay.

Additionally, variable $U$ is supposed to modify the sympathetic tone, which acts both on HR and TPR. This is implemented by dividing the regulation loop into two branches, and defining the auxiliary variables $U_1$ and $U_2$ (analogous to $U$). Further equilibrium delays between changes in $U$ and changes in $U_1$ and $U_2$ are characterized respectively by the time constants $\tau_1$ and $\tau_2$, (analogous to $\tau$). Changes in $U_1$ and $U_2$ induce proportional changes in HR and TPR (through parameters $\alpha$ and $\beta$, respectively). Finally, the model defines cardiac output (CO) as the product of HR by stroke volume (SV), and the difference (MAP − RAP) as the product of TPR by CO. These features complete the mechanistic model.

## 17.3 LITERATURE STUDIES AND DATA FOR MODEL CHECKING

In the paper by Francheteau et al. [2], computer simulations were performed to assess the output characteristics of the mathematical model and to explore the sensitivity of predicted hemodynamic responses to some crucial model parameters. Fitting of experimental data was also achieved with nicardipine data obtained in healthy volunteers under different modes of infusion (constant and exponentially decreasing rate). The suggested mathematical model demonstrated its value in fitting experimental CO and MAP data. Furthermore, the simulations strongly suggested that a graded variation of a few model parameters—including baseline levels of TPR, MAP, and CO, and an additional parameter $a_c$, supposed to reflect and quantify the efficiency of the feedback loop such as the baroreceptor reflex, would also be able to account for the contrasting patterns of hemodynamic response in hypertensive patients and in normotensive subjects (3–7). This assumption was based solely on simulations and thus remained to be tested on actual data in hypertensive patients.

### 17.3.1 Actual Data from Studies in Patients

We analyzed retrospectively several sets of literature data arising from studies with intravenous nicardipine in various subpopulations, including mild to severe hypertensive patients and elderly hypertensives. These data and results are presented first, as a means to strengthen the readers' confidence into the model prior to going into the predictive mode. In the real practice of new drug development, the forecasting would have to rely upon data from healthy volunteers only, without the opportunity to perform model checking as we did here.

Literature data include two studies in mild to moderate hypertensive patients and referred to as studies MH1 (8) and MH2 (9); one study in severe hypertensives referred to as SEV (10); and one study in elderly subjects with elevated blood pressure, noted ELD (11). The designs, material, and methods of these studies are briefly summarized in the upper part of Table 1.

The pharmacokinetic and hemodynamic data of studies MH2, SEV, and ELD were available on file in house. For study MH1, the mean pharmacokinetic and hemodynamic data, for the doses ranging from 1 to 7 mg, were visually recovered from the figures of the original paper. Then, since the hemodynamic data were expressed as differences from baseline, the recovered data were converted to absolute values by assuming baseline values for MAP and HR equal to 124 mmHg and 75 bpm, respectively. In all cases, whenever the MAP was not measured, its value was recovered from the DBP and SBP values according to the standard formula:

$$\text{MAP} = \frac{2 \times \text{DBP} + \text{SBP}}{3} \qquad (2)$$

### 17.3.2 Fitting of Experimental Data

Fitting and optimal parameter estimation were achieved using the modeling and simulation software, SimuSolv (12). Optimal parameter estimates were obtained through maximization of the logarithm of the likelihood function of the data, assuming a proportional error model for nicardipine plasma concentration data, and a constant error model for pharmacodynamic responses.

**Table 1** Pharmacodynamic Parameter Estimates Derived from Mean and Individual Data Fittings of Hypertensive Patients and Elderly Patients Receiving Intravenous Administration of Nicardipine

| | | Mild to moderate hypertension | Mild to moderate hypertension | Severe hypertension[a] | Elderly patients |
|---|---|---|---|---|---|
| **Summarized designs of studies** | **Study** | MH1 | MH2 | SEV | ELD |
| | **Reference** | 8 | 9 | 10 | 11 |
| | **Number of patients** | 11 | 37 | 18 | 28 (71–93 yr) |
| | **Baseline supine DBP (mmHg)** | 95–114 | 95–114 | >120 | $\geq$ 100 |
| | **Dosage regimen** | Single dose | Single dose | Individual titration to a DBP (95 mmHg or a DBP decrease) 25 mmHg | Three successive increasing doses |
| | **Doses** | 0.125, 0.25, 1, 3, 5, and 7 mg (2 min infusion) | 0.5, 1, 2 and 4 mg/h (48 h infusion) | 4 to 15 mg/h (1, 6, or 24 h infusion) | 1.25, then 2.5, then 5 mg (6 min infusions) |
| | **Pharmacodynamic measures** | BP and HR over 30 min | SBP and HR over 72 h | DBP and HR over up to 24 h | SBP and HR over 120 min |
| **Final PD parameter estimates** | $k_{e0}$ **($h^{-1}$)** | 41 | 0.87 | 2.3–∞[b] | 16 |
| | $TPR_{min}$ **(mm Hg min $liter^{-1}$)** | 10.3 | 9.9 | 9.8 (8.3–10.9) | 9.8 |
| | $I_{50}$ **(ng $ml^{-1}$)** | 30 | 47 | 64 (50–76) | 42 |
| | $\gamma$ | Set to 1 | Set to 1 | Set to 1 | 1.4 |
| | $a_c$ **($10^{-3}$ $mmHg^{-1}$)** | 10 | 4.6 | 1.7(0.0 to 5.2) | 1.2 |

[a] Mean values (and range) of three sets of individual parameter estimates
[b] Direct relationship between plasma concentration and effect
BP, blood pressure; DBP, diastolic blood pressure; SBP, systolic blood pressure; HR, heart rate; PD, pharmacodynamic.

A two- or three-compartment open pharmacokinetic model with first-order elimination was fitted to the plasma nicardipine concentration data. It was found that the MH1 study data was best described by a two-compartment model, whereas the other studies supported the more complex three-compartment model.

In a second step, the pharmacokinetic parameters were kept constant and the hemodynamic data (HR and MAP) were used for a global fitting of the PK/PD model. Since stroke volume and right atrial pressure were not measured in any of the selected studies, these parameters were assumed to be constant and were given the standard values of 0.097 liter and 5 mmHg, respectively. Those experiments also did not allow to discriminate, within the TPR kinetics, between the direct effect of the drug and the potential indirect influence of the arterial pressure control. Accordingly, parameters $\alpha$ and $\beta$ were arbitrarily set to one and zero, respectively. In other words, this constrained the blood pressure regulation loop to act solely on heart rate.

For study MH1, the mean pharmacokinetic and hemodynamic data obtained at 1, 3, 5 and 7 mg were modeled. A delay was observed between the end of the 2 min infusion and the time of occurrence of the maximal concentration (frequently 3 min postdosing). In order to account for this phenomenon, the drug was assumed to be delivered to a depot compartment linked to the central compartment via a first-order rate constant.

In study MH2, the mean pharmacokinetic and hemodynamic data of the placebo administration and of the four administrations of active drug were fitted simultaneously. Due to the period of observation (72 h), the model was slightly modified in order to account for the spontaneous circadian variation of the hemodynamic parameters, by adding a sinusoidal term (13–15) to the equation describing the time course of heart rate in the original paper (2),

$$\mathrm{HR(t)} = \mathrm{HR_{eq}} \times (1 - \alpha U_1(t)) + \mathrm{S1} \times \sin\left(\frac{2\pi \mathrm{t}}{24}\right) \tag{3}$$

where S1 is an amplitude of the circadian rhythm.

Note that the argument of the sine function in Eq. (3) does not incorporate any time-shift parameter $\tau$. For studies MH1 and MH2, the pharmacodynamic parameters, with the exception of equilibrium values $MAP_{eq}$ and $HR_{eq}$ which were specific of each experiment, were assumed to be common to the different doses.

In study SEV, the pharmacokinetic and hemodynamic models were fitted to the individual data of three patients, each of them representative of one particular dosage regimen group. In study ELD, the mean hemodynamic data was modeled. The unknown mean pharmacokinetic profile was predicted using typical values of pharmacokinetic parameters for nicardipine disposition estimated in patients with mild to moderate hypertension (mean age ± SD: 52.3 ± 18.5 years) in a separate study (16).

## 17.4 MODEL-BASED PREDICTIONS—ASSUMPTIONS

Model-based predictions were performed to give an overview of the capabilities of the model to *extrapolate* drug responses in hypertensive patients from responses observed in healthy volunteers, as a means to mimic the transition from clinical phase 1 to phase 2. Healthy volunteer data described in the work by Francheteau et al. (2) provided the starting values for pharmacokinetic and pharmacodynamic parameters. We actually extracted three initial sets of PD parameters, either derived from mean of individual parameters, or median of individual parameters, or from the fit of mean data.

The first exercise was to try and predict the hemodynamic response (time course of BP and HR) in a fictive subject thought to be a "patient with *mild to moderate* hypertension." To account for the expected decreased efficiency of the feedback regulation in hypertensive patients in contrast with healthy subjects, the value of $a_c$ was set to 0.005 $mmHg^{-1}$, i.e., 10 times lower than the typical value of healthy volunteers, as suggested by previous simulations (2). In addition, the equilibrium values of the parameters $DBP_{eq}$, $SBP_{eq}$, $MAP_{eq}$, and $TPR_{eq}$ in this fictive subject were set to values consistent with mild to moderate hypertension (i.e., 100, 160, 120, and 15.8 mmHg min $liter^{-1}$, respectively). Moreover, to account for a potential decrease in arterial vascular reactivity to drug administration in hypertensive patients vs. healthy subjects, we performed additional computer simulations with reference values of $I_{50}$ either unchanged or doubled, and/or with values of $\gamma$ either unchanged or set to 1.

The combination of the three initial sets of parameters with the four assumptions on $I_{50}$ and $\gamma$ led to 12 different scenarios, corresponding to 12 different sets of parameter values, as listed in Table 2. For the computer simulations, drug was assumed to be intravenously infused for 24 h with rates ranging from 0.25 to 8 mg/h. This period of time is adequate to reach steady-state blood pressure. Time profiles of diastolic blood pressure in the fictive patient following drug administration were derived from the simulated profiles of mean arterial pressure (MAP) with the following equation:

$$DBP(t) = MAP(t) \times \frac{DBP_{eq}}{MAP_{eq}} \quad (5)$$

For further exploratory purposes, the previous simulation exercise was repeated for another fictive subject thought to be a "patient with *severe* hypertension" with values of the parameters $DBP_{eq}$, $SBP_{eq}$, $MAP_{eq}$, and $TPR_{eq}$ set to 125, 200, 150, and 19.9 mmHg min $liter^{-1}$ respectively. The value of $a_c$ was again set to 0.005

**TABLE 2** Sets of Pharmacodynamic Parameters Used for Predicting Hemodynamic Responses in Hypertensive Patients After Intravenous Administration of Nicardipine Derived from Data of Healthy Volunteers, Together with Associated Scenario Numbers[a]

| | Initial sets of parameters from data in healthy volunteers | | | | | Derived sets of parameters and associated changes in parameters $I_{50}$ and $\gamma$ | | |
|---|---|---|---|---|---|---|---|---|
| Source of parameters | Scenario number | $k_{e0}$ ($h^{-1}$) | $TPR_{min}$ (mmHg min $liter^{-1}$) | $I_{50}$ (ng $ml^{-1}$) | $\gamma$ | Scenario number | $I_{50}$ | $\gamma$ |
| | | | | | | 2 | Doubled | Same |
| Mean values | 1 | 12 | 7.8 | 38 | 2.9 | 3 | Same | 1.0 |
| | | | | | | 4 | Doubled | 1.0 |
| | | | | | | 6 | Doubled | Same |
| Median values | 5 | 12 | 7.9 | 36 | 2.7 | 7 | Same | 1.0 |
| | | | | | | 8 | Doubled | 1.0 |
| | | | | | | 10 | Doubled | Same |
| Fit of mean data | 9 | 26 | 9.0 | 29 | 4.5 | 11 | Same | 1.0 |
| | | | | | | 12 | Doubled | 1.0 |

[a] With regard to derived sets of parameters (right-hand side of the table), changes in parameters $I_{50}$ and $\gamma$ refer to either scenario 1, 5, or 9.

mmHg$^{-1}$ as for the first fictive subject. All computer simulations were done using the modeling and simulation software SimuSolv (12).

## 17.5 RESULTS

### 17.5.1 Fitting of Experimental Patient Data

#### 17.5.1.1 Pharmacokinetics

The PK model was adequately fitted to the mean nicardipine PK data for studies MH1 and MH2 and to the individual PK data for study SEV as illustrated by the upper panels of Figures 2 to 4.

#### 17.5.1.2 Pharmacodynamics

The pharmacodynamic parameters estimated through modeling of the hemodynamic data for all studies are reported in the lower part of Table 1. Some other pharmacodynamic parameters not reported in Table 1 turned out to be nonestimable and were subsequently fixed to predefined constant values for final fits, namely, the parameters $\tau$ and $\tau_1$. Both time constants were set to 0.01 h for studies MH2, SEV, and ELD as in the original paper with healthy volunteer data (2). Given the short observation time in study MH1 (only 20 min postdosing), $\tau$ and $\tau_1$ had to be set to lower values, such as 0.01 min.

Typical patterns of nicardipine-induced hemodynamic changes in hypertensive patients were adequately reproduced by the mathematical model, namely:

- Acute reduction in MAP and the transient increase in HR (study MH1) after bolus dosing of high-dose nicardipine in moderately hypertensive patients (Figure 2)
- Time course of MAP and HR up to 72 h (study MH2), including circadian changes, in mild to moderate hypertensive patients treated with 48 h nicardipine infusion (Figure 3). From this figure, it is clear that intravenous nicardipine administration significantly lowers the arterial pressure, without suppressing the circadian changes in blood pressure and heart rate.
- Pronounced reduction in MAP (up to 40 mmHg) in severe hypertensive patients (study SEV) following 1–24 h infusion (Figure 4)
- Acute reduction in MAP and virtual lack of change in HR (study ELD) after short-term infusion of nicardipine in elderly hypertensive subjects (Figure 5)

The value of the parameter $k_{e0}$ was found to be high (range 16–41 h$^{-1}$), indicating a virtual lack of delay between the kinetics of nicardipine plasma concentration and the time course of TPR reduction for all studies, except for the 48 h infusion study MH2 where the final estimate was equal to 0.87 h$^{-1}$. The

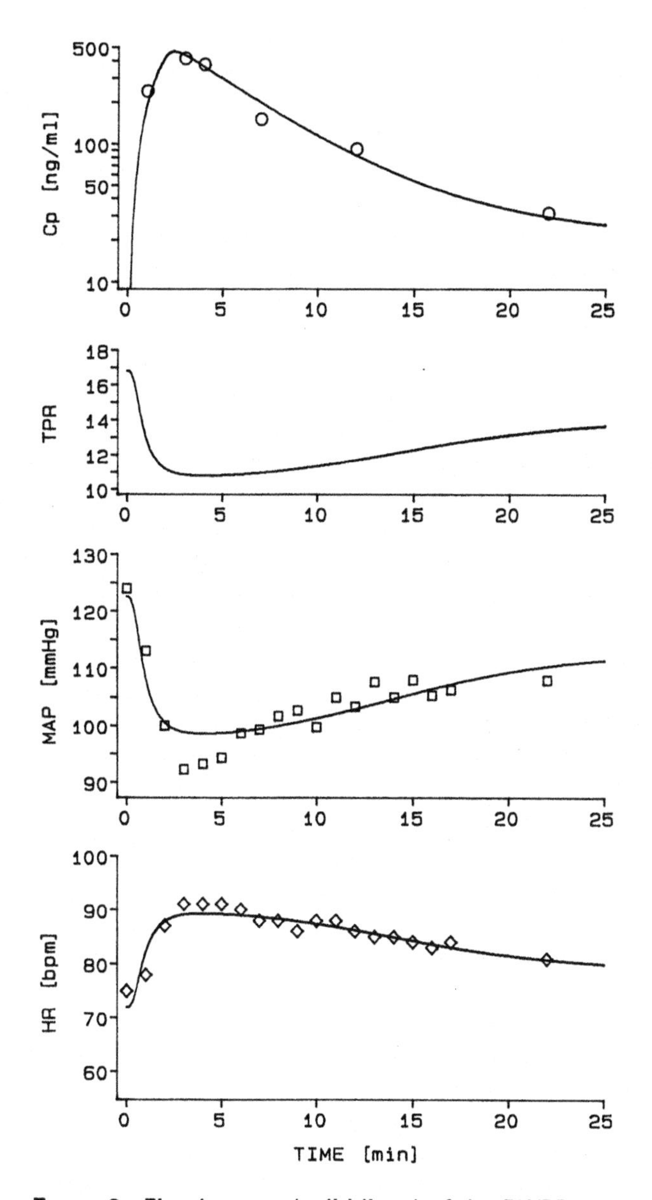


**FIGURE 2** Fitted curves (solid lines) of the PK/PD model with mean plasma concentrations of nicardipine (○) and mean hemodynamic data [heart rate (◇) and mean arterial pressure (□)] of study MH1 after an intravenous bolus dose of 7 mg nicardipine. For clarity, pharmacokinetic and pharmacodynamic profiles were truncated at 25 minutes. Typical standard deviations were 7 mmHg and 14 bpm for mean arterial pressure and heart rate, respectively. TPR is expressed in mmHg min liter$^{-1}$.

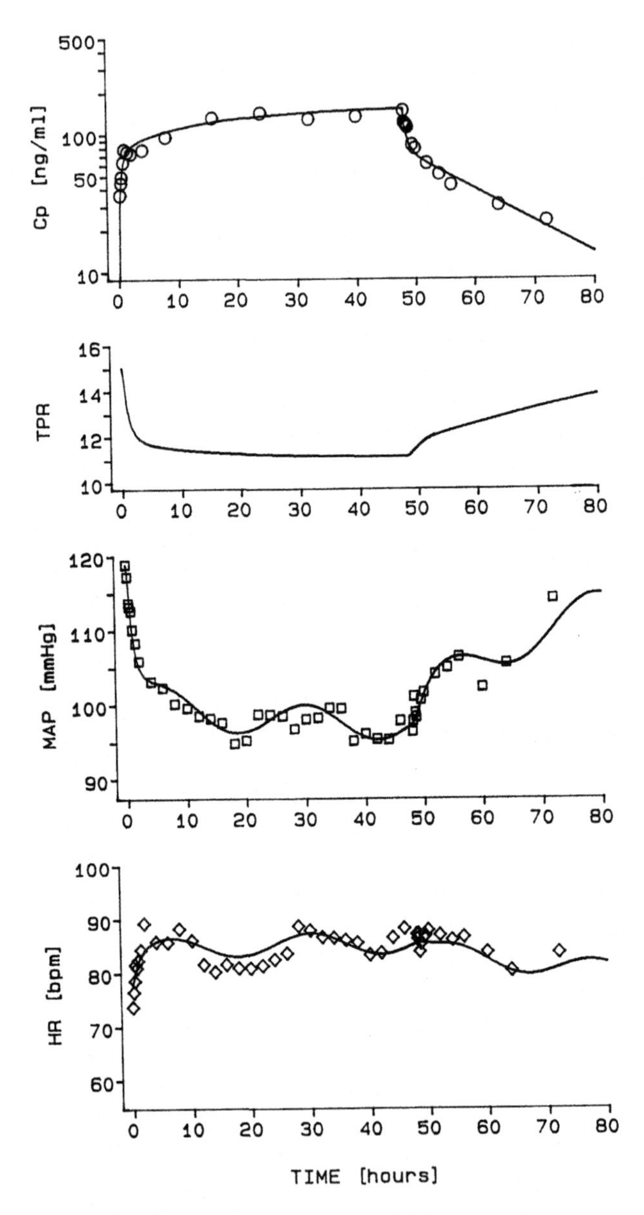


**Figure 3** Fitted curves (solid lines) of the PK/PD model with mean plasma concentrations of nicardipine (○) and mean hemodynamic data [heart rate (◇) and mean arterial pressure (□)] of study MH2 following the 4 mg/h intravenous nicardipine dosage regimen. For clarity error bars were omitted, typical standard deviations were 7 mmHg and 9 bpm for mean arterial pressure and heart rate respectively. TPR is expressed in mmHg min liter$^{-1}$.

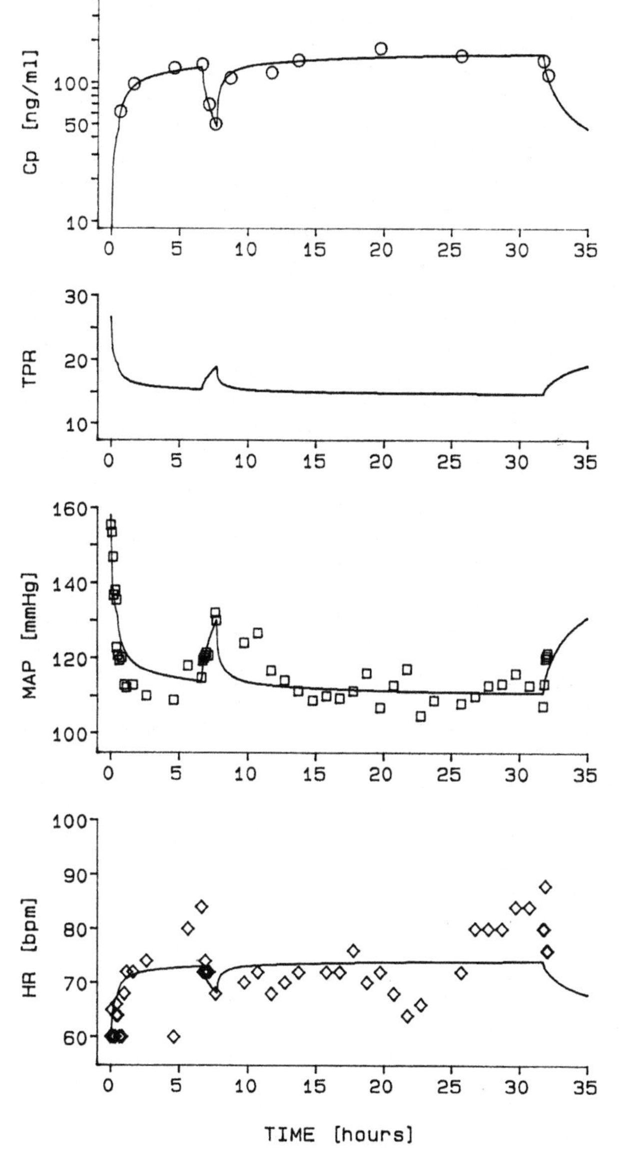


**FIGURE 4** Fitted curves (solid lines) of the PK/PD model with individual plasma concentrations of nicardipine (○) and hemodynamic data [heart rate (◇) and mean arterial pressure (□)] of one subject (group II) of study SEV following a 7 mg/h intravenous nicardipine maintenance dosage regimen. TPR is expressed in mmHg min liter$^{-1}$.

**TABLE 3** Model-Predicted Range of Diastolic Blood Pressure Reduction (DBP, mmHg) After 24 h Constant Rate Intravenous Infusion of Nicardipine in Two Fictive Patients, One with Mild to Moderate Hypertension (left-hand side), and One with Severe Hypertension (right-hand side)[a]

| Infusion rate (mg/h) | Simulated range of DBP reduction (mmHg) in a fictive, *mild to moderate* hypertensive patient (DBP baseline = 100 mmHg) | | | | | | Simulated range of DBP reduction (mmHg) in a fictive, *severe* hypertensive patient (DBP baseline = 125 mmHg) | | | | | |
|---|---|---|---|---|---|---|---|---|---|---|---|---|
| | 0.25 | 0.5 | 1 | 2 | 4 | 8 | 0.25 | 0.5 | 1 | 2 | 4 | 8 |
| **Scenario** | | | | | | | | | | | | |
| **5** | 0–5 | 0–5 | 5–10 | 15–20 | 30–35 | 35–40 | 0–5 | 0–5 | 5–10 | 25–30 | 45–50 | 55–60 |
| **6** | 0–5 | 0–5 | 0–5 | 5–10 | 15–20 | 30–35 | 0–5 | 0–5 | 0–5 | 5–10 | 25–30 | 45–50 |
| **7** | 0–5 | 5–10 | 10–15 | 15–20 | 20–25 | 25–30 | 5–10 | 10–15 | 15–20 | 25–30 | 35–40 | 40–45 |
| **8** | 0–5 | 0–5 | 5–10 | 10–15 | 15–20 | 20–25 | 0–5 | 5–10 | 10–15 | 15–20 | 25–30 | 35–40 |

[a] Pharmacodynamic parameters derive from scenarios 5–8 (see Table 2 for details).

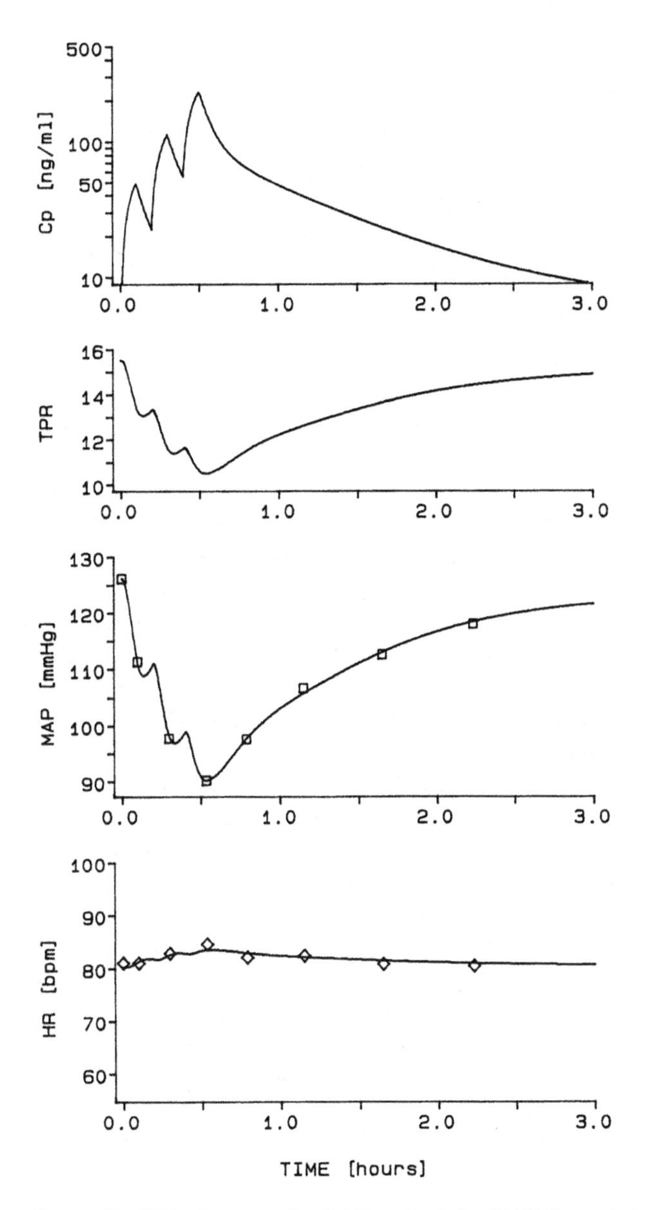


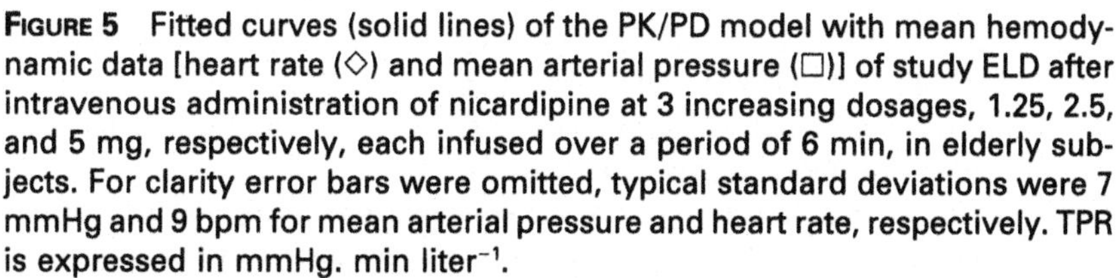
**FIGURE 5** Fitted curves (solid lines) of the PK/PD model with mean hemodynamic data [heart rate (◇) and mean arterial pressure (□)] of study ELD after intravenous administration of nicardipine at 3 increasing dosages, 1.25, 2.5, and 5 mg, respectively, each infused over a period of 6 min, in elderly subjects. For clarity error bars were omitted, typical standard deviations were 7 mmHg and 9 bpm for mean arterial pressure and heart rate, respectively. TPR is expressed in mmHg. min $liter^{-1}$.

consistently low values of $TPR_{min}$ across studies (range 9.8–10.3 mmHg min liter$^{-1}$) vs. values of $TPR_{eq}$ (range 15.1–26.5 mmHg min liter$^{-1}$; not reported in Table 3) reflected the efficacy of the vasodilating action of nicardipine in hypertensive patients.

The estimates of $I_{50}$ were found to be remarkably consistent, ranging from 30 to 64 ng ml$^{-1}$. The parameter of sigmoidicity $\gamma$ was fixed to 1.0, i.e., hyperbolic relationship, for the final fitting of all studies except for study ELD (elderly subjects) where its final estimate was found equal to 1.4. The estimates of the control parameter $a_c$ were found to decrease with both the degree of hypertension and age, ranging from 10.1 and 4.6 ($10^{-3}$ mmHg$^{-1}$) for studies MH1 and MH2 (patients with mild to moderate hypertension) to 1.7 and 1.2 ($10^{-3}$ mmHg$^{-1}$) for studies SEV (patients with severe hypertension) and ELD (elderly subjects), respectively.

### 17.5.2 Model-Based Predictions: Extrapolating to Patients from the Healthy Volunteer Data

This section illustrates the use of the PK/PD model as a tool for predicting hemodynamic responses in hypertensive patients through extrapolation from phase 1 healthy volunteer data. From the model simulations, for all 12 scenarios (described in Table 2), we mimicked one key aspect of the transition from clinical phase 1 to phase 2, namely, the attempt to define the range of doses to be tested in patient trials. In the present case, the established goal was to design a dose-finding study (of intravenously infused nicardipine) in mild to moderate hypertensive patients whose outcomes should be (a) the *maximum noneffective dosage regimen*, i.e., the highest dosage regimen associated with a predicted reduction in diastolic blood pressure less than 10 mmHg (lack-of-efficacy criterion), and (b) the *maximum effective and safe dosage regimen*, i.e., the highest dosage regimen associated with a predicted reduction in diastolic blood pressure greater than 10 mmHg (efficacy criterion) and less than 25 mmHg (safety criterion).

The model-based computer predicted reductions in diastolic blood pressure for the fictive subject thought to be a "patient with mild to moderate hypertension," with the median-based sets of parameters (i.e., scenarios 5–8), are presented for illustrative purposes in Table 3 (left panel) and depicted in Figure 6. (Similar results for the other scenarios are not presented). The target dose ranges to be tested defined for all sets of parameters are summarized in Table 4.

It can be seen in Figure 6 that according to the simulation scenarios, the doses to be tested in a dose-finding study with mild to moderate hypertensive patients should range between 1 mg/h, the highest noneffective dosage regimen for most of the considered scenarios, and 4 mg/h, the highest effective and safe dosage regimen for most of these scenarios.

For further illustrative purposes, the computer predicted reduction of the diastolic blood pressure for the second fictive subject thought to be a "patient

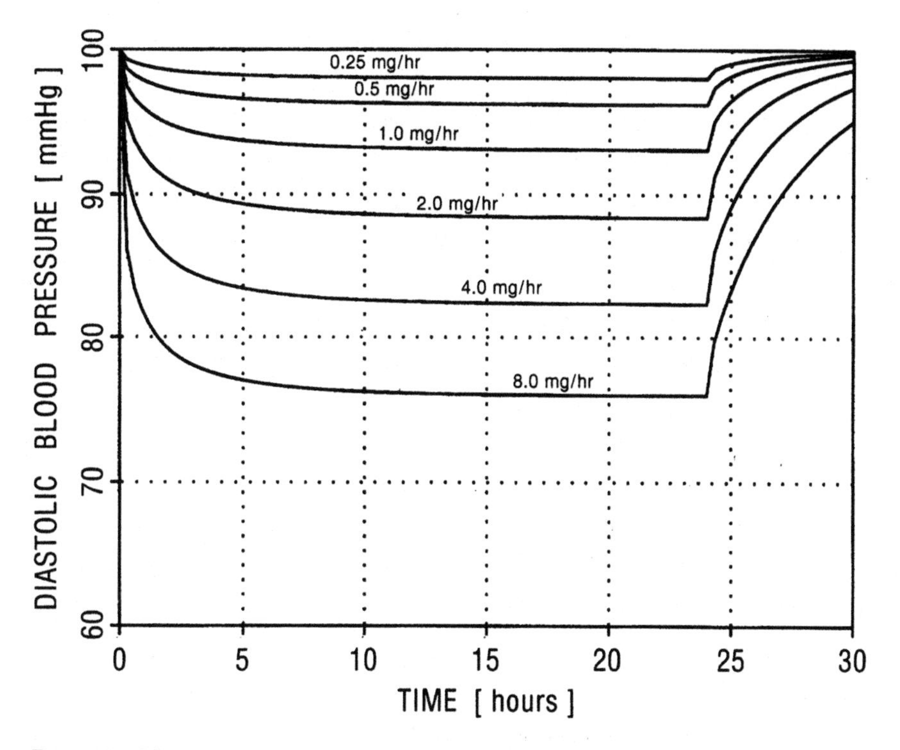


**FIGURE 6** Model-predicted diastolic blood pressure after nicardipine infusion at different rates ranging from 0.25 mg/h (upper curve) to 0.5, 1.0, 2.0, 4.0, and 8.0 mg/h (lower curve) in a fictive subject, thought to be a "typical patient with mild to moderate hypertension," with median-based parameters (scenario 8, see text and Table 2 for explanation).

with *severe* hypertension," for the same scenarios [5–8], are presented in Table 3 (right panel). Examination of Table 3 suggests that the dose range is shifted to higher doses in order to achieve the necessary reduction in DBP. Based on the PK/PD model, a dose-finding study of nicardipine in patients with severe hypertension should include drug regimens ranging between 2 and 8 mg/h.

## 17.6 DISCUSSION

### 17.6.1 A Patho-Physiologically Based, Mechanistic PK/PD Model

In this chapter, we addressed a problem of wide concern in new drug development and clinical evaluation: the transfer of PK/PD knowledge gathered in healthy volunteer trials (typically phase 1) into assumptions for dose-finding in patients

**TABLE 4** Model-Predicted Range of Candidate Nicardipine Infusion Rates to be Tested (mg/h) in a Dose-Finding Study in Mild to Moderate Hypertensive Patients for the Various Scenarios Chosen for Computer Simulations[a]

| Source of parameters | Scenario number | Range of infusion rates (mg/h) |
|---|---|---|
| | 1 | 1–2 |
| Mean values | 2 | 2–4 |
| | 3 | 0.5–4 |
| | 4 | 1–8 |
| | 5 | 1–2 |
| Median values | 6 | 2–4 |
| | 7 | 0.5–4 |
| | 8 | 1–8 |
| | 9 | 1 |
| Fit of mean data | 10 | 2 |
| | 11 | 0.5–4 |
| | 12 | 1–8 |

[a] See Table 2 for definition of scenarios 1–12.

(typically phase 2). It was shown that, in cardiovascular clinical pharmacology, mathematical modeling can be a useful technique for such extrapolation. The principles underlying the suggested approach in terms of the mechanistic model are the following: (a) the basic physiological mechanisms, which are accounted for by the model's structure as established from healthy volunteer data, are qualitatively the same in healthy subjects and in patients; (b) pathology can be expressed quantitatively in terms of predictable changes in some model parameters.

A key feature of the suggested cardiovascular model is the inclusion of one, and indeed only one, fundamental physiological process, i.e., the linkage of total peripheral resistance, mean arterial pressure, and heart rate (or cardiac output) through a feedback control mechanism which regulates arterial pressure. The basal state (of the model) is supposed to reflect a pseudo-steady-state situation. The corresponding combination of $TPR_{eq}$, $MAP_{eq}$, and $HR_{eq}$ is derived from the characteristics of the regulation. This linkage is the crucial sophistication in regard to the classical PK/PD models where sigmoidal $E_{max}$ models are used for *separate* description of the measured hemodynamic variables (see, e.g., Refs. 17, 18). Information about the system's behavior is present in those responses which display a change in presence of the stimulus (the drug). Through the model, information gathered in one operating condition (the healthy subject) can be transposed to another operating condition (the patient).

### 17.6.2 Experimental Fitting of Patient Data

Under normal physiological conditions, the baroreceptors, which are an essential component of the feedback process after acute drug administration, deliver both deviation-sensitive and rate-sensitive signals to the solitary tract nucleus located in the brainstem, in response to the stretching of the arterial wall of the aortic and carotidian vessels. The model indeed incorporates a derivative part in the arterial pressure control which is sensitive to the instantaneous rate of change of MAP (through parameter $b_c$). Ideally, experiments should be specifically designed for the purpose of characterizing a PK/PD model (19, 20), and to get the best chance to properly estimate model parameters. As an example, markedly different programs of intravenous infusion were required to detect an influence of the infusion rate on the intensity of cardiovascular effects of nifedipine (21) and felodipine (22). Because the present PK/PD analysis was retrospective, the PK/PD model could not be fully characterized with the data at hand, and some simplifying assumptions had to be made a priori. For instance, some aspects of the PK/PD model were omitted (parameters $b_c$ and $\beta$ were set to zero), and some model parameters were assigned plausible, constant values (e.g., stroke volume and right atrial pressure).

Despite these simplifications, the PK/PD model was shown to be able to reflect experimental data in a variety of different dosing conditions in several populations of patients. The changes in parameter estimates for $a_c$, $TPR_{min}$, and $I_{50}$ are consistent with the expected patho-physiological changes across populations:

- The mathematical model includes a proportional component which generates, through parameters $a_c$, a "control signal" in response to deviations of MAP from its equilibrium value: the higher the value of $a_c$, the larger the reflex tachycardia and the better the regulation of MAP. This parameter is thought to be related to the elasticity of the arterial wall and to the efficiency with which a change is detected and converted into a control signal. The proportional controller operates less efficiently when the arteries are rigid, a phenomenon which is reported to occur in the elderly (23, 24). This produces a larger change in MAP in response to a given change in TPR. Indeed, the estimates of parameter $a_c$ show a dramatic and monotonic decrease with both degree of hypertension and age, ranging from 10.1 and 4.6 ($10^{-3}$ $mmHg^{-1}$) in patients with mild to moderate hypertension to 1.7 and 1.2 ($10^{-3}$ $mmHg^{-1}$) in patients with severe hypertension and in elderly subjects, respectively. These values are far below the mean estimate of 46.0 ($10^{-3}$ $mmHg^{-1}$) reported in healthy subjects (2).
- Minimal vascular resistance can be derived from cardiac output measurement (5, 25, 26) or estimated locally by various hemodynamic methods after application of maximal vasodilator stimuli, mostly to the forearm or hand (6, 7, 27–29). The elevated minimal vascular resistance

in hypertensive subjects as compared to normotensive subjects is reflecting the decreased vasodilator capacity in hypertension. These abnormalities can be attributed either to an increase in vascular smooth muscle activity or to some structural changes in the vasculature causing vascular hypertrophy (27). Our estimates of $TPR_{min}$, which was found to be higher in hypertensives than in normotensive subjects, are in agreement with the above-mentioned changes.

The estimates of the parameter $I_{50}$ are remarkably consistent across populations, ranging from 30 to 64 ng $ml^{-1}$. These values are in good agreement with the mean values reported by Modi et al. (30), in a separate analysis, for the blood pressure lowering effect (35.1 ng $ml^{-1}$) and for the tachycardic effect (58.0 ng $ml^{-1}$) of intravenous nicardipine in a sample of healthy male volunteers.

Consistently high $k_{e0}$ values were found across studies, except in study SEV where the estimate of $k_{e0}$ was 0.87 $h^{-1}$. In that particular study, interpretation of $k_{e0}$ is likely to be different, because this parameter might have contributed a role normally played by a time-shift parameter in the sine function model of circadian variability, actually omitted in the sine function in Eq. (3).

The suggested model, although being physiologically based, is not intended to be fully comprehensive, nor exhaustive. Although parsimonious, it provides an adequate description of data under different nicardipine inputs, and up to 3 days of drug administration. Of course, extrapolation beyond the limits of the range in which the model was built and validated should be done with caution. This note of caution especially applies to long-term therapy, where other regulatory mechanisms, e.g., cardiovascular remodeling, may be triggered.

### 17.6.3 Extrapolations to Patients from the Healthy Volunteer Data

We have carried out model-based computer steady-state predictions of diastolic blood pressure after intravenous infusion of nicardipine over 24 h, under various scenarios regarding model parameter values. The rationale behind the choice for investigating unchanged and doubled values of $I_{50}$ was that these upper and lower limits allowed, roughly, to cover the range of estimates of this parameter, as defined from fits in healthy volunteers. Expectedly, the time course of DBP is sensitive to the choice of parameter values (e.g., $I_{50}$, $\gamma$) and, therefore, the predictions of the magnitude of the antihypertensive response are quantitatively different. Most importantly, however, the definition of doses to be selected for a phase 2 study aimed at assessing the antihypertensive effect in patients is remarkably robust across scenarios.

Regulatory authorities are increasingly interested in the shape of the dose-(concentration)-response relationship as an outcome from dose-finding studies.

Based on our computer simulations, the suggestion of dosing rates for intravenous nicardipine (placebo, 1, 2 and 4 mg/h) should allow a "trend" to be detected in the decrease in blood pressure at steady state. Indeed, the model-suggested dosing rates for a 24 h infusion in moderate hypertensives are very close to the actual design chosen by Cook et al. [9] (infusion rates from 0.5 to 4 mg/h) for the dose-ranging experiment that these investigators had carried out separately (and earlier than) the present modeling and simulation work (see Figure 7). There is some discrepancy across simulations in the time required to lower blood pressure down

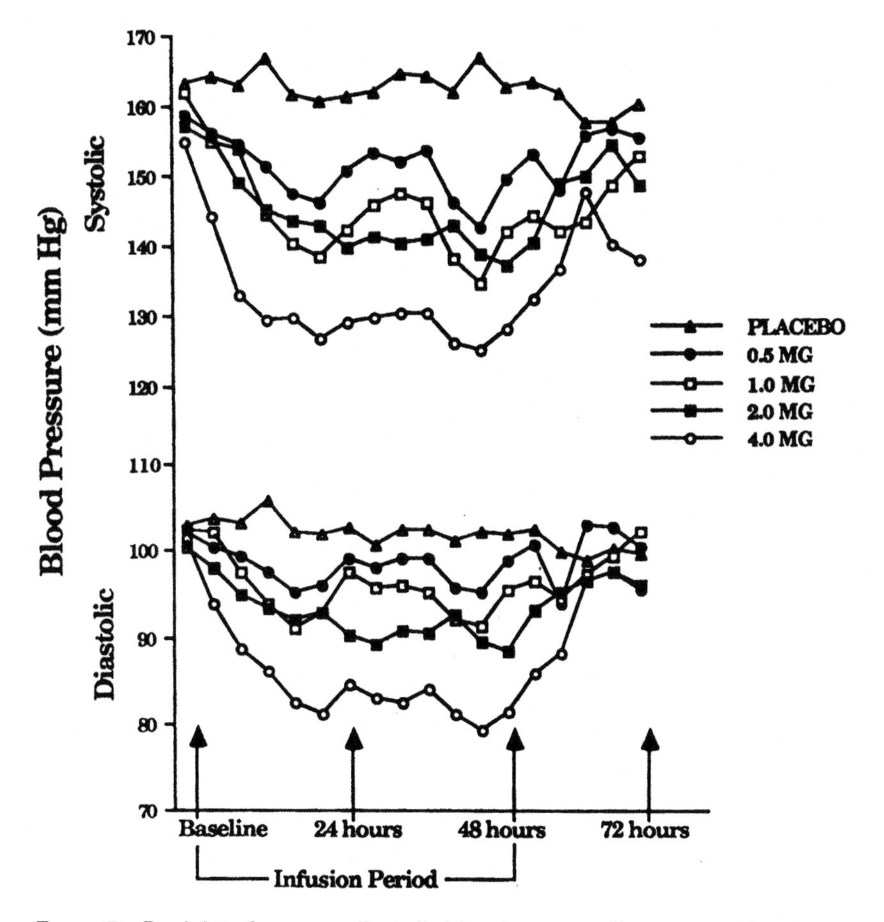


**FIGURE 7** Real data for mean diastolic blood pressure (lower panel) and mean systolic blood pressure (upper panel) after nicardipine infusion at different rates in mild to moderate hypertensive patients. [from Cook et al. [9] with permission.]

to 90 mmHg. This calls for caution and careful monitoring of the patient when infusing the drug in an actual experiment performed at a constant infusion rate; this also illustrates the potential value of using alternate, more sophisticated input functions, e.g., exponential infusions which would achieve effective concentrations more rapidly (21, 22).

The computer simulations highlight that PK/PD modeling can provide useful support for the planning of a phase 2 dose-finding study. We have shown this for a complex system such as the cardiovascular system, where, at a first glance, the nature of the PD response is different in the healthy subject and the hypertensive patient. However, the definite answer can only come from the experiment itself. In our example, the actual response data after the 2 mg/h dosing rate in a phase 2 study would especially be helpful for discriminating between models. The improved characterization of the PK/PD model will inevitably increase the reliability of predictions for further study planning, e.g., for phase 3 confirmatory trials. In addition, although such PK/PD model-based simulations provide a very useful tool for dose and regimen selection, the final decision should be made by the responsible clinician, based on the usual criteria for efficacy and, of course, careful consideration of patients' safety.

## 17.7 CONCLUSION

The incorporation of PK/PD modeling in new drug development has been advocated by many authors (1, 31–34) and the number of successful applications has been steadily expanding in the past decade (35–37). Consensus on how to best take advantage of modeling in early development is developing (38). In the present example, the challenge was to use heart rate data in healthy volunteers to try and predict a qualitatively different response in hypertensive patients, namely a decrease in blood pressure. This problem was solved by the modeling of the regulation loop itself. This kind of prediction, which uses a patho-physiologically based mechanistic PK/PD model, is potentially applicable to other therapeutic areas where qualitative differences between physiological and pathological conditions can be translated into quantitative differences in a few, rationally selected model parameters. Actually, this kind of approach has also been successfully transposed to the regulation of blood glucose and insulin in response to an oral antidiabetic drug, both in healthy volunteers and type 2 diabetic patients (39).

Between-subject variability in both PD and physiological parameters would inevitably be present in practice. This was actually not considered in the present investigation. An anticipated sophistication of the proposed approach would be to estimate model parameters together with their interindividual variability, and to use this variability for Monte Carlo (stochastic) simulation, so as to provide a confidence band prediction.

## ACKNOWLEDGMENTS

The present modeling work was carried out at a period of time where the co-authors were all members of Sandoz Pharmaceuticals, prior to the merge with Ciba which created Novartis. The contents of the present chapter had been released by Sandoz for publication. Yamanouchi Pharmaceutical Co. Ltd. (Tokyo, Japan) and Dr. J.D. Wallin are kindly acknowledged for their help in providing the raw data of the nicardipine studies MH2 and SEV. Ms. R. Benghozi, Messrs. C. Dubray, A. Tarral (all former colleagues at Sandoz) and Dr. S. Vozeh are gratefully acknowledged for their critical review and input to an earlier version of the present manuscript.

## REFERENCES

1. PD Kroboth, VD Schmith, RB Smith. Pharmacodynamic modelling: application to new drug development. Clin Pharmacokinet 20:91–98, 1991.
2. P Francheteau, JL Steimer, H Merdjan, M Guerret, C Dubray. A mathematical model for dynamics of cardiovascular drug action: application to intravenous dihydropyridines in healthy volunteers. J Pharmacokinet Biopharm 21:489–514, 1993.
3. O Lederballe-Pedersen, NJ Christensen, KD Rämsch. Comparison of acute effects of nifedipine in normotensive and hypertensive man. J Cardiovasc Pharmacol 2: 357–366, 1980.
4. GA McGregor, C Rotellar, ND Markandu, SJ Smith, GA Sagnella. Contrasting effects of nifedipine, captopril, and propranolol in normotensive and hypertensive subjects. J Cardiovasc Pharmacol 4(Suppl):358–362, 1982.
5. S Julius, K Jamerson, A Mejia, L Krause, N Schork, K Jones. The association of borderline hypertension with target organ changes and higher coronary risk. Tecumseh Blood Pressure study. JAMA 264:354–358, 1990.
6. J Cleroux, N Kouame, A Nadeau, D Coulombe, Y Lacourciere. Baroreflex regulation of forearm vascular resistance after exercise in hypertensive and normotensive humans. Am J Physiol 263(Suppl H):1523–1531, 1992.
7. M Leschke, FC Schoebel, M Vogt, M Heintzen, M Kelm, W Motz, BE Strauer. Reduced peripheral and coronary vasomotion in systemic hypertension. Eur Heart J 13(Suppl D):96–99, 1992.
8. DG Cheung, JL Gasster, JM Neutel, MA Weber. Acute pharmacokinetic and hemodynamic effect of intravenous bolus dosing of nicardipine. Am Heart J 119:438–442, 1990.
9. E Cook, GG Clifton, R Vargas, G Bienvenu, R Williams, N Sambol, G McMahon, S Grandy, CM Lai, C Quon, CR Anderson, P Turlapaty, JD Wallin. Pharmacokinetics, pharmacodynamics, and minimum effective clinical dose of intravenous nicardipine. Clin Pharmacol Ther 47:706–718, 1990.
10. JD Wallin, ME Cook, L Blanski, GS Bienvenu, GG Clifton, H Langford, P Turlapaty, A Laddu. Intravenous nicardipine for the treatment of severe hypertension. Am J Med 85:331–338, 1988.

11. M Escande, D David, B Diadema. Effets antihypertenseurs de la nicardipine intraveineuse dans l'hypertension arterielle du sujet âgé. Thérapie 44:161–165, 1989.
12. EC Steiner, TD Rey, PS McCroskey. Reference Guide for SimuSolv: Modeling and Simulation Software. Midland, MI: Dow Chemical Company, 1990.
13. AN Kong, EA Ludwig, RL Slaughter, PM DiStefano, J DeMasi, E Middleton Jr., WJ Jusko. Pharmacokinetics and pharmacodynamic modeling of direct suppression effects of methylprednisolone on serum cortisol and blood histamine in human subjects. Clin Pharmacol Ther 46:616–628, 1989.
14. JA Wald, WJ Jusko. Corticosteroid pharmacodynamic modeling: osteocalcin suppression by prednisolone. Pharm Res 9:1096–1098, 1992.
15. A Chakraborty, W Krzyzanski, WJ Jusko. Mathematical modeling of circadian cortisol concentrations using indirect response models: comparison of several methods. J Pharmacokinet Biopharm 27:23–43, 1999.
16. JP Fillastre, M Godin, M Guerret, P Bernadet, D Lavène. Cinétique d'une nouvelle forme de nicardipine. Arch Mal Coeur Vaisseaux 81(Suppl):312, 1988.
17. G Mikus, C Zekorn, T Brecht, M Eichelbaum. Acute haemodynamic effects of i.v. nitrendipine in healthy subjects. Eur J Clin Pharmacol 41:99–103, 1991.
18. KH Graefe, R Ziegler, W Wingender, KD Rämsch, H Schmitz. Plasma concentration-response relationships for some cardiovascular effects of dihydropyridines in healthy subjects. Clin Pharmacol Ther 43:16–22, 1988.
19. WA Colburn, MA Eldon. Models of drug action: experimental design issues. In: CJ van Boxtel, NHG Holford, M Danhof, eds. The in vivo study of drug action. Amsterdam: Elsevier, 1992, pp 17–29.
20. E Bellissant, V Sébille, G Paintaud. Methodological issues in pharmacokinetic-pharmacodynamic modelling. Clin Pharmacokinet 35:151–166, 1998.
21. CH Kleinbloesem, P van Brummelen, M Danhof, H Faber, J Urquhart, DD Breimer. Rate of increase in the plasma concentration of nifedipine as a major determinant of its hemodynamic effects in humans. Clin Pharmacol Ther 41:26–30, 1987.
22. AF Cohen, MA van Hall, J van Harten, RC Schoemaker, P Johansson, DD Breimer, R Visser, B Edgar. The influence of infusion rate on the hemodynamic effects of felodipine. Clin Pharmacol Ther 48:309–317, 1990.
23. JD Bristow, AJ Honour, GW Pickering, P Sleight, HS Smyth. Diminished baroreflex sensitivity in high blood pressure. Circulation 39:48–54, 1969.
24. AC Simon, J Levenson, J Bouthier, ME Safar, AP Avolio. Evidence of early degenerative changes in large arteries in human essential hypertension. Hypertension 7: 675–680, 1985.
25. SJ Montain, SM Jilka, AA Ehsani, JM Hagberg. Altered hemodynamics during exercise in older essential hypertensive subjects. Hypertension 12:479–484, 1988.
26. GE McVeigh, DE Burns, SM Finkelstein, KM McDonald, JE Mock, W Feske, PF Carlyle, J Flack, R Grimm, JN Cohn. Reduced vascular compliance as a marker for essential hypertension. Am J Hypertens 4:245–251, 1991.
27. PA Carberry, AMM Shepherd, JM Johnson. Resting and maximal forearm skin blood flows are reduced in hypertension. Hypertension 20:349–355, 1992.
28. Y Hirooka, T Imaizumi, H Masaki, S Ando, S Harada, M Momohara, A Takeshita. Captopril improves impaired endothelium-dependent vasodilation in hypertensive patients. Hypertension 20:175–180, 1992.

29. DJ Patel, RN Vaishnav, BR Coleman, RJ Tearney, LN Cothran, CL Currey. A theoretical method for estimating small vessel distensibility in humans. Circ Res 63: 572–576, 1988.
30. NB Modi, P Veng-Pedersen, DJ Graham, RJ Dow. Application of a system analysis approach to population pharmacokinetics and pharmacodynamics of nicardipine hydrochloride in healthy males. J Pharm Sci 82:705–713, 1993.
31. CC Peck, WH Barr, LZ Benet, J Collins, RE Desjardins, DE Furst, JG Harter, G Levy, T Ludden, JH Rodman, L Sanathanan, JJ Schentag, VP Shah, LB Sheiner, JP Skelly, DR Stanski, RJ Temple, CT Viswanathan, J Weissinger, A Yacobi. Opportunities for integration of pharmacokinetics, pharmacodynamics and toxicokinetics in rational drug development. Int J Pharm 82:9–19, 1992.
32. ME Sale, TF Blaschke. Incorporating pharmacokinetic/pharmacodynamic modeling in drug development–Are we ready? Drug Inf J 26:119–124, 1992.
33. H Derendorf, LJ Lesko, P Chaikin, WA Colburn, P Lee, R Miller, R Powell, G Rhodes, D Stanski, J Venitz. Pharmacokinetic/Pharmacodynamic modeling in drug research and development. J Clin Pharmacol 40:1399–1418, 2000.
34. P Chaikin, GR Rhodes, R Bruno, S Rohatagi, C Natarajan. Pharmacokinetics/Pharmacodynamics in drug development: an industrial perspective. J Clin Pharmacol 40: 1428–1438, 2000.
35. JPP Heykants, E Snoeck, F Awouters, A Van Peer. Antihistamines. In: CJ van Boxtel, NHG Holford, M Danhof, eds. The in vivo study of drug action. Amsterdam: Elsevier, 1992, pp 337–356.
36. BG Reigner, PEO Williams, IH Patel, JL Steimer, C Peck, P van Brummelen. An evaluation of the integration of pharmacokinetic and pharmacodynamic principles in clinical drug development. Experience with Hoffmann La Roche. Clin Pharmacokinet 33:142–152, 1997.
37. LB Sheiner, JL Steimer. Pharmacokinetic/Pharmacodynamic Modeling in Drug Development. Annu Rev Pharmacol Toxicol 40:67–95, 2000.
38. L Aarons, MO Karlsson, F Mentré, F Rombout, JL Steimer, A van Peer, and invited COST B15 experts. Role of modelling and simulation in Phase I drug development. Eur J Pharm Sci 13:115–122, 2001.
39. D Martin, F Bouzom, H Merdjan. Building of a physiologically based mathematical model for dynamics of an antidiabetic agent. Proceedings Third International Symposium on Measurement and Kinetics of In Vivo Drug Effects, Noordwijkerhout, 1998, pp 168–170.

# 18

## Assessment of QTc Interval Prolongation in a Phase I Study Using Monte Carlo Simulation

**Peter L. Bonate**
Quintiles Transnational Corp, Kansas City, Missouri*, U.S.A.

### 18.1 INTRODUCTION

It has become apparent that some drugs adversely prolong the QT interval on an electrocardiogram (ECG) (1). Regulatory authorities are also becoming increasingly sensitive to this potential safety issue as evidenced by recent interactions between pharmaceutical companies and the Food and Drug Administration. The difficulty in assessing QT interval prolongation is that QT intervals have large intrasubject and between-subject variability and are affected by heart rate (2, 3), sex (4, 5), food (6), chest lead placement (7), and diurnal variation (8), to name a few. Since QT intervals are dependent on heart rate, raw QT intervals are rarely analyzed. Instead, corrected QT intervals (QTc), which control for heart rate, are used. Bazett's correction (QTcB) is the most commonly used and is defined as

$$\text{QTcB} = \frac{\text{QT}}{\sqrt{\text{RR}}} = \frac{\text{QT}}{\sqrt{60/\text{HR}}} \tag{1}$$

* *Current affiliation*: ILEX Oncology, Inc., San Antonio, Texas, U.S.A.

where QT is the QT interval in msec, RR is the RR interval in msec, and HR is heart rate in bpm (9).

A phase I study was conducted to assess the effect of drug concentration on QTcB intervals in healthy volunteers. In this study, 10, 30, and 60 mg doses were used as treatments. Of particular interest was the nature of the QTcB interval response surface as a function of the relevant covariates identified from a population analysis of the data. Nonlinear mixed effect modeling and Monte Carlo simulation were used to complete this task. The goal of the analysis was to identify any covariates that might be of particular importance in the pharmacokinetic and pharmacodynamic response. Specifically, weight was thought to be an important covariate in explaining the pharmacokinetics of the drug, whereas it was hoped that dose or concentration would not be an important covariate in explaining the pharmacodynamic response. One goal of this analysis was to confirm the weight effect and to determine if the weight effect correlates with the pharmacodynamic effect. That is, if the weight effect was confirmed as a significant covariate in the pharmacodynamic model, would an actual study be necessary to confirm it clinically? The approach to answer this question via modeling and simulation is discussed in this chapter.

## 18.2 METHODS

### 18.2.1 Study Design

The study was a single-center, randomized, double-blind, placebo-controlled, four-period crossover trial. Within each period, subjects received either active drug (10, 30, or 60 mg) or matching placebo once daily for 7 days. Prior to each treatment, there was a single-blind placebo acclimation day (day −1). A washout period of at least 6 days separated treatments. Each subject was to receive each treatment once. However, during the first period, some subjects experienced severe nausea to the drug, which after unblinding was found to occur with the 60 mg dose. Also, some subjects showed mild nausea and light-headedness to the drug at the 30 mg dose, but not at the same level of severity seen with the 60 mg dose. Therefore, in those subjects that were to receive 60 mg in an upcoming period and who showed previous sensitivity to the 30 mg dose, an unblinded monitor had the option of having these subjects receive the 30 mg dose a second time, instead of receiving the 60 mg dose. The 10 and 30 mg treatments consisted of encapsulated tablets containing two tablets (placebo, 10 mg, or 20 mg), whereas the 60 mg treatment consisted of three 20 mg overencapsulated tablets. Blood samples were collected and assayed for drug concentrations at time 0 (predose), 1.5, 3, 6, 9, 12, and 24 h after dosing on days −1, 1 and 7 and at predose on days 5 and 6. A full pharmacokinetic sampling day was included on day −1 to maintain the blind.

### 18.2.2 Subjects

Forty healthy subjects (20 male and 20 female) were enrolled. Subjects were to be in good health, nonsmokers for 6 months prior to the screening visit, between 18 and 50 years old (inclusive), body weight within 20% of the desirable weight for adults at the screening visit, able to provide informed written consent, had a normal ECG based on 24 h Holter monitoring, willing to abstain from alcohol and xanthine-containing food and beverages for the duration of each treatment period, and willing to remain in the clinic for the inpatient portions of the study. Female subjects were to be receiving a medically accepted contraceptive regimen, be postmenopausal for a period of at least 1 year, or have been surgically sterilized. Informed consent was obtained for all subjects before enrollment in the study. Dropouts were not replaced.

Subjects were excluded from the study if any of the following were present at the time of screening: abnormal preadmission vital signs or clinical laboratory evaluations considered clinically significant by the clinical investigator; significant organ abnormality or disease; history of seizures, or significant renal, hepatic, gastrointestinal, cardiovascular, or hematologic illness; a positive test for hepatitis B surface antigen, hepatitis C antibody, or HIV antibody; any female subject with a positive serum pregnancy test or who was breastfeeding; or any subject with a mental capacity limited to the extent that the subject could not provide legal consent or understand information regarding the side effects or tolerance of the study drug. Subjects were also excluded from the study if any of the following were present at the time of the first treatment period: history of alcohol and/or drug abuse within the past year; serious mental or physical illness within the past year; history of donated plasma or blood within the last 30 days; history of taking medication, either prescription or nonprescription, within the last 14 days; ingestion of any investigational medication within the last 60 days; or acute illness within the last 5 days.

### 18.2.3 ECG Assessments

Readings from 12-lead ECGs were collected at time 0 (predose), 1.5, 3, 6, 9, 12, and 24 h postdose on days −1, 1, and 7 and at predose on days 5 to 6. All original ECGs were re-read by a single cardiologist who was independent of the sponsor and study site and blinded to period, treatment, and time after dose. ECGs were analyzed for PR, QRS, RR, and QT intervals. QTcB intervals were computed using Bazett's formula (9).

### 18.2.4 Population Analysis of Pharmacokinetic Data

Plasma samples were assayed for drug of interest using a GLP-validated LC/MS assay with a linear range of 0.1–250 ng/ml. Plasma concentrations were modeled using nonlinear mixed effect models with NONMEM (University of California, San Francisco, version 5). The following covariates were tested: age, sex, and body surface area calculated using Gehan and George (10). The general modeling approach to be taken followed the guidelines set forth by Ette and Ludden (11) and Bruno et al. (12). Model selection was based on physiological and pharmacological rationale and the principle of parsimony—simpler models were chosen over more complex models when statistically justified (13). Model development was performed using forward selection. Nested models were compared using the likelihood ratio test with a statistical criteria of 0.05 (14). Random effects were modeled as a log-normal distribution for both between-subject and interoccasion variability. Residual errors were modeled as a constant coefficient of variation. Covariate screening was done by visual inspection of the scatter plots of the empirical Bayes estimate under the base model for the pharmacokinetic parameter of interest vs. covariate. All models were developed using first-order conditional estimation (FOCE), except the final model which was estimated using FOCE with interaction.

### 18.2.5 Population Analysis of Pharmacodynamic Data During Placebo Administration

QTcB intervals for subjects in the placebo arm of the study were modeled using nonlinear mixed effects models (NONMEM). The following covariates were tested: day, period, gender, race, food, baseline serum potassium concentration, baseline serum calcium concentration, and chest lead. Model terms were entered into the model in a linear manner, but the food effect was modeled as an exponential decline from maximal effect. Model selection was based on criteria described in the previous subsection. Random effects and residual error were treated as normally distributed.

### 18.2.6 Population Analysis of Combined Pharmacokinetic and Pharmacodynamic Data

Once the pharmacokinetic and placebo pharmacodynamic model were developed, a combined pharmacokinetic/pharmacodynamic model of QTcB intervals following active drug and placebo administration was developed. The effect of drug concentration on observed QTcB intervals was tested using the following models:

$$\mathrm{QTcB}(t) = P(t) \tag{2}$$

$$\mathrm{QTcB}(t) = P(t) + \theta_1 \hat{C}(t) \tag{3}$$

$$\mathrm{QTcB}(t) = P(t) + \theta_1 \hat{C}(t)^{\theta_2} \tag{4}$$

$$\mathrm{QTcB}(t) = P(t) + \theta_1 \hat{C} + \theta_2 \hat{C}(t)^2 \tag{5}$$

$$\mathrm{QTcB}(t) = P(t) + \frac{\theta_1 \hat{C}(t)}{\theta_2 + \hat{C}(t)} \tag{6}$$

where QTcB($t$) is the observed QTcB interval at time $t$ after dosing, $P(t)$ is the placebo effect at time $t$ after dosing, $\hat{C}(t)$ is the individual predicted drug concentration based on the pharmacokinetic model at time $t$, and $\theta$ is the set of parameters to be estimated. The models were not centered. Model selection was based on the criteria previously described.

### 18.2.7 Monte Carlo Simulation and Statistical Analysis

Once the final pharmacokinetic and pharmacodynamic model (after placebo administration) was chosen, QTcB intervals were simulated using a response surface approach where male and female subjects were simulated using NONMEM with dose varying over the interval 10–60 mg by 10 mg increments, and weight varying from 40 to 110 kg by 10 kg increments. Although patient weight and dose were treated as fixed effects, the within- and between-subject variability estimates obtained from the modeling of the QTcB interval data were used to simulate "observed" QTcB intervals for a particular subject at a particular time. One hundred subjects in one virtual trial were simulated with dosing occurring once a day for 7 days with multiple ECGs collected every hour on day 7. The mean, maximal, and maximal change from baseline QTcB interval on day 7 for each subject was assessed. For both males and females, the mean and maximal change from baseline QTcB intervals were then normalized by dose and weight and fit to a response surface consisting of a quadratic polynomial (dose and weight as primary factors) with interaction (15, 16). Nonsignificant model terms were removed sequentially and the model refit until only statistically significant ($p < 0.05$) model terms remained.

## 18.3 RESULTS

### 18.3.1 Population Analysis of Pharmacokinetic Data

Base model analysis showed that a two-compartment oral model with lag time was the superior structural model. Since the 60 mg dose consisted of three 20 mg tablets that were overencapsulated, separate absorption related parameters (F, Ka, and lag time) were needed for this treatment. The only covariate that was shown to be of importance was body surface area (BSA), which was shown to be linearly related to intercompartmental clearance $Q$. Residual error was best modeled using a proportional error model. Table 1 summarizes the population parameter estimates under the final model. Figure 1 shows a plot of individual

TABLE 1 Summary of Final Pharmacokinetic Model

| Parameter | Units | Estimate | SE (estimate) | BSV (%) | BOV (%) |
|---|---|---|---|---|---|
| Common to all doses | | | | | |
| CL | liters/h | 153 | 15.6 | 65.6 | 14.4 |
| V2 | liters | 15.4 | 4.35 | 75.2 | 76.4 |
| V3 | liters | 1240 | 104 | 42.8 | 23.6 |
| Q | liters/h | 48.6 * BSA | 5.24 | 44.0 | 15.7 |
| For 10 and 30 mg treatments | | | | | |
| Ka | Per hour | 0.192 | 0.00964 | 14.2 | 16.5 |
| Lag time | hours | 0.909 | 0.0214 | Fixed | |
| For 60 mg (3 × 20 mg) treatment | | | | | |
| Relative F | | 1.23 | 0.0410 | Fixed | |
| Ka | Per hour | 0.221 | 0.0126 | Fixed | |
| Lag time | hours | 0.813 | 0.0512 | Fixed | |
| Residual variability: 17.5% | | | | | |

CL, clearance; V2, central volume; V3, peripheral volume; Ka, absorption rate constant; Q, intercompartmental clearance; F, bioavailability; BSV, between-subject variability; BOV, between-occasion variability; SE, standard error.

predicted concentrations against observed concentrations. Figure 2 shows a scatter plot of individual weighted residuals against individual predicted concentrations.

### 18.3.2 Population Analysis of Pharmacodynamic Data During Placebo Administration

Exploratory data analysis revealed that food, gender, chest lead, and day of ECG collection were all significant covariates. Baseline QTcB intervals tended to increase as the number of days in the clinic increased, suggesting that the model may not be valid beyond a 7 day period as this would lie outside the bounds of the time frame studied. The major sources of variability were between subject and within subject. Figure 3 plots the mean QTcB intervals over time against the model predicted QTcB intervals over time following placebo administration. Table 2 presents a summary of the final model parameter estimates for QTcB intervals after placebo administration. The model did not partition the total variability well since residual variability was larger than between-subject variability. Between- and within-subject variability was estimated at 3.1 and 4.1%, respec-

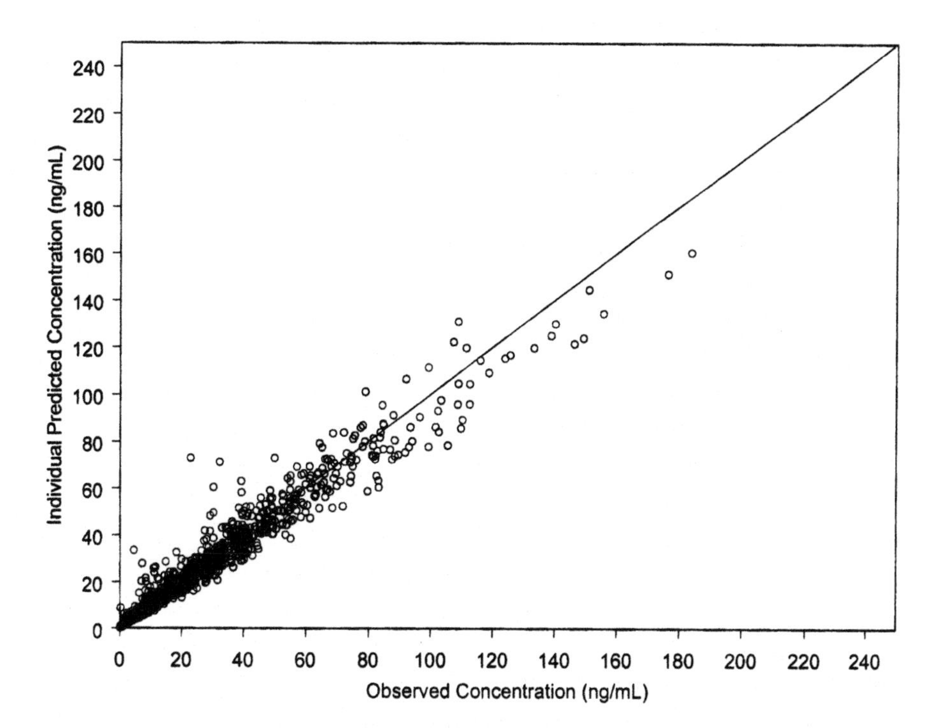


**FIGURE 1** Individual predicted plasma concentrations under final population pharmacokinetic model vs. observed plasma concentrations. Solid line is the line of unity.

tively. The condition number, which was calculated as the square root of the ratio of the largest to smallest eigenvalue of the variance-covariance matrix, for the final model was 4.5, indicating that the degree of collinearity in the model was negligible (17).

### 18.3.3 Population Analysis of Combined Pharmacokinetic and Pharmacodynamic Data

The final pharmacokinetic and pharmacodynamic models were combined and the individual predicted concentrations used as the input concentration for the pharmacodynamic model. Table 3 presents the model development summary. A sigmoid $E_{max}$ model was not attempted since the $E_{max}$ model indicated that it was overparameterized. A linear model was deemed the superior model. The slope of the relationship between drug concentration and QTcB intervals was 0.198 ms $ng^{-1}ml^{-1}$ (95% confidence interval: 0.137, 0.259). Figure 4 presents a scatter

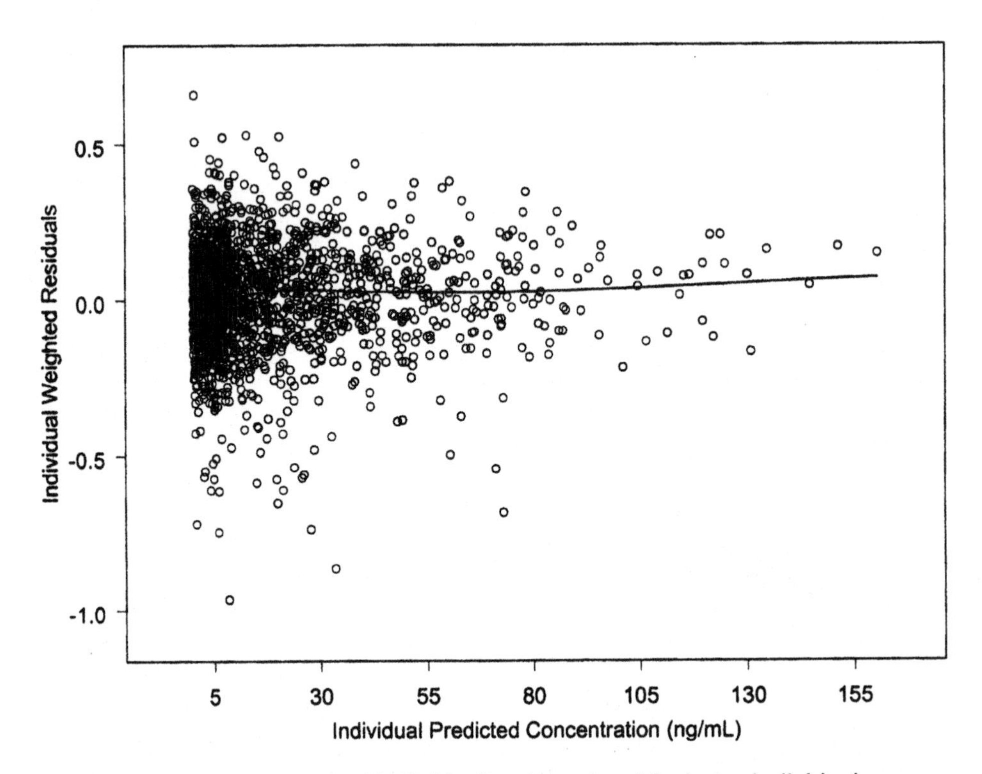


**Figure 2** Residual plot of individual weighted residuals vs. individual predicted concentrations. Solid line is a locally weighted scatter plot smooth (LOESS) to the data.

plot of placebo corrected QTcB intervals against the individual predicted plasma concentrations under the final pharmacokinetic model. The condition number for the final model was 4.5, indicating that the degree of collinearity in the combined model was almost nonexistent (17).

### 18.3.4 Monte Carlo Simulation and Analysis

The response surface plots for males and females are shown in Figures 5 and 6, respectively. Simulated mean maximal QTcB intervals for females at all doses and weights were greater than for males, a not unexpected result. Although weight (by virtue of its influence of BSA) had an effect on the pharmacokinetics of the drug, weight had a smaller overall effect than did dose. In both males and females, the quadratic response fit to the data revealed that for all three metrics (maximal QTcB interval, mean QTcB interval, and maximal change from baseline QTcB

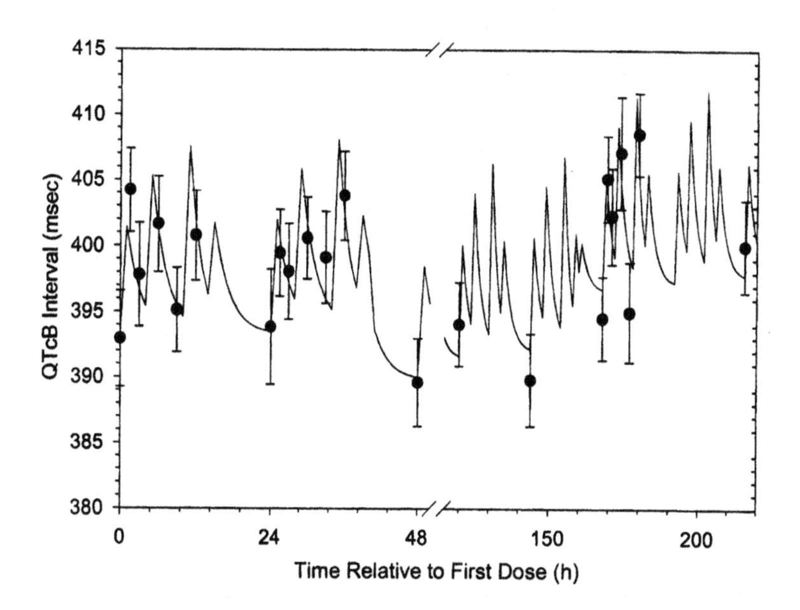


**Figure 3** Plot of model predicted QTcB intervals against mean QTcB intervals over time after placebo administration. Error bars are the standard error of the mean.

interval) the only significant factor was dose ($p < 0.05$). Table 4 presents a summary of the reduced models using dose as a factor for both males and females.

## 18.4 DISCUSSION

A two-compartment model with lag time resulted in an adequate fit to the data. In order to adequately characterize the pharmacokinetics of the 60 mg dose, which consisted of three 20 mg tablets overencapsulated for blinding purposes, separate absorption parameters were needed. Allowing a separate relative bioavailability term for the 60 mg dose group revealed that the 60 mg group had 23% greater bioavailability than the other dose groups, an effect that, if uncorrected, would seriously confound any analysis based on dose alone. Between-subject variability (BSV) in the pharmacokinetic parameters was large (more than 40% for all parameters except Ka). Also, between-occasion variability (BOV) in the pharmacokinetic parameters was modest to large. Clearance (CL) showed the smallest BOV, whereas central volume (V2) had as large as BOV as BSV. Body surface areas was only covariate which showed a significant relationship with any of the pharmacokinetic parameters, which was characterized by a linear relationship

**TABLE 2** Summary of Final Model for QTcB Intervals Following Placebo Administration

| Parameter | Model component | Estimate | SE | BSV ($ms^2$) |
|---|---|---|---|---|
| Baseline | $E0 = \theta(1) + \eta(1)$ | 389 | 3.21 | 147 |
| Trend effect | $K = \theta(2) + \eta(2)$ | 0.0225 | 9.34E-3 | 8.7E-4 |
| Female sex effect (0 = male, 1 = female) | $SX = \theta(3) * SEX$ | 7.57 | 4.31 | |
| Chest lead IV effect (1 = lead IV, 0 = other) | $CE = \theta(8) * LEAD$ | 9.54 | 4.33 | |
| Rate of decline in food effect | $KFE = \theta(7) * \exp(\eta 3)$ | 0.400 | 0.208 | 0.0117 |
| Food effect | B: $FE = \theta(4) * \exp(-KFE * t)$ | $\theta(4)$: 10.6 | $\theta(4)$: 2.62 | |
| | L: $FE = \theta(5) * \exp(-KFE * t)$ | $\theta(5)$: 12.5 | $\theta(5)$: 4.52 | |
| | D: $FE = \theta(6) * \exp(-KFE * t)$ | $\theta(6)$: 14.7 | $\theta(6)$: 3.90 | |
| Day effect, if day = {2, 5, 6} | $DE = \theta(9)$ | −4.02 | 2.01 | |

Final model: $QTcB = E0 + K * Time + SX + FE + CE + DE + \varepsilon$

Residual variability $\sigma^2$: 252 $ms^2$

$\theta$, population estimate; $\eta$, subject-specific deviation from the population estimate where $\eta \sim N(0, \omega^2)$; $\varepsilon$, randomly distributed residual error where $\varepsilon \sim N(0, \sigma^2)$; SE, standard error; BSV, between-subject variability ($\omega^2$); B, breakfast; L, lunch; D, dinner.

**TABLE 3** Summary of Pharmacokinetic/Pharmacodynamic Models for QTcB Intervals After Active Drug Administration

| Model | OFV | Parameter | θ | SE(θ) | BSV |
|---|---|---|---|---|---|
| Placebo model | 20846.2 | | | | |
| $QTcB(t) = P(t) + \theta_1 \hat{C}(t)$ | 20773.4 | $\theta_1$ | 0.198 | 0.0311 | 9.00E-3 |
| $QTcB(t) = P(t) + \theta_1 \hat{C}(t)^{\theta_2}$ | 20773.4 | $\theta_1$ | 0.183 | 0.181 | 7.68E-3 |
| | | $\theta_2$ | 1.02 | 0.226 | |
| $QTcB(t) = P(t) + \theta_1 \hat{C} + \theta_2 \hat{C}(t)^2$ | 20773.0 | $\theta_1$ | 0.171 | 0.0572 | 9.72E-3 |
| | | $\theta_2$ | 4.06E-4 | 5.77E-4 | |
| $QTcB(t) = P(t) + \dfrac{\theta_1 \hat{C}(t)}{\theta_2 + \hat{C}(t)}$ | 20773.4, R | $\theta_1$ | 1820 | | |
| | | $\theta_2$ | 9140 | | |

BSV, between-subject variability; R, R matrix singular; $\theta_1$ has units ms $ng^{-1}$ $ml^{-1}$; BSV has units of (ms $ng^{-1}$ $ml^{-1})^2$.

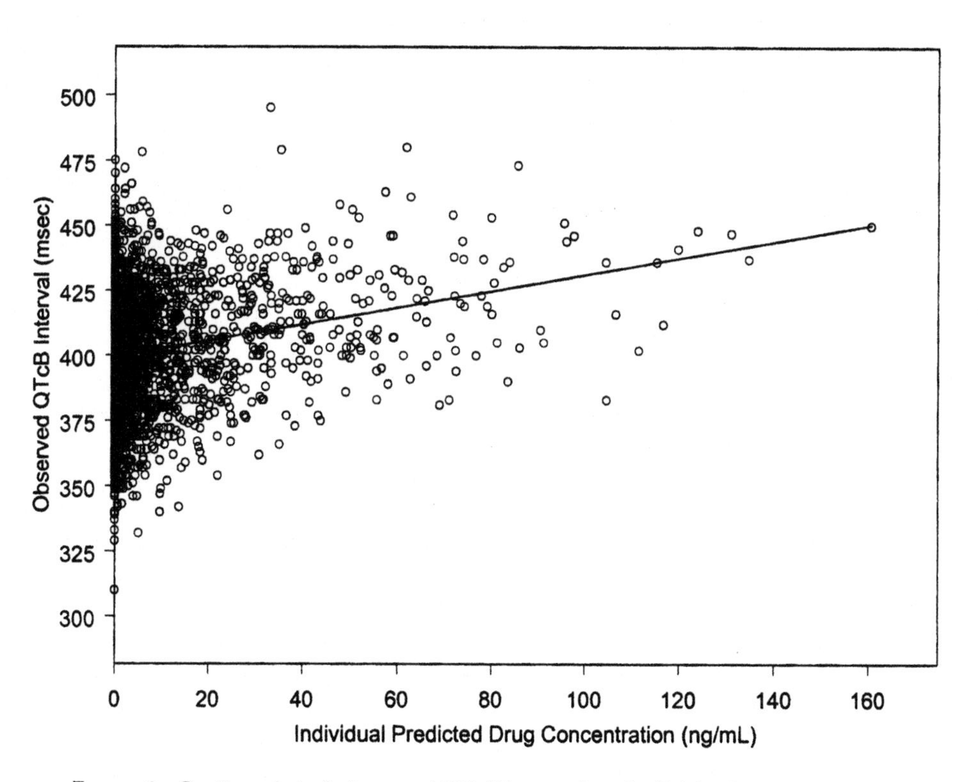


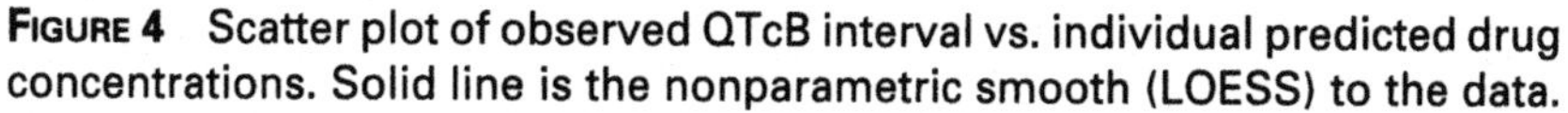

**FIGURE 4** Scatter plot of observed QTcB interval vs. individual predicted drug concentrations. Solid line is the nonparametric smooth (LOESS) to the data.

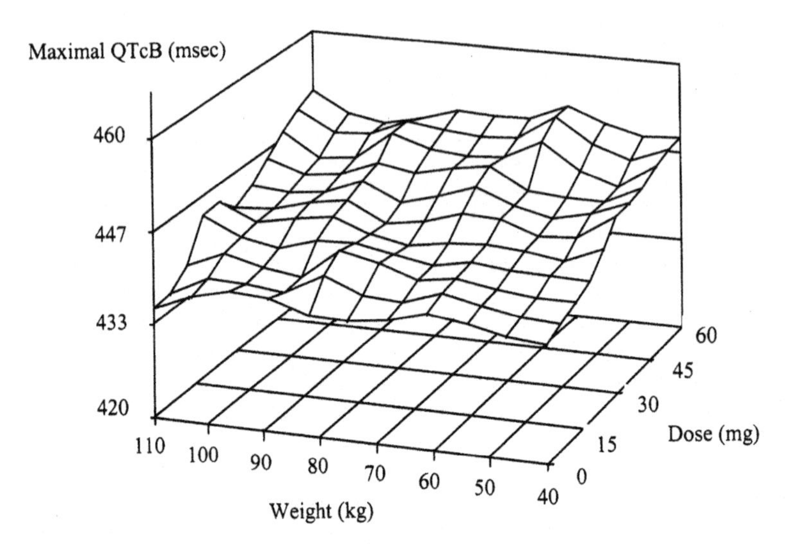


**FIGURE 5** Response surface of simulated mean maximal QTcB intervals in males who were dosed once a day with drug for 7 days. ECGs were serially collected on day 7.

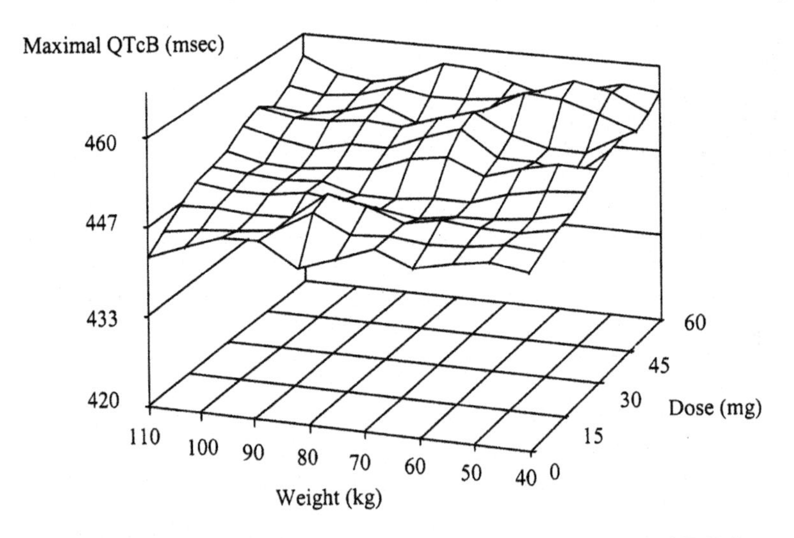


**FIGURE 6** Response surface of simulated mean maximal QTcB intervals in females who were dosed once a day with drug for 7 days. ECGs were serially collected on day 7.

**TABLE 4** Summary of Response Surface Mapping to Simulated QTcB Intervals

| Metric | Sex | Parameter | Estimate | Standard error | $R^2$ |
|---|---|---|---|---|---|
| Maximal QTcB interval | Female | Intercept | 444.98 | 0.422 | 0.8039 |
| | | β(1) * Dose | 0.1742 | 0.0117 | |
| | Male | Intercept | 437.27 | 0.385 | 0.8643 |
| | | β(1) * Dose | 0.1979 | 0.0107 | |
| Mean QTcB interval | Female | Intercept | 413.02 | 0.342 | 0.7125 |
| | | β(1) * Dose | 0.1097 | 9.47E-3 | |
| | Male | Intercept | 405.4 | 0.316 | 0.7994 |
| | | β(1) * Dose | 0.1286 | 8.77E-3 | |
| Maximal change from baseline QTcB interval | Female | Intercept | 4.17 | 0.407 | 0.4849 |
| | | β(1) * Dose | 0.0804 | 0.0113 | |
| | Male | Intercept | 3.15 | 0.401 | 0.6396 |
| | | β(1) * dose | 0.1088 | 0.0111 | |

$R^2$, coefficient of determination; β(1) has units ms $mg^{-1}$.

with intercompartmental clearance (Q). Hence, subjects with smaller weights will tend to have higher drug concentrations and possibly higher QTcB intervals than others.

Although the mean fit for the QTcB interval data from the placebo arms was adequate, the model did not partition the overall variability well. The final placebo model had larger residual variability than between-subject variability in baseline QTcB intervals. Still, the population mean and between-subject variance estimates were consistent with the literature (6, 18, 19). The results of this analysis confirm previous reports where sex, chest lead, and food all affect QTcB intervals. Two items of interest were noted in this analysis. The QTcB intervals increased by about 4 ms in relation to the number of procedures performed on each subject. On days where there was a lot of activity, e.g., serial blood draws and ECG collections, QTcB intervals were higher than normal. Second, as subjects remained in the clinic over time, there was a tendency for their baseline QTcB interval to increase over time with QTcB intervals increasing about 0.5 ms on average per day. It should be noted that the study period was 8 days in length. It is unknown what happens beyond the study period. It is likely that a new steady state will eventually be reached but how long this may take could not be determined.

Including drug concentrations in the pharmacodynamic model significantly improved the goodness of fit for all the models tested. The best-fit model was a linear one between predicted drug concentration and observed QTcB intervals. The model suggests that for every 10 ng/ml increase in drug concentration there was a corresponding increase in QTcB intervals by 1.98 ms. Fitting the simulated data to a response surface indicated that despite individuals with smaller weights having higher drug concentrations, weight (or BSA) had no independent effect on the QTcB interval response surface. The only significant factor was dose. Under the reduced response surface, for every 10 mg increase in dose, there was a corresponding increase in maximal and mean QTcB intervals of 1.7 and 1.1 ms, respectively, in females and 2.0 and 1.3 ms, respectively, in males.

The goal of this study was to confirm the weight effect on the pharmacokinetics of the drug, which was confirmed. The second question, Did weight directly influence the pharmacodynamic effect? was not supported in this case. When weight (or its surrogate BSA) was systematically varied there was no apparent effect on QTcB intervals. Hence it was concluded that a clinical study in obese patients was probably not necessary.

In summary, a population approach was used to characterize the pharmacokinetics and pharmacodynamics of the drug of interest. Using Monte Carlo simulation, QTcB intervals were simulated to characterize the response surface as a function of each of the covariates identified from the population analysis. From the response surface, the relevant factors that affect QTcB intervals were identified. The data suggest that after controlling for sex, only the dose of medication was an important factor in QTcB interval prolongation. One outcome of this conclusion may be the alleviation of performing a confirmatory study assessing the role of weight as a factor in the pharmacodynamics of the drug.

## REFERENCES

1. PL Bonate, T Russell. Assessment of QTc prolongation for non-cardiac drugs from a drug development perspective. J Clin Pharmacol 39:349–358, 1999.
2. S Ahnve. Correction of the QT interval for heart rate: review of different formulas and the use of Bazett's formula in myocardial infarction. Am Heart J 109:568–574, 1985.
3. C Funck-Bretano, P Jaillon. Rate-corrected QT interval: techniques and limitations. Am J Cardiol 72:17B–22B, 1993.
4. European Agency for the Evaluation of Medicinal Products (EMEA): Committee for Proprietary Medicinal Products (CPMP). Points to Consider: The assessment of the potential for QT interval prolongation by non-cardiovascular medicinal products, London, 17 December 1997. CPMP/986/96.
5. AJ Moss. Measurement of the QT interval and the risk associated with QTc interval prolongation: a review. Am J Cardiol 72:23B–25B, 1993.
6. D Nagy, R DeMeersman, D Gallagher, A Pietrobelli, AS Zion, D Daly, SB Heymsfield. QTc Interval (cardiac repolarization): lengthening after meals. Obesity Res 5: 531–537, 1997.

7. DM Mirvis. Spatial variation of QT intervals in normal persons and patients with myocardial infarction. J Am Coll Cardiol 5:625–631, 1985.
8. RS Bexton, HO Vallin, AJ Camm. Diurnal variation of the QT interval—influence of the autonomic nervous system. Br Heart J 55:253–258, 1986.
9. HC Bazett. An analysis of time-relations of electrocardiograms. Heart 7:353–370, 1920.
10. EA Gehan, SL George. Estimation of human body surface area from height and weight. Cancer Chem Rep 54:225–235, 1970.
11. EI Ette, TM Ludden. Population pharmacokinetic modeling: the importance of informative graphics. Pharm Res 12:1845–1855, 1995.
12. R Bruno, N Vivier, JC Vergniol, SL De Phillips, G Montay, LB Shiener. A population pharmacokinetic model for docetaxel (Taxotere): model building and validation. J Pharmacokin Biopharm 24:153–172, 1996.
13. JR Wade, SL Beal, NC Sambol. Interaction between structural, statistical, and covariate models in population pharmacokinetic analysis. J Pharmacokin Biopharm 22: 165–177, 1994.
14. EI Ette. Comparing non-hierarchical models: application to non-linear mixed effects modeling. Comp Biol Med 6:505–512, 1996.
15. RL Iman, JC Helton, JE Campbell. An approach to sensitivity analysis of computer models: part II—ranking of input variables, response surface validation, distribution effect, and technique synopsis. J Qual Technol 13:232–240, 1981.
16. DJ Downing, RH Gardner, FO Hoffman. An examination of response surface methodologies for uncertainty analysis in assessment models. Technometrics 27:151–163, 1985.
17. PL Bonate. The effect of collinearity on parameter estimates in nonlinear mixed effect models. Pharm Res 16:709–717, 1999.
18. J Morganroth, AM Brown, S Critz, WJ Crumb, DL Kunze, AE Lacerda, H Lopez. Variability of the QTc interval: impact on defining drug effect and low-frequency cardiac event. Am J Cardiol 72:26B–31B, 1993.
19. J Morganroth, FV Brozovich, JT McDonald, RA Jacobs. Variability of the QT measurement in healthy men, with implications for selection of an abnormal QT value to predict drug toxicity and proarrhythmia. Am J Cardiol 67:774–776, 1991.

# 19

# Optimizing a Bayesian Dose-Adjustment Scheme for a Pediatric Trial: A Simulation Study

**Marc R. Gastonguay**
University of Connecticut, Farmington, Connecticut, U.S.A.

**Ekaterina Gibiansky**
GloboMax LLC, Hanover, Maryland, U.S.A.

**Leonid Gibiansky**
The Emmes Corporation, Rockville, Maryland, U.S.A.

**Jeffrey S. Barrett**
Aventis Pharmaceuticals, Bridgewater, New Jersey, U.S.A.

## 19.1 INTRODUCTION

### 19.1.1 Clinical Problem and Goals

Limited pharmacokinetic (PK) and pharmacodynamic (PD) data exist in pediatric populations for the majority of medicines approved for adult indications. Despite the lack of dosing and PK/PD information available, physicians treating children are often left with the difficult task of adjusting adult dosages or simply avoid prescribing the available medications. Seeking to improve on this situation, the

U.S. Food and Drug Administration has requested that drug companies study drugs likely to be administered to pediatric populations in prospective dose-finding trials. While actual efficacy trials in pediatric populations are seldom conducted, these guidelines have increased the number of studies available to define pediatric dosing regimens. The FDA Modernization Act of 1997 [21 U.S.C. 355a(b)] required the FDA to develop, prioritize, and publish an initial list (Docket No. 98N-0056) of approved drugs for which pediatric information may produce benefits in the pediatric population. While the list does not constitute a written request under 21 U.S.C. 355(a) or infer that the drug is entitled to pediatric exclusivity, it does reveal the indications and compound classes for which additional experience in the pediatric population is warranted. The regulatory appreciation for the necessity of studying pediatric populations has been further codified by 21 CFR Parts 201, 312, 314, and 601 which require drug manufacturers to provide sufficient data and information to support directions for pediatric use for claimed indications. These regulations have been reauthorized with the "Best Pharmaceuticals for Children Act," which was signed into law in January 2002.

The difficulty with such studies, of course, revolves around the availability and willingness of patients (and/or their parents/guardians) to participate in such studies and the concern with the collection of venous blood for PK and/or PD characterization. More specifically, the number of samples typically required to characterize individual PK/PD relationships cannot often be accommodated, particularly in very young children. One solution to this dilemma has been the adoption of mixed-effects modeling techniques to assess population PK/PD. These techniques offer numerous benefits to such studies including: less intensive sampling, accommodation of imbalanced data, permission of variations in dosing regimen and sample collection, the ability to study a broader (less restrictive) patient population, the ability to screen for drug interactions, estimation of drug exposure correlation with safety and efficacy parameters, and the ability to pool data across studies. This approach has been advocated by FDA in both the Population Pharmacokinetics Guidance (1999) and the Draft Guidance: General Considerations for Pediatric Pharmacokinetic Studies for Drugs and Biological Products (1998).

Heparin and enoxaparin, a low-molecular weight heparin (LMWH), make the recent list of agents included in the Pediatric Priority List (Docket No. 98N-0056, May 19, 2000). Both heparin and LMWHs are polydisperse polysaccharides which include biologically inactive species. LMWHs are polycomponent moieties with multiple biological actions, each with a distinct time course which confounds the pharmacokinetic characterization of these agents. Hence, assays developed for a single pharmacologic activity more appropriately describe the anticoagulant pharmacodynamics and not the pharmacokinetics. Plasma anti-Xa activity is an accepted surrogate for the concentration of molecules that contain the high-affinity binding site for antithrombin III and correlates with molecules greater than 5400 Da (1). While useful as a biomarker for LMWH activity, anti-Xa levels have not been closely related with antithrombotic efficacy in most studies. Accordingly, monitoring patients based on anti-Xa activity is generally not advised. In the absence of a better surrogate and given the little data in pediatric patients, monitoring anti-Xa activity has become a prerequisite for judging the therapeutic exposure of LMWH in dose-finding trials.

Recent studies in adults suggest that most of the problems with heparin and oral anticoagulant therapy in children can likely be minimized by the use of LMWH. Compared to heparin, LMWH causes less bleeding, minimizes the need for monitoring, and is injected subcutaneously rather than by continuous intravenous administration. The risk of heparin-induced thrombocytopenia (HIT) and osteoporosis appears to be less with LMWH compared to heparin (2). Compared to oral anticoagulants, doses of LMWH are not influenced by other medications or dietary intake of vitamin K; monitoring can be considerably reduced, and safety may be enhanced due to stable and predictable pharmacokinetics. The discomfort of daily injections can be minimized by the use of a topical anesthetic and a subcutaneous catheter (3) that can remain in place for up to 7 days. A list of some recent trials of various LMWH in pediatric patients is provided in Table 1. All were guided by anti-Xa monitoring with dosing to therapeutic levels defined by experience in adult populations.

Tinzaparin sodium is a low molecular weight heparin produced by controlled enzymatic depolymerization of conventional, unfractionated porcine heparin. Indications for which tinzaparin is currently administered include treatment of deep vein thrombosis (9) and pulmonary embolism (10), and perioperative thromboembolism prophylaxis in orthopedic (hip/knee replacement) (11) or general surgery (12) settings, and for extracorporeal anticoagulation in hemodialysis (13). A dose-finding study of tinzaparin, (Innohep) in children with venous or arterial thromboembolic diseases has been planned in order to address the lack of available information on dosing tinzaparin in pediatric patients.

The primary objective of this dose-finding study is to determine the dose of tinzaparin required to achieve a therapeutic plasma anti-Xa level in children with venous or arterial thromboembolic diseases (TEs). The secondary objectives are to determine the pharmacokinetic profile of tinzaparin and to describe the long-term safety of tinzaparin over 90 days administration. Thirty-five patients, between the ages of newborn (>36 weeks of gestational age) and 16 years are planned for enrolment. The primary response criterion is the dose of tinzaparin required to achieve a satisfactory plasma therapeutic anti-Xa level of activity. Patients will receive an initial dose of 175 IU/kg, which is based on adult pharmacodynamic data (11). Blood will be drawn at 4 h after administration of study medication. On assessment, if the anti-Xa levels are within the therapeutic range (0.5–1.0 anti-Xa IU/ml), no dose adjustment is required and the patient enters the maintenance phase. If the therapeutic range for tinzaparin is not achieved

**Table 1** Clinical Trial Experience with LMWHs in Pediatric Populations

| LMWH (Reference) | Population | Design | Conclusion |
|---|---|---|---|
| Enoxaparin (4) (Massicotte, et al, 1996) | Newborn to 17 years of age<br>DVT (treatment and prophylaxis), PE, CNS thrombosis, congenital heart disease patients | Prophylaxis patients received 0.5 mg/kg SC twice daily.<br>Treatment patients received 1 mg/kg SC every 12 h with subsequent doses to achieve a 4 h anti-Xa level between 0.5–1.0 IU/ml.<br>Median duration 14 days. | Newborn infants required an average of 1.6 units/kg to achieve therapeutic heparin levels.<br>Dose is age dependent. |
| Enoxaparin (5) (Punzalan, et al, 2000) | 18 days to 19 years of age<br>DVT (treatment and prophylaxis) | 1 mg/kg SC every 12 h with subsequent doses adjusted to achieved 3 h anti-Xa level between 0.5 and 1.2 IU/ml | Trend toward higher dose requirements to achieve levels in the younger children.<br>Neonates needed higher doses.<br>Older and heavier patients required less enoxaparin. |
| Fraxiparin (6) (Laporte, et al, 1999) | 15 days to 8 years of age<br>Congenital and coronary heart disease patients | Initial dose 330–3300 IU aXa (33–447 IU anti-Xa/kg), twice daily in all but 6 patients.<br>Follow-up from 2 h to 30 days. | Higher dosage needed for children than adults in order to reach similar anti-Xa levels. |
| Dalteparin (7) (Nohe et al, 1999) | DVT (treatment and prophylaxis) | 95 ± 52 IU/kg required for prophylaxis.<br>129 ± 43 IU/kg required for treatment. | Dose requirement was higher in younger children and decreased with age. |
| Dalteparin (8) (Fijnvandraat et al, 1993) | 8 to 16 years of age<br>Chronic stable hemodialysis patients | 4-way crossover comparing 3 doses of Dalteparin to heparin. | The bolus 24 IU/kg followed by 15 IU/kg/h dose was safe and at least as effective as heparin. |

with the initial dose at 4 h, the following day, the dose is adjusted on the basis of the anti-Xa levels achieved on the previous day. Dose adjustment can continue until day 7 in order to achieve satisfactory therapeutic anti-Xa levels. Once achieved, the patient enters the maintenance phase. If the anti-Xa level is not achieved by seventh day following entry, the patient is withdrawn from the study.

During the maintenance phase, the patient receives the new adjusted dose (the dose which achieved the therapeutic range for tinzaparin of 0.5–1.0 anti-Xa IU/mL) for a minimum of 5 days. Blood samples will be drawn daily at different time intervals (2, 4, 6, 12, and 24 h) for pharmacokinetic assessments. Collections at other time points may be made at the discretion of the investigator. Two assessments for each time interval will be performed. After this phase, the patient enters the follow-up phase. If the assessments for different time intervals are not completed within the 5 days, the maintenance phase may be extended for an additional 3 days. Children will be followed up to 90 days with the fixed dose of tinzaparin or until the duration of the anticoagulation treatment, whichever comes first. The dose of tinzaparin may be adjusted monthly based on anti-Xa levels to ensure therapeutic anti-Xa levels are maintained. After discontinuation of the study medication, the subsequent therapy for the patient will be the decision of the investigator.

In order to guide the investigator in the adjustment of doses to maintain anti-Xa activity within the acceptable target activity window, a tabular nomogram was constructed based on the anti-Xa response to a dose of 175 IU/kg. As the table was created from an adult patient study (11), the utility of the anti-Xa prediction in pediatric patients is based on the assumption that the tinzaparin disposition kinetics and pharmacodynamic response in pediatrics will be similar to adults. Moreover, this information is static and does not improve with the knowledge from real-time data generated from the study. The development of the Bayesian forecasting routine described herein (and the associated simulation-based testing) was initiated to provide an additional resource to the investigator.

### 19.1.2 Theoretical Background and Rationale for the Use of Simulation

A population pharmacokinetic model for tinzaparin was previously developed from clinical trials in adults (14). In the adult population analysis, the tinzaparin PK data were described by a linear two-compartment open model with first-order absorption and a baseline endogenous anti-Xa activity parameter. PK parameters in the adult model were weight-normalized due to the fact that tinzaparin is dosed clinically on a body weight basis. This parameterization allowed scaling of the model to predict PK in subjects with a wide range of body weights.

It has been shown, however, that scaling the PK parameter clearance on the basis of body weight usually results in underdosing when the model is extrapolated from adults to the pediatric population (15). For other LMWHs, pediatric patients exhibit a higher weight-normalized clearance when compared with adults, with an even greater increase in those patients less than 2 months old (4, 6, 16, 17). Therefore, the adult model may be inadequate for dose prediction and dose adjustment of tinzaparin in a pediatric population. To this end, a Bayesian population model capable of predicting pediatric tinzaparin PK based on prior information gathered from adults and new, sparse information from pediatric patients was developed and tested. The model-based dose-adjustment algorithm was developed to provide additional guidance for the clinical investigator. The estimation algorithm was designed to adapt as new information in pediatrics was gained so that both individual-specific doses and the best current estimate of the typical starting dose for new pediatric patients could be provided (Figure 1).

The estimation algorithm implemented in the population model is a maximum-likelihood method and is comprised of two Bayesian components: (a) individual parameter estimation and (b) population parameter estimation. Individual parameters are estimated based on mean (typical value) and variance (interindividual and residual) estimates of pediatric population PK parameters, using a maximum a posteriori probability (MAP) Bayesian estimation method (18). This method is also known as the POSTHOC estimation step in the NONMEM soft-

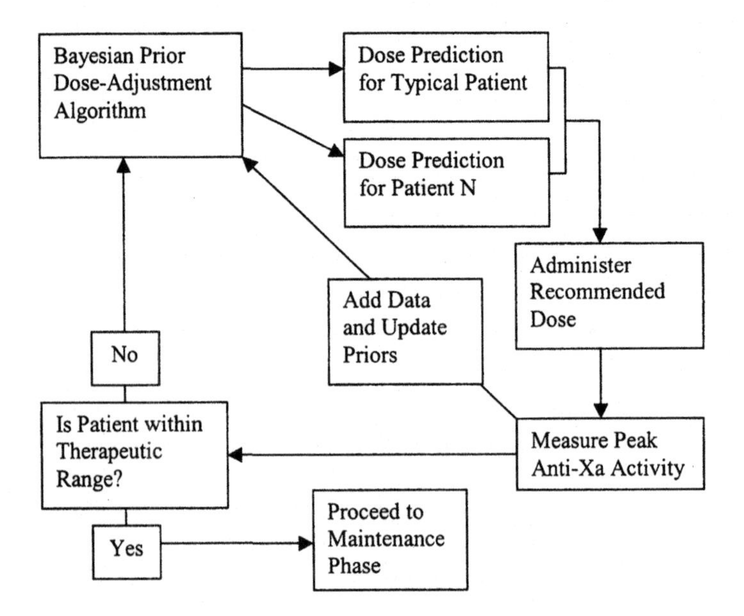


**FIGURE 1** Dose-adjustment algorithm flowchart.

ware (19). Given the sparse pharmacokinetic sampling strategy of the current clinical protocol and the lack of knowledge about the true pediatric pharmacokinetic model parameters, the MAP Bayesian estimation method alone may not be sufficient for accurate and stable estimation of pediatric tinzaparin PK parameters. Therefore, the second component of the Bayesian estimation algorithm combines information about the PK model structure and parameter estimates from adults with new data from the pediatric population to estimate the pediatric population mean and variance parameters that are necessary for the individual MAP Bayesian estimation step. The population parameters can also be used to make informed decisions about starting doses for new, tinzaparin-naïve pediatric patients.

The population Bayesian estimation method requires an additional level of random effects in the nonlinear mixed-effects model so that prior information can be incorporated in the estimation of population-level PK parameters. This method, which includes the addition of a Bayesian penalty function to the typical log-likelihood approximation(s) [Eq. (1)], has been implemented as an experimental, undocumented subroutine in the NONMEM software (S. Beal, personal communication, 1999) and will likely be available as a standard component of future versions of NONMEM. For the purpose of clarity in this document, we will adopt a convention used in the Bayesian hierarchical modeling literature and refer to the parameters defining the prior distribution of population fixed and random effect parameters as hyper-priors (20). With the NONMEM population Bayesian prior implementation, the estimates of the population PK fixed-effects parameter are constrained by a multivariate-normal hyper-prior distribution with mean $\theta_p$ and covariance matrix $\Psi$. In the pediatric tinzaparin application, weight-normalized adult population parameter values were used as the hyper-prior means and a measure of uncertainty in the mean (which is proportional to the square of the standard error of the estimate from the adult data analysis) served as the hyper-prior variance. A similar hierarchy exists for the population interindividual random effect covariance matrix $\Omega$, except that an Inverse-Wishart hyper-prior distribution is assumed, with a mode $\Omega_p$ (obtained from the adult model $\Omega$ estimate) and degrees of freedom $\nu$. Increasing the degrees of freedom in the Inverse-Wishart distribution has the effect of creating a more informative hyper-prior distribution (20). Given that the number of pediatric subjects in the dose-adjustment application is very small, an informative hyper-prior based on the adult model was assumed and $\nu$ was set to a value that was approximately equal to the number of subjects in the adult population analysis. Based on the new pediatric data and prior information, the Bayesian prior objective function allows estimation of new pediatric population parameters, which are simply the modes of the posterior distributions for each parameter.

$$\begin{aligned} OF_{BP} &= OF_{FO} + (\theta - \theta_p)'\Psi^{-1}(\theta - \theta_p) \\ &\quad + (\nu + k + 1) \cdot (\log|\Omega| + \mathrm{tr}(\Omega_p\Omega^{-1})) \end{aligned} \tag{1}$$

where

$OF_{BP}$ = first-order Bayesian prior objective function in NONMEM
$OF_{FO}$ = first-order objective function in NONMEM
$\theta$ = vector of population fixed effects parameters
$\theta_p$ = vector of mean values of the hyper-prior distributions of the elements of $\theta$
$\Psi$ = covariance matrix for the hyper-prior distribution of the elements of $\theta$
$\Omega$ = covariance matrix for the interindividual random effects
$\Omega_p$ = mode of the hyper-prior distribution of $\Omega$
$v$ = degrees of freedom in the Inverse-Wishart hyper-prior distribution of $\Omega$
$k$ = $\Omega$ matrix dimension

The dispersion of the hyper-prior distribution on a population parameter can be thought of as a measure of the uncertainty in the population parameter. The magnitude of this uncertainty relative to the uncertainty in the new data (also known as the residual variance $\sigma^2$) influences the estimation of the posterior mode of the population model parameters. If uncertainty in a parameter is set at a very low level relative to uncertainty in the new data (a so-called informative prior), the population parameter estimates will tend to regress toward the hyper-prior mean or mode. This effect is probably insignificant when the population contributing prior information is the same (or similar to) the population contributing new data. An informative hyper-prior may be too restrictive, however, if the new study population is significantly different from the population used to derive the hyper-prior model. On the other hand, if population parameter uncertainty is very high relative to the uncertainty in the data (so-called diffuse prior), the new data, although sparse, will dominate the population parameter estimation. The dilemma, therefore, is in making careful decisions about the magnitude of prior uncertainties on population parameters. One solution is to explore, through simulation, values for hyper-priors that allow a balance between prior knowledge and new data for the particular model, study population, and study design.

Various options for the pediatric tinzaparin Bayesian dose-adjustment strategy, trial design, and estimation algorithm were explored via computer simulation (Table 2). The impacts of these choices on both population and individual parameter estimates were evaluated. A thorough description of the methods and results for all simulation scenarios is beyond the scope of this chapter. For the purpose of illustration, a few components of the simulation study are presented in greater detail. These are: design scenario 2, "Updating scheme for population hyper-priors," and design scenario 6, "Effect of number and timing of PK samples on individual parameter estimates."

**TABLE 2** Trial and Algorithm Design Issues Explored by Simulation

1. Estimation and a posteriori identifiability of population PK parameters
2. Updating scheme for population hyper-priors
3. Impact of the magnitude of the hyper-prior variance on population parameter estimates
4. Impact of the magnitude of the hyper-prior variance on sample size
5. Effect of informative hyper-prior distributions
6. Effect of number and timing of PK samples on individual parameter estimates
7. Influence of outlier observations on individual parameter estimates

## 19.2 OBJECTIVES

The objective of this work was to design and optimize a Bayesian dose-adjustment strategy for a PK/PD trial of tinzaparin, a new low molecular weight heparin (LMWH) administered for the first time in a pediatric population. The scope of the clinical tinzaparin dose-adjustment problem was well defined and included the following specifications:

- All prior knowledge about tinzaparin PD has been obtained in adults.
- Target activity of 0.5–1.0 anti-Xa IU/ml and starting dose of 175 IU/kg/day have been identified from results of adult studies.
- Target levels must be achieved within the 7 day dose-adjustment period of the trial and maintained throughout the study.
- Dose adjustment will be based on a daily sample at the time of expected peak activity (4 h post-dose) and an additional blood sample taken on day 1 or 2 at a random time, 0–8 h post-dose.
- The algorithm should provide a better starting dose for new patients as patients are enrolled in the study, with a target of achieving accurate population parameter estimates within 5 patients.
- The algorithm must be sensitive to differences in PK between the adult (prior) and pediatric populations, while avoiding unrealistic dose predictions based on one or two outlier blood levels.

## 19.3 METHODS

### 19.3.1 Simulation Plan

Due to the large number of possible trial and algorithm design scenarios, a simulation plan was developed to facilitate the execution and completion of the simulation studies and to ensure that the project objectives were achieved. The simulation plan defined various elements of the simulation experiments, such as the simulation model, trial design options, dose-adjustment algorithm settings, simulation assumptions, and methods of assessment of simulation results (21).

### 19.3.2 Simulation Model

The pediatric population PK simulations were based on the population PK model and parameters (population means and variances) obtained from a previous study in adults (14). Monte Carlo simulations with nested random effects were conducted using the $SIMULATION block in the NONMEM software with the ONLYSIMULATION and SUBPROBLEMS options. The random effects hierarchy was defined according to the adult population model's interindividual variance-covariance matrix $\Omega$ and the residual variance matrix $\Sigma$. For each of the simulation scenarios, the structure of the simulated dosing and sampling times, as well as individual covariate values, were conveyed to the simulation software via a template simulation data set, which was created as a comma-separated ASCII text file.

A two-compartment model with first-order absorption (22) and nonzero endogenous anti-Xa level was used for simulation of pediatric PK data with the parameter values defined in Table 3. The model was parameterized in terms of

**TABLE 3** Base Model Simulation Parameters

| Fixed-effects parameters $\theta$ | | |
|---|---|---|
| Clearance | CL (liters/h/kg) | 0.0173 |
| Central volume of distribution | V2 (liters/kg) | 0.103 |
| Absorption rate constant | KA ($h^{-1}$) | 0.212 |
| Intercompartmental clearance | Q (liters/h/kg) | 0.0064 |
| Peripheral volume of distribution | V3 (liters/kg) | 0.6 |
| Endogenous anti-Xa activity | ENDO (IU/ml) | 0.0984 |

| Interindividual covariance matrix $\Omega$ | | | | | |
|---|---|---|---|---|---|
| | $\eta_{CL}$ | $\eta_{KA}$ | $\eta_{Q}$ | $\eta_{V3}$ | $\eta_{ENDO}$ |
| $\eta_{CL}$ | 0.228 | — | — | — | — |
| $\eta_{KA}$ | −0.0937 | 0.186 | — | — | — |
| $\eta_{Q}$ | 0 | 0 | 0.988 | — | — |
| $\eta_{V3}$ | 0 | 0 | −1.2 | 1.47 | |
| $\eta_{ENDO}$ | 0 | 0 | 0 | 0 | 0.402 |

| Residual variance matrix $\Sigma$ | | |
|---|---|---|
| | $\varepsilon_1$ | $\varepsilon_2$ |
| $\varepsilon_1$(additive) | 0.0103 | — |
| $\varepsilon_2$(proportional) | 0 | 0.0416 |

weight-normalized clearances and volumes, with exponential interindividual variance models and a combined additive and proportional residual variance model. No covariate factors other than weight were included in the simulation model. Variations on this basic simulation model were created in order to address different objectives of the simulation study.

### 19.3.3 Estimation Model

The estimation model was generally the same as the simulation model, except for the specific estimation methods employed. One difference included the addition of a separate weight-normalized CL parameter to be estimated for patients less than or equal to 2 months old, a modeling decision that was based on prior experience with other LMWH in pediatrics (4, 6, 16). Population parameter estimates were obtained using the Bayesian prior objective function described in Eq. (1). These estimates were equivalent to the mode of the posterior distribution for population PK parameters. Individual parameters were estimated using a maximum a posteriori probability (MAP) Bayes method (NONMEM POSTHOC Step). Individual pharmacokinetic parameter estimates reflected a balance between the individual pediatric data and the population priors (specifically, the population mean, interindividual variance, and residual variance estimates).

### 19.3.4 Assumptions

It was necessary to make several assumptions about the trial design and simulation/estimation models. A few of these are listed here:

- It is unlikely that any of the concentration observations will be below the assay quantitation limit. Most of the planned concentration observations will be near $C_{max}$ (4 h post-dose) while the rest will be sampled at random times between 1 and 8 h. No pre-dose samples are scheduled.
- The typical weight-normalized pediatric tinzaparin clearance may be up to two-fold greater than typical weight-normalized adult clearance values.
- Weight-normalized PK parameters in the youngest patients (less than 2 months old) may be different than other pediatric patients.
- Random effect distributions for the pediatric population model are similar to the adult distribution models.
- In the clinical use of the algorithm, outlier or potentially erroneous anti-Xa observations could be observed at values that were 10 times greater or less than the typical expected value.

### 19.3.5 Model Evaluation

When the use of a model is entirely for extrapolation, as was the case in this pediatric application of an adult model, a priori evaluation of the predictive performance of the model is not possible. The goodness of fit of the adult tinzaparin model had already been addressed (14), but there were no available observed data for the evaluation of this model in a pediatric population. Under such circumstances, the validation or evaluation of the model is not as important as assessing the impact of certain model or trial-design assumptions on the desired simulation outcome. With this in mind, ranges of values for a variety of PK model parameters, hyper-priors, and trial design features were utilized in the trial simulations.

### 19.3.6 Simulation Design

#### 19.3.6.1 General Simulation Study Design

Hypothetical pediatric population data were simulated according to a general sampling scheme that was defined in the pediatric study protocol. Subjects received once-daily subcutaneous tinzaparin doses for 8 days. PK samples were obtained at 4 h after each dose. An additional sample for each individual was taken on each of the first 2 days of therapy. These samples were drawn randomly from a different block of time (0–4 or 4–8 h post-dose) on each day.

#### 19.3.6.2 Updating Scheme for Population Hyper-Priors

In the adaptive Bayesian estimation problem, various schemes for updating the prior information could have been employed. Two updating schemes were tested by simulation (a) "all at once" and (b) "step by step." The population hyper-priors for the all-at-once method were maintained at the same values throughout the adaptive dose-adjustment period. Instead of updating the hyper-prior distributions directly, the entire pediatric data set was included in each dose-adjustment estimation step. As the pediatric database grew, the influence of the data relative to the hyper-prior also increased. In contrast to this method, the step-by-step method allowed for updating the values of the hyper-prior distribution parameters after each new pediatric subject's data became available. The hyper-prior means (or modes) were modified to reflect the resulting population parameter (posterior mode) estimates from the previous parameter estimation step. The precision of the hyper-prior parameter estimates were not updated—hence the "informativeness" of the prior was not altered. Thus, only the new individual pediatric data were included with each dose-adjustment step.

The impact of the hyper-prior variance on population parameter estimates was also investigated in these simulations. Values for the hyper-prior variance were defined as the uncertainty level, which was based on multiples of the squared standard error from the adult model parameter estimates. Therefore, an uncertainty level of 10 would be equivalent to 10 times the squared standard error (adult model) for a particular parameter. Thus, a larger uncertainty level resulted in a more diffuse hyper-prior distribution.

The performance of the prior updating methods was evaluated by simulation. Population pediatric anti-Xa activities were simulated following the per-protocol tinzaparin dosing and sampling regimen. All simulations were conducted with a typical pediatric clearance of 0.04 L/h/kg, which was more than two-fold greater than the weight-normalized adult clearance (and also two-fold greater than the hyper-prior mean for the population clearance estimate). This value was believed to be near the extreme of expected clearance values for pediatric patients, as determined in previous studies with other LMWH (4, 6, 16). The result of this choice creates the plausible condition whereby the prior information (hyper-prior distribution) and the new data (simulated with a larger population clearance estimate) provide conflicting information in the Bayesian estimation scheme. For both the all-at-once and step-by-step methods, population parameter estimates and anti-Xa activity predictions were assessed in scenarios including varying numbers of pediatric patients and different magnitudes of the hyper-prior variances.

#### 19.3.6.3 Effect of PK Sampling Times on Individual Parameter Estimates

Given the necessity to estimate individual PK parameters as accurately as possible with a minimum number of PK samples, it was useful to explore the impact of different PK sampling designs on individual parameter estimates. Pediatric tinzaparin PK data were simulated under different scenarios reflecting a range of assumptions about the values of typical pediatric PK parameters relative to adults, as well as the number and timing of PK samples. The timing of PK samples was defined by the per-protocol design: one 4 h point every day, plus 1 additional point on each of days 1 and 2 (0–4 or 4–8 h post-dose), or by a modified design: one 4 h point every day plus 1 additional point on day 1 at 8 h plus 1 additional point on day 2 at 1 h post-dose. Individual parameters were estimated (empirical MAP Bayes estimate) based on adult model priors and new observations of anti-Xa activities. Given the individual parameter estimates, anti-Xa activities at 4 h post-dose on the first day of dosing were predicted for each subject based on different numbers of PK samples and the specified sampling designs (Table 4). These scenarios reflected the clinical dose-adjustment problem in the case when hyper-prior distributions had not been updated.

### 19.3.7 Assessment of Simulation Results

For population parameter estimation, the typical value estimates of population PK parameters (CL, V2, and ENDO) were compared to the true simulation parameter values. The accuracy of the population prediction of the peak anti-Xa activity (more precisely, activity at 4 h post-dose) was also reported for each of the simulation/estimation scenarios. The average $PR/PR_0$ ratio (where PR is the population model prediction at 4 h post single dose, and $PR_0$ is the same prediction for

**TABLE 4** Various Tinzaparin Sampling Designs Explored by Simulation

| Description of sampling design | Samples per patient |
|---|---|
| D1. One sample at 4 h on day 1 | 1 |
| D2. One sample at 4 h on day 1 and an additional point on day 1 | 2 |
| D3. Two samples on day 1 and one sample at 4 h on day 2 | 3 |
| D4. Two samples on days 1 and 2 | 4 |
| D5. Two samples on days 1 and 2, and one sample on day 3 | 5 |
| D6. One sample at 4 h each day with an additional sample on day 1 and day 2 | 10 |

the true simulation parameters) was calculated. Individual parameter estimation results were summarized by comparing the individual model-estimated ($C_{i,\text{est}}$) and the true simulated ($C_{i,\text{sim}}$) anti-Xa activity at 4 h post-dose on the first day for each subject. The median percent prediction error [MPPE, Eq. (2)] and the median percent absolute error [MPAE, Eq. (3)] were calculated as measures of bias and precision, respectively.

$$\text{MPPE} = \text{median}\left(\frac{C_{i,\text{est}} - C_{i,\text{sim}}}{C_{i,\text{sim}}} * 100\right) \tag{2}$$

$$\text{MPAE} = \text{median}\left(\frac{\text{abs}\,(C_{i,\text{est}} - C_{i,\text{sim}})}{C_{i,\text{sim}}} * 100\right) \tag{3}$$

## 19.4 RESULTS

### 19.4.1 Updating Scheme for Population Priors

Results of the simulations for the prior updating schemes were summarized as average parameter estimates and $PR/PR_0$ ratios for approximately 100 trial replicates. Summaries of simulation outcomes are presented as three-dimensional surface plots, where various $PR/PR_0$ values or parameter estimates are presented as functions of the uncertainty level and number of patients $N$ (Figures 2 to 5). The optimal surface for the dose-adjustment algorithm is represented by the light-gray plane at a $PR/PR_0$ value of 1.0 or a clearance estimate of 0.04 liter/h/kg.

For the all-at-once method, a relatively diffuse hyper-prior distribution (uncertainty level of approximately 10 or more), coupled with sample sizes of at

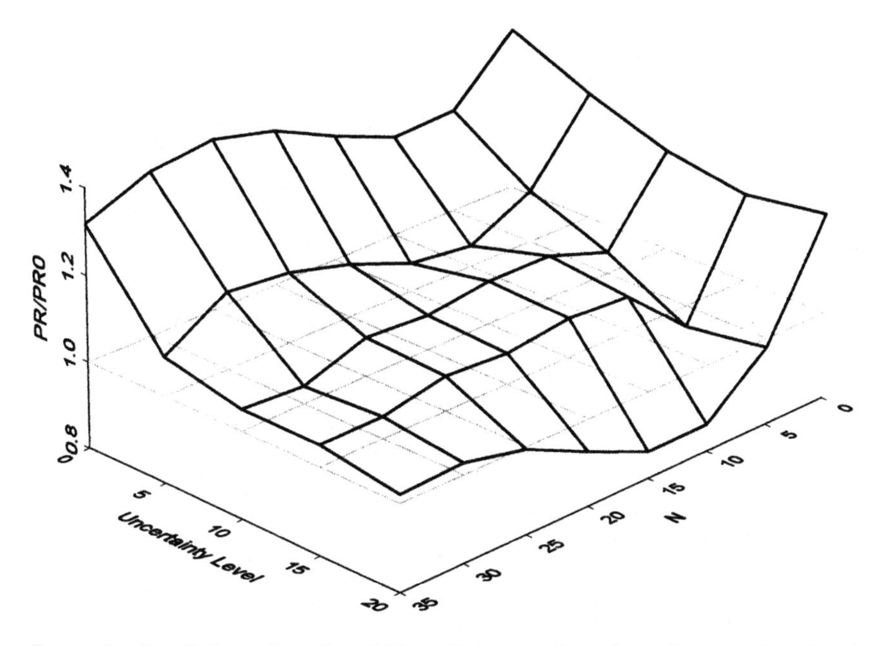


**FIGURE 2** Prediction of peak anti-Xa activity as a function of uncertainty level and sample size: All-at-once method. The ratio of model predicted to true simulated anti-Xa activity at 4 h post-dose ($PR/PR_0$) is plotted as a function of uncertainty level and sample size $N$, using a three-dimensional surface plot. The dark solid lines represent the $PR/PR_0$ resulting from the simulation studies, while a reference plane at $PR/PR_0 = 1$ is presented in light gray.

least five patients, resulted in anti-Xa predictions that were similar to the simulation values (Figure 2). The ability of the all-at-once algorithm to overcome an incorrect hyper-prior assumption is demonstrated in Figure 3, where population clearance estimates move away from the hyper-prior and toward the true simulation value when uncertainty levels approach 10 and the number of patients is at least 5 (Figure 3). When a more informative hyper–prior was employed (uncertainty level less than 10), the prior information dominated the population parameter estimation, even with the complete study size ($N = 35$) (Figure 3). With a target of accurate estimation within five patients, it was evident that an uncertainty level of at least 10 was necessary for adequate peak anti-Xa predictions and accurate population estimates of clearance.

With the same prior uncertainty level, parameter estimates from the step-by-step method reflected the true parameters for the new data after inclusion of data from fewer individuals than the all-at-once method (Figure 4 and 5). For example, $PR/PR_0$ is near 1 when only three subjects have entered the study and

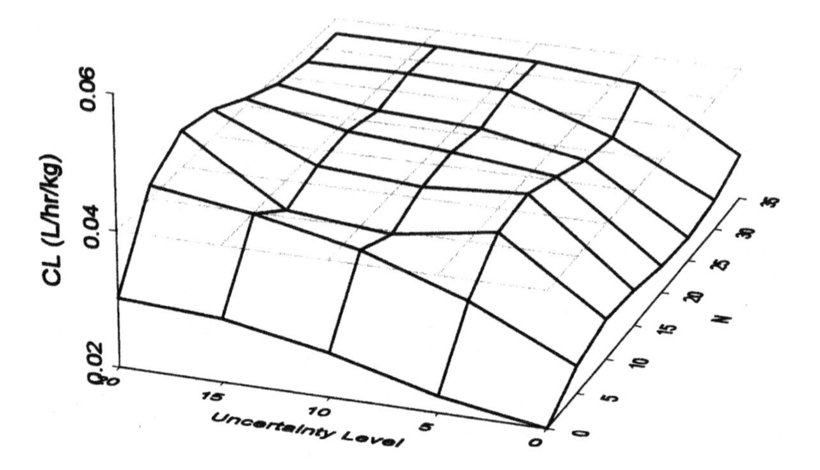


**FIGURE 3** Clearance as a function of uncertainty level and sample size: All-at-once method. The population estimate of clearance (CL) is plotted as a function of uncertainty level and sample size $N$, using a three-dimensional surface plot. The dark solid lines represent the population CL estimates resulting from the simulation studies, while a reference plane at the simulation value of CL = 0.04 L/h/kg is presented in light gray.

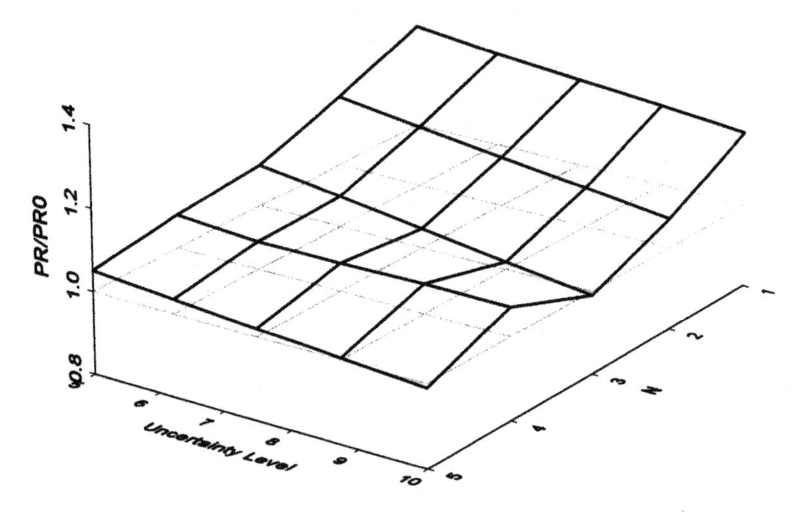


**FIGURE 4** Prediction of peak anti-Xa activity as a function of uncertainty level and sample size: Step-by-step method. The ratio of model predicted to true simulated anti-Xa activity at 4 h post-dose ($PR/PR_0$) is plotted as a function of uncertainty level and sample size $N$, using a three-dimensional surface plot. The dark solid lines represent the $PR/PR_0$ resulting from the simulation studies, while a reference plane at $PR/PR_0 = 1$ is presented in light gray.

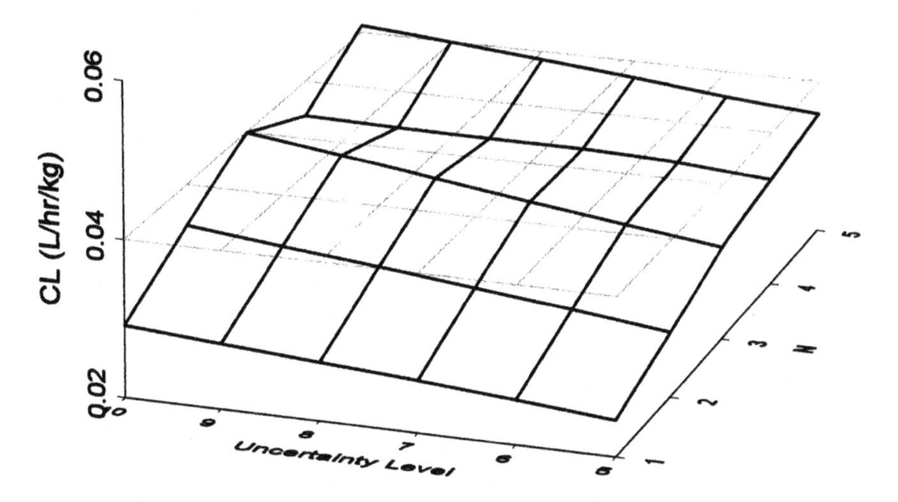


FIGURE 5 Clearance as a function of uncertainty level and sample size: Step-by-step method. The population estimate of clearance (CL) is plotted as a function of uncertainty level and sample size $N$, using a three-dimensional surface plot. The dark solid lines represent the population CL estimates resulting from the simulation studies, while a reference plane at the simulation value of CL = 0.04 L/h/kg is presented in light gray.

an uncertainty level of 10 is used (Figure 4). Likewise, the population estimate of clearance reflects the true simulation value of 0.04 liter/h/kg with a study size of 4 and an uncertainty level greater than 5 (Figure 5). This increased sensitivity to the new data was expected because the hyper-prior was shifted toward the data-driven estimate with each step. The limited range of scenarios displayed in Figures 4 and 5 (relative to Figures 3 and 4) demonstrates the fact that these simulation studies were conducted in a real-world drug development context, where efforts were focused on those scenarios that were thought to be most relevant.

### 19.4.2 Effect of PK Sampling Times on Individual Parameter Estimates

Approximately 100 trial replicates of 35 patients each were simulated for each of the sampling design scenarios and results were reported as average MPPE and MPAE for predicted anti-Xa activity at 4 h post-dose. As shown in Table 5, predictions were not biased and were acceptably accurate when the population parameters were correct (MPAE = 14, 12, and 9.1% after 1, 2, and 3 measurements, respectively). There was a tendency to overestimate the individual pre-

**TABLE 5** Impact of PK Sampling Scheme on Individual Predictions of Peak Anti-Xa Activity: Per-Protocol Design

| | PK sampling design (from Table 4) | | | | | |
|---|---|---|---|---|---|---|
| | D1 | D2 | D3 | D4 | D5 | D6 |
| CL = adult CL, V2 = adult V2 | | | | | | |
| MPPE | 3.2 | 0.92 | 1.6 | 3.8 | 2.2 | 0.97 |
| MPAE | 14 | 12 | 9.1 | 7.8 | 9.0 | 7.0 |
| CL = 1.5 * adult CL | | | | | | |
| MPPE | 4 | 10 | 3.9 | 2.7 | 1.2 | 0.9 |
| MPAE | 15 | 14 | 11 | 9.8 | 9.8 | 6.5 |
| CL > 2 * adult CL | | | | | | |
| MPPE | 13 | 18 | 9 | 7.1 | 5.3 | 1.9 |
| MPAE | 18 | 19 | 14 | 12 | 11 | 6.9 |
| V2 = 1.5 * adult V2 | | | | | | |
| MPPE | 14 | 11 | 9.9 | 8.4 | 8.8 | 8.9 |
| MPAE | 18 | 13 | 12 | 11 | 11 | 11 |
| V2 = 2 * adult V2 | | | | | | |
| MPPE | 18 | 13 | 17 | 18 | 14 | 14 |
| MPAE | 19 | 16 | 17 | 18 | 15 | 14 |
| V2 = 2 * adult V2, CL > 2 * adult CL | | | | | | |
| MPPE | 34 | 25 | 17 | 14 | 12 | 5.7 |
| MPAE | 34 | 25 | 18 | 16 | 13 | 8.5 |

PK parameters are as defined in Table 3.
MPPE = median percent prediction error
MPAE = median percent absolute error

dicted concentration when the population estimate of clearance or volume was less than the simulated value. The degree of overestimation increased as the difference between the estimation and simulation parameters increased. The overestimation remained under 20% except for the most extreme case when both the population clearance and volume parameters used in the individual estimation were half the population clearance and volume values used for simulation. The second measurement point on the first day was very important in the extreme case, where it decreased the prediction error from 34% (D1) to 25% (D2). In less extreme cases the second measurement point improved the prediction by 3–5%. The third measurement point, at 4 h on the second day, further improved the prediction: by 8% in the extreme case, and by approximately 3% in other cases. Gains from the additional samples were not important. When the second measurements on days 1 and 2 were taken at fixed nominal times (modified design) instead of at random-block times, prediction errors increased moderately (results not shown).

## 19.5 CONCLUSIONS AND DISCUSSION

The goal of this work was to optimize the Bayesian pediatric dose-adjustment algorithm with a series of simulation studies depicting potential clinical scenarios. An important feature was to allow updating of the pediatric population parameters (priors) during the trial, based on a weight-scaled adult model and new, sparse data from pediatric patients. Other factors examined by simulation included the impact of the magnitude of the prior uncertainty in the population parameters, stability of parameter estimates to outliers or data errors, and the impact of increasing the number of PK samples per patient. The simulations identified the necessary algorithm settings and desirable trial design options for robust dose-adjustment recommendations under a variety of conditions.

Final decisions about the pediatric trial design and dose-adjustment algorithm were guided by the results of simulation studies and by the clinical investigator's prior knowledge and practical experience. With regard to the specific simulation results presented here for the population parameters, a decision was made to set the hyper-prior uncertainty to a level of 10 and to use the all-at-once prior updating scheme. It was obvious that the sensitivity to new data for each method could be "tuned" by adjusting the level of uncertainty, so this decision was based, in part, on practical implementation considerations. For the individual predictions, the simulation studies identified the importance of obtaining two PK samples on day 1, with the second sample taken randomly, and a subsequent sample on day 2. These results supported the proposed per-protocol design.

Other design factors (Table 2) were studied by simulation, although results are not presented here. For example, the effect of changing the variance of the hyper-prior distributions for peripheral compartment PK parameters was investigated. Because of the sparse sampling design, a posteriori parameter identifiability was poor when diffuse hyper-prior distributions were used for the peripheral compartment parameters. Fixed parameter values and informative hyper-priors performed equally well, but in keeping with the Bayesian nature of the algorithm, it was preferable to use the informative hyper-prior distributions. The potential impact of outlier or erroneous observations on individual dose recommendations was also a major concern. Additional simulation studies revealed that because of the assumed prior distributions, the probability of underdosing due to one aberrant observation was greater than the probability of overdosing. Guidelines, based on diagnostic plots and the necessity for additional samples, were developed to assist the investigator in recognizing unrealistic dose-adjustment recommendations. Undoubtedly, many more simulation scenarios and possible clinical situations could be imagined and tested in the context of the current trial design. It is important to note, however, that the ideas and simulations described in this chapter were conceived and carried out within the practical restrictions and constraints of a real-time drug development process.

The Bayesian dose-adjustment strategy was particularly well suited to the investigation of LMWH in pediatric patients. Current opinion suggests that dosing of LMWH in pediatric patients should be guided by anti-Xa monitoring to therapeutic levels extrapolated from experience in adult populations (23). This notion has primarily been based on knowledge gained with a single agent (enoxaparin) and, consequently, may result in suboptimal or even inappropriate dosing recommendations for other LMWHs. Studies in adults have revealed that anti-Xa and anti-IIa profiles as well as safety and efficacy across LMWHs are often unrelated and/or unpredictable (24). Thus, the large degree of prior uncertainty associated with first time in pediatric studies of LMWHs is expected and ideally managed with an adaptive dosing algorithm.

The design of the pediatric tinzaparin dose-finding trial was characterized by a multitude of potential design options, and significant uncertainty. The use of an adaptive dose-adjustment algorithm is helpful under conditions such as these, but the formal incorporation of prior knowledge in the estimation method can itself be a subjective undertaking. A useful method of evaluating these subjective choices and prior assumptions is through the use of carefully designed Monte Carlo simulations. As shown in this pediatric tinzaparin example, simulations were necessary to assess the impact of multiple design options and were subsequently a useful resource for making the final trial and algorithm design decisions.

## ACKNOWLEDGMENTS

The authors thank Dr. Stuart Beal for making the NONMEM Bayesian prior subroutine code available and for his guidance in its application. The clinical trials and simulation studies were supported by DuPont Pharmaceuticals Company (Newark, DE) as part of the tinzaparin sodium (Innohep) development program and the simulation work was performed at GloboMax LLC (Hanover, MD).

## REFERENCES

1. HB Nader, JM Walenga, SD Berkowitz, F Ofosu, DA Hoppensteadt, G Cella. Preclinical differentiation of low molecular weight heparins. Semin Thromb Hemost 25(3):63–72, 1999.
2. I Murdoch, R Beattie, D Silver. Heparin-induced thrombocytopenia in children. Acta Paediatr 82:495–497, 1993.
3. S Siragusa, B Cosmi, F Piovella, J Hirsh, JS Ginsberg. Low molecular weight heparins and unfractionated heparin in the treatment of patients with acute venous thromboembolism: results of a meta-analysis. Am J Med 100:1–9, 1996.

4. P Massicotte, M Adams, VL Marzinotto, LA Brooker, M Andrew. Low molecular weight heparin in paediatric patients with thrombotic disease: a dose finding study. J Pediatr 128:313–318, 1996.
5. RC Punzalan, CA Hillery, RR Montgomery, JP Scott, JC Gill. Low-molecular-weight heparin in thrombotic disease in children and adolescents. J Pediatrics Hematology/Oncology 22(2):137–142, 2000.
6. S Laporte, P Mismetti, P Piquet, S Doubine, A Touchot, H Decousus. Population pharmacokinetic of nadroparin calcium (Fraxiparine®) in children hospitalised for open heart surgery. Eur J Pharm Sci 8:119–125, 1999.
7. N Nohe, A Flemmer, R Rumler, M Praun, K Auberger. The low molecular weight heparin dalteparin for prophylaxis and therapy of thrombosis in childhood: a report on 48 cases. Eur J Pediatr 158 Suppl 3(2–3):S134–S139, 1999.
8. K Fijnvandraat, MT Nurmohamed, M Peters. A crossover-over dose finding study investigating a low molecular weight heparin (Fragmin®) in six children on chronic hemodialysis. Thromb Haemost 69:649, 1993.
9. RD Hull, GE Raskob, GF Pineo, D Green, AA Towbridge, CG Elliott, RG Lerner, J Hall, T Sparling, HR Brettell, J Norton, CJ Carter, R George, G Merli, J Ward, W Mayo, D Rosenbloom, R Brant. Subcutaneous low-molecular weight heparin compared with continuous intravenous heparin in the treatment of proximal-vein thrombosis. N Engl J Med 326:975–982, 1992.
10. G Simonneau, H Sors, B Charbonnier, Y Page, J-P Laaban, J-L Bosson, D Mottier, B Beau. A comparison of low-molecular-weight heparin with unfractionated heparin for acute pulmonary embolism. N Engl J Med 337:663–669, 1997.
11. RD Hull, GE Raskob, GF Pineo, D Rosenbloom, W Evans, T Mallory, K Anquist, F Smith, G Hughes, D Green, CG Elliott, A Panju, R Brant. A comparison of subcutaneous low-molecular weight heparin with warfarin sodium for prophylaxis against deep-vein thrombosis after hip or knee implantation. N Engl J Med 329:1370–1376, 1993.
12. A Leizorovicz, H Picolet, JC Peyrieux, JP Boissel, HBPM research group. Prevention of perioperative deep vein thrombosis in general surgery: a multicenter double blind study comparing two doses of logiparin and standard heparin. Br J Surgery 78:412–416, 1991.
13. KE Ryan, DA Lane, A Flynn, J Shepperd, HA Ireland, JR Curtis. Dose finding study of a low molecular weight heparin, Innohep, in haemodialysis. Thromb Haemost 66(3):277–282, 1991.
14. JS Barrett, E Gibiansky, RD Hull, A Planes, H Pentikis, JW Hainer, TA Hua, MR Gastonguay. Population pharmacodynamics in patients receiving tinzaparin for the prevention and treatment of deep vein thrombosis. Int J Clin Pharmacol Ther 39(10): 431–46, 2001.
15. BJ Anderson, AD McKee, NHG Holford. Size, myths and the clinical pharmacokinetics of analgesia in paediatric patients. Clin Pharmacokinet 33:313–327, 1997.
16. A Sutor, P Massicotte, M Leaker, M Andrew. Heparin therapy in pediatric patients. Semin Thromb Hemost 23(3):303–319, 1997.
17. MM Samama, GT Gerotziafas. Comparative pharmacokinetics of LMWHs. Semin Thromb Hemost 26 Suppl 1:31–38, 2000.

18. DZ D'Argenio, A Schumitzky. ADAPT II User's Guide, Biomedical Simulations Resource, University of Southern California, Los Angeles, CA.
19. SL Beal, AJ Boeckmann, LB Sheiner. NONMEM Users Guides Parts I-VIII, NONMEM Project Group, University of California, San Francisco, CA.
20. A Gelman, JB Carlin, HS Stern, DB Rubin. Bayesian Data Analysis. London: Chapman & Hall, 1995.
21. NHG Holford, M Hale, HC Ko, J-L Steimer, CC Peck, P Bonate, WR Gillespie, T Ludden, DB Rubin, D Stanski. Simulation in drug development: good practices. Draft Publication of the Center for Drug Development Science. Draft version 1.0, 1999. http://cdds.georgetown.edu/research/sddgp723.html
22. JG Wagner. Fundamentals of Clinical Pharmacokinetics. Hamilton, IL: Drug Intelligence Publications, Inc., 1975.
23. M Andrew, AD Michelson, E Bovill, M Leaker, MP Massicotte. Guidelines for antithrombotic therapy in pediatric patients. J Pediatr 132:575–588, 1998.
24. L Bara, A Planes, and M-M Samama. Occurrence of thrombosis and haemorrhage, relationship with anti-Xa, anti-IIa activities, and D-dimer plasma levels in patients receiving low molecular weight heparin, enoxaparin or tinzaparin, to prevent deep vein thrombosis after hip surgery. Br J Haematology 104:230–240, 1999.

# Index